The early years of radio astronomy

This volume is dedicated to Karl Guthe Jansky (1905—1950)
and the spirit of inquiry which he exemplifies

The early years of
RADIO ASTRONOMY

REFLECTIONS FIFTY YEARS AFTER JANSKY'S DISCOVERY

Edited by

W. T. SULLIVAN, III

Associate Professor of Astronomy, University of Washington, Seattle

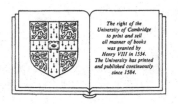

The right of the
University of Cambridge
to print and sell
all manner of books
was granted by
Henry VIII in 1534.
The University has printed
and published continuously
since 1584.

CAMBRIDGE UNIVERSITY PRESS

Cambridge

London   New York   New Rochelle

Melbourne   Sydney

PUBLISHED BY THE PRESS SYNDICATE OF THE UNIVERSITY OF CAMBRIDGE
The Pitt Building, Trumpington Street, Cambridge, United Kingdom

CAMBRIDGE UNIVERSITY PRESS
The Edinburgh Building, Cambridge CB2 2RU, UK
40 West 20th Street, New York NY 10011–4211, USA
477 Williamstown Road, Port Melbourne, VIC 3207, Australia
Ruiz de Alarcón 13, 28014 Madrid, Spain
Dock House, The Waterfront, Cape Town 8001, South Africa

http://www.cambridge.org

First published 1984
First paperback edition 2004

A catalogue record for this book is available from the British Library

Library of Congress catalogue card number: 83–23227

ISBN 0 521 25485 X hardback
ISBN 0 521 61602 6 paperback

CONTENTS

Contents

PREFACE

Fifty years ago Karl Jansky accidentally discovered that the Milky Way is a copious source of radio waves. This eventually led to detailed study of extraterrestrial radio waves, which for three reasons has unquestionably been one of the key developments in the astronomy of our century. First, there are of course the startling results themselves, revealing a panoply of unexpected phenomena. After radio galaxies, quasars, pulsars, the cosmic background radiation, and complex interstellar molecules, the Universe would never again be the same. Second, the style of research of the radio researchers eventually also changed the way that traditional astronomy was done; here I refer to the use of electronics and the attitude that primary training in astronomy was not necessary for success. Third, the achievements of radio astronomy provided much of the basis for the desire to scan the skies at every electromagnetic frequency possible, a program which has dominated much of subsequent observational astronomy. When an historian of the distant future characterizes the astronomy of our own era, major emphasis will surely be placed on this continual opening of the electromagnetic window and our resultant expanded view of the Universe, a process which began with radio astronomy.

It is thus both important and fitting that we pause after fifty years and collect the reflections of the pioneers of radio astronomy. To this end a score of major participants have contributed to the present volume their recollections and analyses of how the field developed. While recollection by itself is not history, these articles yield invaluable glimpses, not otherwise obtainable, of the spirit of the times, and also serve as starting points for delving further into the history.

This collection is the outgrowth of three half-day sessions, each connected to the golden anniversary of Jansky's discovery and held at a recent major conference: American Association for the Advancement of Science (San Francisco, January 1980), International Union of Radio Science (URSI) (Washington, D.C., August 1981), and International Astronomical Union (Patras, Greece, August 1982). As a general rule the papers deal with the development of radio astronomy only before 1960, a point at which the nature of the field was distinctively changing from "little" to "big" science. A major goal has been to achieve a reasonably fair, worldwide coverage of all important aspects of the field prior to 1960, but this has not been completely possible. The following gaps in the coverage are notable: the development of radio astronomy in the 1950s both in the United States and in the Netherlands, theoretical interpretations of radio emission and radio sources in the West, the Cantabrigian view of the controversy over radio source counts, and the early work of the post–World War II group led by James S. Hey (although see his 1973 book *The Evolution of Radio Astronomy*).

Three further publications in which radio astronomers describe their early work are also recommended. Bernard Lovell has written *The Story of Jodrell Bank* (1968), while John Bolton has given insights into the early years at the Radiophyics Laboratory in Sydney in his reminiscence "Radio Astronomy at Dover Heights" (*Proc. Astronomical Soc. of Australia*, **4**, 349–58 (1982)). Finally, the proceedings of a workshop held in May 1983 at the National Radio Astronomy Observatory in Green Bank, West Virginia, in honor of Jansky and entitled "Serendipitous Discoveries in Radio Astronomy" (edited by K.I. Kellermann), also contain numerous accounts of the early days of radio astronomy.

Many persons besides the twenty–one names listed in the Table of Contents have contributed to this volume. Aid in the organization of the original sessions at conferences was given by Haruo Tanaka, John Findlay, Michael Hoskin, Govind Swarup, and Ken Kellermann. Simon Mitton has been very helpful as editor at Cambridge University Press. I thank Vladimir Chalupka for his help with refinement of the translations of the two Soviet papers. I am grateful to the individual contributors for their willingness to put up with my editorial suggestions, and to the numerous typists around the world, and in particular Charlotte Arthur at the University of Washington, for skill and patience in dealing with the

infernal mechanics of producing camera-ready copy. Finally, I thank
Barbara Sullivan for her strong support of all my historical endeavors.

August 1983   W.T. SULLIVAN, III
                                                   Department of Astronomy
                                                   University of Washington
                                                   Seattle, Washington 98195
                                                   U.S.A.

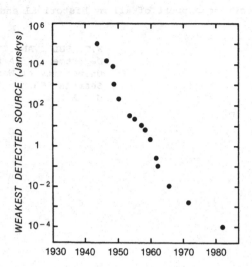

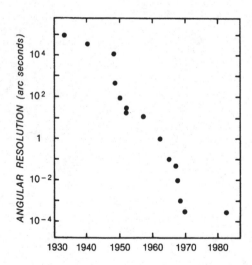

*The increasing sensitivity and angular resolution of radio telescopes
during the half century since Jansky's discovery (courtesy Ken
Kellermann)*

SECTION ONE: *The Earliest Years*

The beginning of radio astronomy was remarkable -- who does not find the stories of Karl Jansky and Grote Reber appealing? In the first case we have a man working on the very practical business of bettering telephone communications, but in the end, through a combination of skill, persistence and fortune, revealing a phenomenon which eventually would transform the way astronomy had been done for centuries. And then while the scientific establishment ignores Jansky's discovery, in the second instance a man devotes his spare time for almost a decade to the task of studying the waves of "cosmic static" falling in the back yard of his suburban home.

Following World War II the study of extraterrestrial noise grew rapidly, but it was another decade before the traditional community of astronomers fully recognized the value of the new radio techniques. Jesse Greenstein was one of the very few in that community who paid any attention to radio results before 1950 -- witness his co-authorship of the first theoretical paper dealing with radio noise (1937) and of the first review article on the subject (1947). He thus closes this section with an authoritative account of how early relationships evolved among theorists, optical astronomers, and radio observers.

Karl Jansky on his 23rd birthday, 23 October 1928, shortly after beginning work at the Bell Telephone Laboratories (courtesy Alice Jansky)

KARL JANSKY AND THE DISCOVERY OF EXTRATERRESTRIAL RADIO WAVES

Woodruff T. Sullivan, III
Department of Astronomy
University of Washington, Seattle

On 27 April 1933 a small audience in Washington, D.C. heard a talk entitled "Electrical Disturbances of Extraterrrestrial Origin." Today we view Karl Jansky's paper as the beginning of radio astronomy, but at that time it was neither the birth of a new science nor greatly acclaimed by Jansky's engineering and scientific colleagues. A week after his talk, Jansky wrote to his father:

I presented my paper in Washington before the U.R.S.I., or International Scientific Radio Union, an almost defunct organization. It was not my wish that my paper was presented there, but at Mr. Friis's [Jansky's supervisor] insistence.... The U.R.S.I. meetings in Washington are attended by a mere handful of old college professors and a few Bureau of Standards engineers. The meeting was conducted in such a manner...that not a word was said about my paper except for a few congratulations that I received afterwards. Besides this, Friis would not let me give the paper a title that would attract attention, but made me give it one that meant nothing to anybody but a few who were familiar with my work. So apparently my paper attracted very little attention in Washington.[1]

To understand this lukewarm reception, we must examine Jansky and his work in the context of contemporary research into radio communications and astronomy. Such an examination reveals that this basic discovery was a misfit. Neither fish nor fowl, it was unable to be appreciated by either the scientists or engineers, and therefore lay untouched as an isolated curiosity.

JANSKY'S EARLY YEARS

Karl Guthe Jansky[2] was born on 22 October 1905 in the Territory of Oklahoma. His father's parents had emigrated from Czechoslovakia in the 1860s, while his mother had a French and English background. His father, Professor of Electrical Engineering at several Midwestern schools, finally settled at the University of Wisconsin where he served on the faculty for 32 years. Karl grew up in a competitive academic environment, attended the local university,[3] and in 1927 obtained his B.A. in Physics (Phi Beta Kappa) with an undergraduate thesis, under the direction of E.M. Terry, involving experimental work with vacuum tubes.

Following an extra year of graduate study in physics, Jansky applied for a job with the Bell Telephone Laboratories in New York City. A physical exam, however, confirmed the kidney disease (Bright's disease) which Jansky had already had for several years and which would eventually lead to his early death. The Labs were reluctant to assume such a risk in a new employee, and it was only through the intervention of his older brother C. Moreau[4] that the Company was finally persuaded that his professional promise outweighed any medical problem. Partly for reasons of health, Jansky was not kept at the main laboratories in New York City, but rather assigned to a radio field station in rural Cliffwood, New Jersey. And so in August 1928 he found himself working on the problem of shortwave static under Harald Friis, for $33 per week.

THE SETTING FOR JANSKY'S WORK

Before resuming with Jansky's specific story, I will set its context with respect to the overall field of radio communications and to activities at Bell Labs over the preceding decade.

Radio communications research in 1928

Marconi first sent radio signals across the Atlantic in 1901. This startling feat prompted O. Heaviside and A.E. Kennelly in the following year to propose the existence in the upper atmosphere of an electrified layer which guided radio waves around the curve of the earth. But not until two decades later was strong evidence put forth for this "Kennelly-Heaviside layer" at a height of ~100 km. This came in 1925 in the form of propagation work by E.V. Appleton and M.A. Barnett in England and pulsed radar by G. Breit and M.A. Tuve in America.

Before 1920 intercontinental radio circuits were typically at frequencies less than 100 kHz, but the increasing requirements for message channels and reliability steadily pushed radio communications to higher frequencies, which led to exploitation of the high frequency, or long wave, region up to 1.5 MHz. After 1920 the amateur operators of the day found that even the frequencies above 1.5 MHz (wavelengths of "200 m down," or "shortwaves"), previously thought useless, worked surprisingly well for long-distance contacts. Although at first only a curiosity, the technology of these shortwaves was intensively developed, and their use became dominant after the introduction of the high-vacuum, oxide-coated triode, valuable both as a power tube in transmitters and as a sensitive amplifying tube for reception.

Besides the push to higher frequencies, another major trend in radio communications of the 1920s was from radiotelegraphy, which employs Morse code, to radiotelephony, the transmission of voice. The latter is a much more difficult technical proposition, requiring increased bandwidth, simultaneous two-way service, greater reliability and fidelity, and 24-hour service. By 1927 American Telephone & Telegraph (A T & T) could offer its first radiotelephone service between New York and London for $75 per three minute call. But this was at long waves (60 kHz) and it soon became clear that shortwaves were superior. Shortwaves required smaller antennas and smaller transmitters to achieve the same signal levels, allowed many more voice channels, and did not suffer the degree of interference from atmospheric noise which plagued the long waves, especially in the summertime when tropical thunderstorms were devastating.

In 1929 a shortwave, trans-Atlantic radiotelephone service was opened to the public and it quickly superseded the long-wave circuits. It operated at a variety of frequencies between 9 and 21 MHz and employed a 15 kw transmitter with air-cooled power tubes and a quartz crystal oscillator. But there were now new sources of interference such as automobiles, intrinsic noise in the receiving electronics, and magnetic storms which needed investigation. Moreover other phenomena such as rapid fading due to changing signal paths called for wholly new designs in receivers and antennas.

These then were the kinds of problems facing the radio engineers whom Jansky joined at Bell Labs in the late 1920s. The goal always was to maximize, within economic limits, the ratio of wanted signal to unwanted noise in the telephone subscriber's ear.

Bell Telephone Laboratories in 1928

In the first decade of this century several major industrial research laboratories were established in the United States, in particular at General Electric, DuPont, Eastman-Kodak, and A T & T. The last of these evolved in 1925 into the Bell Telephone Laboratories. Its first director, Frank Jewett, placed strong emphasis on step-by-step attacks on problems in communications engineering; in particular, precision of measurement was seen as vital. Bell Labs quickly established an international reputation in a wide variety of scientific fields -- witness the work in the 1920s of C.J. Davisson and L.H. Germer on electron diffraction by crystals and that of H. Nyquist and J.B. Johnson on the noise generated by electronic components. The strength of the Labs in physics can be gauged by the fact that over the period 1925-28 it ranked amongst the top ten institutions in terms of number of articles published in *Physical Review* (Hoddeson 1980). In fact the Labs were very much like a large technical university, but with no students, much better equipment, and a greater concern for achieving practical results on a schedule.[5]

The first A T & T research connected with radio telephony was begun by C.R. Englund in 1914 on long wave signal and static levels. In 1919 R. Bown and Friis joined him in investigating problems of radio propagation, measurement methodology, and receiver and antenna design. As radio communications steadily moved to higher frequencies, research at Bell Labs became more quantitative with detailed, synoptic studies of the level and character of both signals and noise. The Radio Research Division, one of eight in the Labs, was run by W. Wilson in the early 1930s and had five Branches. The Branch of 15 to 20 men which Jansky joined was concerned primarily with problems of reception and was jointly headed by Friis and Englund.

JANSKY'S INVESTIGATIONS

Jansky's work at Bell Labs relevant to the discovery of extraterrestrial radio waves can be divided into four phases. From 1928 to 1930 he was "learning the ropes," recording long-wave static, and beginning to build a rotatable antenna and shortwave receiver. In the midst of several diversions during the following two-year phase, he finished testing of all components and his first shortwave observations contained hints of a new source of static. In the climactic years of 1932 and 1933 the "hiss type static" was recorded and studied in detail and a full astronomical explanation developed. In the last phase, over the next three years, Jansky only occasionally made measurements of his "star static" while he primarily worked on more practical aspects of radio noise.

Phase One (1928-30): orientation and building

After only six weeks on the job at Cliffwood, Jansky was deeply involved in Friis's assignment to build an electric field-strength receiver to record the intensity of shortwave atmospheric interference ("atmospherics"). But it is also clear that the young graduate was spending much time simply adjusting:

> I have been building apparatus for the last few weeks for my new shortwave recorder. It will be several months yet before I get any actual results....When I first came here the language they spoke was almost foreign to me, but I am beginning to get used to it now. At Madison I had never heard of such things as attenuators, T.U.'s, gain controls, double detection, etc., but that is what I get for not taking engineering.[6] [23 Sept 1928, KJ:CJ]

Jansky found that many of the components needed for his study already were available, but others required considerable modification and some simply did not exist. For instance the superheterodyne shortwave receiver (Fig. 1) was basically a Friis design, and had a sensitivity and stability unsurpassed for its day.[7] A novel requirement for the receiver was a long integration time, that is, the need to average the output over a considerable time (30 seconds) in order to be able to detect weaker levels. This recording circuit was developed largely by W.W. Mutch, another newcomer to the Labs.

Jansky's shortwave antenna was also a mixture of old and new ideas. Friis had already built a 115 ft long, manually rotatable platform with a loop antenna at each end, operating at 43 kHz and allowing the rough azimuthal direction of any long-wave static to be determined. Although Jansky's focus was to be on shortwave static, one purpose of his study was to find correlations with the better-known long-wave static. Thus he began recording long-wave static early on, and

Figure 1. The long-wave (left) and shortwave receivers with which Jansky recorded the intensities and times of occurrence of atmospherics on 43 kHz and 20.5 MHz. Two strip chart recorders can also be seen between the receivers. (Jansky 1932; c. IRE (now IEEE))

was able to adapt the idea of a rotating array to shortwaves. The shortwave antenna was of the type known as a Bruce array, recently developed for trans-Atlantic commercial circuits by E. Bruce, also a member of Friis's group. On 24 August 1929 Jansky recorded in his notebook: "Mr. Sykes will start work on the merry-go-round next Monday."[8]

Phase Two (1930-31): diversions and first shortwave observations

At this point Jansky lost several months on the shortwave static project. In February 1930 the entire group at Cliffwood moved to a new field station a few miles away in Holmdel; this meant not only the usual disruptions accompanying any major move, but also that the concrete foundation and track for his antenna had to be re-done. But now the growing group had what it really needed -- the room to carry out properly a wide variety of experiments. The Holmdel site was 440 acres of rolling farmland and woods, complete with a trout stream and skating pond. At the center was a large, wooden frame laboratory nicknamed the "turkey farm" and scattered all over were smaller shacks for individual experimenters.

Figure 2. The 29 m (two wavelengths) long rotating Bruce array at the Holmdel, New Jersey field station of Bell Telephone Laboratories. With this antenna Jansky discovered extraterrestrial radio noise in 1932-33. The active element is the farther from the camera; it was connected through the small white box and an underground copper pipe to the receiver shack about 100 m away. (Jansky 1932; c. IRE (now IEEE))

In the spring Jansky searched for a "quiet" band on which to observe and eventually settled on a wavelength of 14.5 m. With this datum in hand, the final dimensions of his Bruce array were determined and the carpentry shop built it in the summer from 400 ft of 7/8 inch brass piping supported by ordinary glass telephone-wire insulators on 2 by 4 inch fir lumber. The array consisted of two almost identical crenellated curtains of quarter-wavelength sections, spanning in total two wavelengths, or 29 m (Fig. 2). They were separated by one-quarter wavelength and one of them, the reflector or passive element, was about 15% taller. The receiver was connected to the center of the smaller element and the net effect was that the antenna responded only to radiation arriving from a single direction perpendicular to the array. The whole thing was mounted on four Ford Model T front wheels and automatically rotated completely around every twenty minutes through the agency of a 1/4 horsepower motor, speed reducer, and 10 ft sprocket wheel and chain.

By the autumn of 1930 Jansky had carefully tuned the antenna, measured its azimuthal antenna pattern (using an oscillator at a distance of 1000 ft), gotten the bugs out of the motor and chain arrangement, and occasionally recorded shortwave static. But as the winter approached he switched to other projects because he took it for granted that "there is practically no shortwave static in winter." In fact he spent most of that winter using the array to study the direction of arrival of shortwave signals from a transmitter in South America, finding variations of as much as 6°. In the spring of 1931 he was working on something else again, this time an ultra-shortwave (4 m wavelength) receiver for static studies at yet higher frequencies. Only by the summer was he finally back on track, regularly recording shortwave static at 14.6 m wavelength (20.5 MHz)[9] over a bandwidth of 26 kHz. In his August 1931 work report we find what in retrospect appears to be the first recognition of a new, weak component of static. Faintly audible with headphones above "first circuit" noise (intrinsic receiver noise), it followed a daily east-to-west pattern:

> Static was strongest during the month just before, during, or just after an electrical storm; however, nearly every night that the receiver was run, static was received from a source that apparently always follows the same path. Early in the evening, about 6 P.M., this static (it has always been quite weak) comes from the southeast; by about 8 P.M. it has slowly moved to the south; by midnight it comes from the southwest; and by 3 A.M. it comes from the west. The reason for this phenomenon is not yet known, but it is believed that a study...of the known thunderstorm areas of the world will reveal the cause.

In August it was a night-time phenomenon, but when it persisted through the autumn and began to shift to different times of the day, Jansky became intrigued.

Phase Three (1932–33): the astronomical discovery

Hiss type static. As Jansky continued irregular monitoring of the shortwave atmospherics in the fall and winter of 1931-32, he began to isolate a component which he called "hiss type static." At one point it appeared to be interference transmitted from some unmodulated carrier, but he later became convinced of its natural origin. His January 1932 work report talks about a "very steady continuous interference...that changes direction continuously throughout the day, going completely around the compass in 24 hours." In a letter home on 18 January 1932 he not only for the first time describes this static to his father ("Sounds interesting, doesn't it?"), but also reports on his newborn first child. At first Jansky also used the term "sun static," for the direction of arrival seemed to coincide quite closely with the sun's position. But by February he had studied his records closely enough to see that the mysterious static was no longer aligned with the sun, but was preceding it "by as much as an hour." Practical problems intervened: in February two weeks were lost while the motor and chain assembly were repaired and in April another three weeks as a consequence of windstorm damage. But enough data were obtained in the spring to establish that the hiss type static was exhibiting a continual shift in time "in accordance with the approaching summer season and the lengthening day." Although Jansky now could see that the radio waves certainly were not coming directly from the sun, he still felt it likely that they were somehow being controlled by the vernal northerly swing of the sun, perhaps through the changing angle of incidence of sunlight on the atmosphere or through the shifting position of the sub-solar point. He thus fully expected the shift in the daily signal to reverse itself when the sun retraced its path back south after the summer solstice in June. But only more data would settle the point.

    In February Friis told Jansky to write up his results on both the long-wave and shortwave static for publication in the *Proceedings of the Institute of Radio Engineers* (I.R.E.) and for presentation at an April meeting in Washington, D.C. of the U.S. section of U.R.S.I. Jansky spent much of the spring working on this, his first publication, entitled "Directional Studies of Atmospherics at High Frequencies." The paper is largely concerned with a description of his equipment and techniques, as well as with data on thunderstorm static, but he does spend three pages discussing the steady hiss type static along the lines I have indicated above. A paragraph from the paper summarizes his findings:

    From the data obtained it is found that three distinct groups of

static are recorded. The first group is composed of the static
received from local thunderstorms and storm centers. Static in
this group is nearly always of the crash type....The second
group is composed of very steady weak static coming probably by
Heaviside layer [ionosphere] refractions from thunderstorms some
distance away. The third group is composed of a very steady
hiss type static the origin of which is not yet known. [Jansky
1932]

At this point, just when the scientific problem was proving
fascinating, the reality of the Great Depression fell hard on Bell Labs.
In June 1932 the work week was cut back from 5 1/2 to 4 days (with a
corresponding reduction in pay, not to be restored until 1936). The
overall budget was cut to 60 per cent of its peak in 1930 and 20 per cent
of all employees were fired. Morale amongst the engineers at Holmdel
sagged, and in one letter to his father Jansky even worried about the
entire field station being shut down. He asked his father about possible
teaching positions at colleges or high schools in Wisconsin, but also
allowed that "I can't think of a better company to work for."

Nevertheless radio work at Holmdel continued. In the second
half of 1932 Jansky was again distracted from directly studying the
origin of the shortwave static as he began investigating the general
methodology of measuring noiselike signals. What were the effects of
changing the bandwidth? Should one record the effective, peak, or
average voltage? Was a linear or square-law detector better? Through
this he still managed to take shortwave data several times each month,
but the daily 12 ft lengths of strip chart recordings were only cursorily
examined. Even so, by August he had noticed that the pattern on the
daily traces was *not* shifting back to its springtime position, as he had
anticipated, but rather kept steadily arriving earlier. Also then, a
fortuitous partial solar eclipse on 31 August allowed him to test any
effect on the hiss type static attributable to the temporary covering of
the sun. But when the records on three consecutive days centered on the
eclipse proved to be indistinguishable, the enigmatic static seemed even
less likely to be associated, directly or indirectly, with the sun.

Connection to the stars. The idea of the static arriving on a *sidereal*
schedule, and therefore being fixed in celestial coordinates, has an
uncertain genesis. December 1932, however, was the date. Jansky's

correspondence and work reports indicate that George C. Southworth (then working for A T & T in New York City) asked him to plot up his entire year's data in order to see if it correlated with "diurnal changes in the directions of earth currents"[10] which Southworth was then studying. This request apparently caused Jansky for the first time to see the precision in the shift of the overall pattern and the fact that after one year the pattern had slipped exactly one day -- the peak signal was now in the south at the same time of day as it had been the previous December. To an astronomer this kind of shift is a fact of life -- a star or other source fixed in celestial coordinates each day rises four minutes earlier (with respect to the sun) as a result of the earth orbiting the sun; after a year, this slippage amounts to one day. But to a communications engineer, the connection is not at all obvious. It may thus have been A. Melvin Skellett, a good friend and bridge partner of Jansky's, who provided the key suggestion.[11] Skellett was leading a highly unusual schizophrenic life as simultaneously a radio engineer for Bell Labs (at the Deal Beach field station) and a graduate student in astronomy at nearby Princeton University (working on the effect of meteors on the ionosphere). In any case by the end of December Jansky had consulted Skellett and had learned a good bit about astronomical coordinates. His December work report cites a preliminary direction of arrival: right ascension = 18^h, declination = $-4°$. In a letter home on 21 December 1932 he could hardly contain his excitement:

> Since I was home [early November] I have taken more data which
> indicates definitely that the stuff, whatever it is, comes from
> something not only extraterrestrial, but from outside the solar
> system. It comes from a direction that is *fixed in space* and the
> surprising thing is that...[it] is *the direction towards which the
> solar system is moving in space.* According to Skellett (our
> friends in Deal) there are clouds of "cosmic dust" in that
> direction through which the earth travels.
>
> There is plenty to speculate about, isn't there? I've got
> to get busy and write another paper right away before somebody
> else interprets the results in my other paper in the same way
> and steals the thunder from my own data. [italics in original]

In the first two months of 1933 Jansky continued his attack along two fronts. First he tried to analyze the past records more carefully in order to pin down the direction of arrival of the

extraterrestrial radio waves. Further study, however, proved as
confusing as it was helpful, for the daily time of arrival was not
behaving as regularly (on the assumption of a *single* source) as it should
have. Secondly, he made several attempts with another antenna on the
site to determine the *vertical* angle of arrival of the hiss type static.
In particular he varied the height above the ground of a "horizontal
antenna setup" available on site, thus shifting its response in elevation
angle, but found no detectable change in intensity of the source.[12]

Jansky also got busy writing a second paper, although
apparently without much enthusiasm from his boss:

> My records show that the "hiss type static"...comes...from a
> direction fixed in space. The evidence I now have is very
> conclusive and, I think, very startling. When I first suggested
> the idea of publishing something about it to Friis, he was
> somewhat skeptical and wanted more data. Frankly, I think he
> was scared. The results were so very important that he was
> timid about publishing them. However, he mentioned them to
> W. Wilson, the department boss, and Wilson discussed it with
> Arnold, who is in charge of the whole Research Department of the
> Bell Labs (he reports directly to Jewett), and Arnold wanted the
> data published immediately. [KJ:CJ, 15 February 1933]

He worked on the paper all spring and presented it in Washington on 27
April (as described in the opening of this paper) and on 27 June at the
national convention of the I.R.E. in Chicago. A short note was sent to
Nature in May and the full paper was submitted in June for publication in
the *Proc. I.R.E.* Until the end Jansky was wrestling with Friis over the
tone of the scientific claims:

> I haven't the slightest doubt that the original source of these
> waves, whatever it is or wherever it is, is fixed in space. My
> data prove that, conclusively as far as I am concerned. Yet
> Friis will not let me make a definite statement to that effect,
> but says I must use the expressions *"apparently* fixed in space"
> and *"seem* to come from a fixed direction," etc., etc., so that
> in case somebody should find an explanation based upon a
> terrestrial source, I would not have to go back on my
> statement. I am not worried in that respect, but I suppose it
> is safer to do what he says. [KJ:CJ, 10 June 1933; italics in
> original]

Thus it is that Jansky's 1933 paper in *Proc. I.R.E.*, unlike his two talks (or the *Nature* article, entitled "Radio Waves from Outside the Solar System"), has the relatively cautious title "Electrical Disturbances Apparently of Extraterrestrial Origin."

This paper, one of the most important in the astronomy of this century, ironically devotes fully half of its twelve pages to explanations for its engineer audience of elementary astronomical concepts such as right ascension, declination, and sidereal time (Fig. 3); the only citations in the entire article are to introductory astronomy texts of the day and to Jansky's first paper. Once past these fundamentals and a presentation of the observations (Fig.'s 4 and 5), he derives that the direction of arrival of the radio waves has a right ascension of 18^h 00^m ± 30^m and a declination of $-10°$ ± 30°. The right ascension is determined in a rather straighforward manner from the sidereal time when the maximum signal was in the south, on Jansky's meridian (Fig. 6). The determination of the declination, however, is a good bit more complex. It is deduced from a comparison of (a) the observed curves showing azimuth of arrival of peak signal versus time with (b) the curves expected for various declinations *on the supposition that the signal is received even when the source is below the horizon*. Jansky had somehow to reconcile the idea of a single source of emission with his receiving the hiss type static for typically twenty hours every day (Fig. 4). Thus his notion was that when the source was below the horizon, the radio waves struck the limb of the earth (as seen from the source) and travelled to New Jersey along the shortest path on the surface, i.e., in such a way that the azimuth was preserved. The last figure in the paper (Fig. 7) presents the comparison between actual and calculated curves and illustrates, as Jansky discusses, that no *single* value of declination seems to work. Jansky can only suggest that this might be due to effects of refraction and attenuation, effects influencing most seriously those waves arriving at low elevation angles.

Jansky also has a few thoughts about the origin of the hiss type static. In the first place he is not certain that he is observing the actual *primary* phenomenon: "It may very well be that the waves...are secondary radiations caused by some primary rays of unknown character...striking the earth's atmosphere" (Jansky 1933b). These "primary rays" would be high-speed particles, that is, cosmic rays. But whatever the origin, he notes the approximate agreement (within 40°) of

Figure 3. An example from Jansky (1933b) of the author's efforts to explain astronomical fundamentals to his engineering audience. (c. IRE (now IEEE))

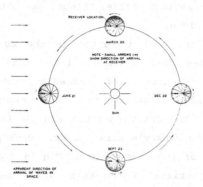

Figure 4. This strip chart recording of 16 September 1932 is all that remains of Jansky's original data. The "intensity" scale refers to measured electric-field strength, with the definition of "decibel" such that 3 db corresponds to a factor of two in received power. The cardinal direction of the rotating antenna is indicated at the top of each half day. Depending on the orientation of the Milky Way with respect to the local horizon, each 20 minute rotation yielded either one or two prominent humps. On this day the galactic center crossed the meridian 21° above the horizon at 5:56 P.M. Eastern Standard Time. (Jansky 1935; c. IRE (now IEEE))

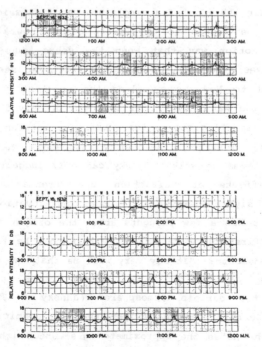

Figure 5. Plot from Jansky (1933b) illustrating the monthly slippage in the time of day when the maximum signal from the "hiss type static" was detected from a given azimuth. (c. IRE (now IEEE))

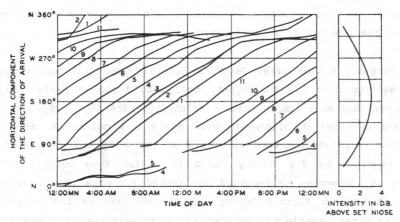

Direction of arrival of waves of extraterrestrial origin.

1. Jan. 21, 1932
2. Feb. 24, 1932
3. March 4, 1932
4. April 9, 1932

5. May 8, 1932
6. June 11, 1932
7. July 15, 1932

8. Aug. 21, 1932
9. Sept. 17, 1932
10. Oct. 8, 1932
11. Dec. 4, 1932

Figure 6. Jansky's determination of the right ascension of the maximum signal. The plotted points represent the (solar) time of transit of the signal through the year. The diagonal lines correspond to different right ascensions -- from these data Jansky derived a value of $18^h \pm 30^m$. (Jansky 1933b; c. IRE (now IEEE))

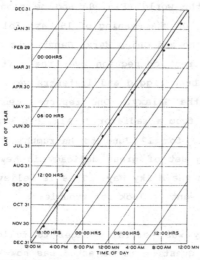

Time of coincidence of the direction of arrival and the meridian of the receiver for the different days of the year.

his derived position with two significant directions in the sky: (1) the direction in Hercules towards which the sun moves with respect to nearby stars, and (2) the direction in Sagittarius towards the center of our Milky Way galaxy. No arguments are given in the paper as to a preference for one or the other of these alternatives.

Popular reaction. A week after Jansky's talk in Washington had garnered little reaction from his colleagues, Bell Labs issued a press release which in contrast made him an instant celebrity. *The New York Times* for 5 May 1933 heralded the discovery with one entire column of its front page: "New radio waves traced to center of the Milky Way -- mysterious static, reported by K.G. Jansky, held to differ from cosmic ray -- direction is unchanging -- recorded and tested for more than a year to identify it as from Earth's Galaxy -- its intensity is low -- only a delicate receiver is able to register it -- no evidence of interstellar signalling." The remainder of that front page reported on other world events which help place Jansky's discovery in the larger historical context: Japan invading China, F.D. Roosevelt inaugurating the policies of his New Deal in the famed "One Hundred Days" (Jansky was a rabid anti-New Dealer), Nazi demonstrations in Germany, and financial troubles for the League of Nations. On the lighter side, that spring the film *King Kong* was released; and in sports, Babe Ruth was to hit the winning home run in the first All-Star game in Chicago. Articles about Jansky's discovery appeared all over the world (Fig. 8) and he soon appeared on the (NBC) Blue radio network. On the evening of 15 May, following Lowell Thomas with the news and Groucho and Chico Marx, Jansky was featured on a weekly science program. A direct connection over the fifty miles from New York City to Holmdel had been rigged and listeners were able to hear for themselves the hiss of the Galaxy. One reporter described it as "sounding like steam escaping from a radiator." All this brouhaha caused the staid *New Yorker* magazine for 17 June to harrumph: "It has been demonstrated that a receiving set of great delicacy in New Jersey will get a new kind of static from the Milky Way. This is believed to be the longest distance anybody ever went to look for trouble."

Follow-up investigations. Jansky was naturally eager to follow up on his discovery. He felt the next question to be asked was how the intensity of the radiation varied with frequency, and so in the second half of 1933

Figure 7. Jansky's determination of the declination of the maximum signal. The long-dashed line corresponds to the measurements and the other curves to expected behavior for various declinations; short dashes indicate that the source of emission is below the horizon. From these data Jansky at first (1933b) derived a declination of -10° ± 30°, later revised (1933c) to -20°. (c. IRE (now IEEE))

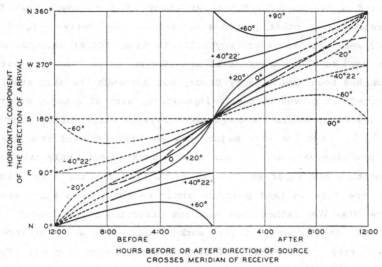

Comparison between the actual curve of the horizontal component of the direction of arrival and the theoretical curves for different declinations.

Figure 8. Karl Guthe Jansky (1905-1950), here posing by his "merry-go-round" for a publicity shot in 1933.

he tried to measure something of the spectrum. His own antenna could not be tuned over a wide range of frequencies, but he did find that there was no detectable change in intensity when the frequency was changed by ±5-10 per cent from its usual 20.5 MHz. He then got a 75 MHz ultra-shortwave receiver running to check the higher frequencies. A.C. Beck had already mounted a "fishbone" antenna (an array of 1 m length rods for 75 MHz operation) for other purposes on top of the merry-go-round, but they never were able to secure any reliable data. Other antennas on the site were used at frequencies of 11 and 19 MHz, in particular a large rhombic which had been developed by Bruce; but although the star static could be detected, it proved virtually impossible, without a major effort, to make the desired quantitative comparisons between different frequencies.

But the next major step in 1933 followed from new thinking, not new observations. In his August work report, only two months after submitting his paper to *Proc. I.R.E.*, Jansky for the first time mentions that the data in hand possibly indicate that the static comes from *the entire Milky Way* rather than just the direction of its center. By October he had this theory fully worked out and was convinced of its superiority. The pattern of bumps obtained over any day (Fig. 4) very nicely fit the concept that an entire band of emission was being detected, and one didn't have now to suppose that the galactic center was being detected even when well below the horizon. When the antenna's sweep along the horizon was also along the Milky Way, a broad bump resulted; when perpendicular, a narrow peak. Sometimes a single sweep even intersected the Milky Way twice, in which case it could be directly seen how much stronger the emission was from the general direction of the galactic center than from the anti-center.

In an effort to communicate with the astronomical community, Jansky submitted an article in September to *Popular Astronomy*, the leading popular astronomy magazine in America. The article is very similar to his *Proc. I.R.E.* paper, but it does briefly mention the above new "fascinating explanation of the data" and says that considerable new data and a detailed analysis will be required to substantiate the idea. Jansky also takes the opportunity to quote the peak intensity as -20° declination, based on a new weighting of the data.

Jansky effectively ceased working on the star static at this point (see below), and not until two years later were these new ideas published. In this short 1935 article, "A Note on the Source of

Interstellar Interference," he sets forth the interpretation outlined above and states that the signal strength seems to be proportional to the number of stars in the beam at any given time. But then he reasons: if the noise somehow is emitted by stars, why is the sun not an overpowering source? Is it "some other [unknown] class of heavenly body" which emits radio waves? Or could the emission be due to "thermal agitation of electric charge"? This latter, prescient suggestion is made because of the remarkable similarity between the audible characteristics of the star static, as monitored on a receiver headset, and the sound of static produced by thermal effects in electronic circuits. Jansky further notes that such hot charged particles are also found in the "very considerable amount of interstellar matter that is distributed throughout the Milky Way." He obviously had been learning a good bit more astronomy in an attempt to explain his results, although he still only cites a popular book of Eddington's rather than the primary literature.

Phase Four (1934-37): Practical work, with occasional star static

After the magnificent results of 1933, Jansky did little more on the star static from 1934 onwards:

> Have I told you that I now have what I think is definite proof that the waves come from the Milky Way? However, I am not working on the interstellar waves anymore. Friis has seen fit to make me work on the problems of methods of measuring noise in general -- a fundamental and interesting work, but not near as interesting as the interstellar waves, nor will it bring me near as much publicity. I am going to do a little theoretical research of my own at home on the interstellar waves, however.
>
> [KJ:CJ, 22 January 1934]

Here begins a period of several years in which the evidence of Jansky's correspondence indicates that he very much wanted to continue work in a major way on the star static, but only sporadically was allowed to make any measurements. His work continued to be focussed on shortwave and ultra-shortwave static of all kinds and the methods for measuring it, without any special attention paid to the hiss from the Milky Way. The little star noise work he was able to do was concentrated in two periods. In the first half of 1935 interest in his work was rekindled (see the letter of 19 February 1935 below) and led to the publication of his 1935 paper discussed in the previous section, and at the end of 1936 a brief period of data collection occurred (see below).

The letters speak for themselves:
Nothing new has developed in the field of interstellar
noise....Friis came out with a surprising remark the other day.
He said, "You don't want to forget about that 'hiss static,'
Jansky..." and yet he keeps me so busy doing other things that
I don't have time to think much about it. [KJ:CJ, 10 December
1934]

I have finally succeeded in stirring up considerable interest
among the men in the New York Laboratories on my work. It all
started late last fall when Mr. Buckley, the present Director of
Research of the Bell Labs, called me in to give him some
pointers on static and noise in general for a speech he was
giving in Toronto in January. Later, at a meeting of the
Colloquium (an organization of the physicists of the Bell Labs
and a few outsiders) he discussed some of the points of his
Toronto speech. At the end he attached considerable importance
to my work. In fact his whole talk was pointed towards a
discussion of the importance and implications of my data. He
concluded his speech with the statement that he thought it was
the most interesting discovery made in recent years!

Well, anyhow, a short time afterwards Friis came around and
suggested (he had heard about Buckley's talk) that I write
another paper for publication setting down my ideas on the
subject, as well as giving certain other deductions I had made
from my data. This I did, and with the consent and approval of
Friis and Bown (Friis's boss). [KJ:CJ, 19 February 1935]

I am going to attempt [to receive the radiations on a wavelength
of one meter] if the powers that be will ever give me time
enough from my other jobs. [KJ:CJ, 9 July 1935]

During the last hour of work this last week I got my
ultra-shortwave apparatus for measuring star static working and
immediately detected the static on 10 meters. I will now make a
study of it in the range of 3.5 to 12 meters.[13] Also they have
discovered that they get it on their new big antenna system with

which they are studying the direction of arrival of signals.[14] In fact it appears that this star static, as I have always contended, puts a definite limit upon the minimum strength signal that can be received from a given direction at a given time, and when a receiver is good enough to receive that minimum signal, it is a waste of money to spend any more on improving the receiver. Friis is really beginning to show a little interest! [KJ:CJ, 20 September 1935]

I have just recently finished a few experiments on the effect of various receiving bandwidths on the peak voltages of static crashes and Friis thinks I am getting data enough to make a real good paper for publication soon. Well, anyhow, it is the end of the work that has kept me away from the star noise, so maybe after I get the paper written I'll be able to go back to it. [KJ:CJ, 29 March 1936]

I am considering the possibility of taking a job at Ames [Iowa State University]....I am at present making $2825 and in spite of the advantages don't think I should accept much less than $3000....What do you know of Ames? Of course I would ask for the time and facilities to carry on my research, which would be more than I have had for the last two years here. [KJ:CJ, 18 May 1936]

At present I have one of my measuring sets set up on Friis's front yard at the edge of the river where I am making a study of the interference caused by motor boat ignition systems....I was given the job of obtaining this information along with all my other jobs. I don't mind, though; it is a very pleasant place to work and the work is interesting, but I would like to put in some time on the star noise problem. [KJ:CJ, 4 August 1936]

Finally, after about two years, I am working on star static again. The reasons why and the manner in which I got back on the job are interesting....[For all] antenna system...sites along the sea shore....I pointed out that [for] the weakest signal that can be picked up, the star noise and not the

ignition noise will be the limiting factor. This, of course,
led to a request for more information on star noise....It begins
to appear that some very definite practical importance can be
attached to my discovery after all. [KJ:CJ, 13 December 1936]

The study mentioned in the last letter led to a few weeks of
measurements of star static which Jansky incorporated into a 1937 paper.
In its abstract he states the basic theme: "On the shorter wavelengths
and in the absence of man-made interference, the usable signal strength
is generally limited by interstellar noise." He emphasizes that in all
his measurements he has never found a noise level less than 6 db above
the intrinsic receiver noise, and that most of this excess is
interstellar in origin. Also reported are some absolute measurements (in
micromicrowatts) of received power at frequencies ranging from 9 to 21
MHz, obtained both with simple half-wave dipoles and a large rhombic
antenna. But there is little directional information available and the
large and variable absorption of the D region of the ionosphere (at this
time of maximum in the 11 year solar sunspot cycle) renders most of his
data inconclusive.

This then was the muted end of Jansky's involvement with his
star static. But almost a thousand miles away and within months of the
submission of this 1937 paper, Grote Reber began construction of a 31 ft
dish in his back yard -- radio studies of the heavens were to stay alive,
and in an even less likely setting (see the following article in this
volume).

JANSKY'S LATER YEARS

Until his death in 1950 Jansky's career at Holmdel continued
on the same theme of understanding and minimizing the sources of noise in
radio communications, whether they were internal to the electronics or
external to the antenna, manmade or natural. He published only three
more papers: in the Proc. I.R.E. in 1939 on intermittent investigations
over the years of the best methods for measuring noise and in 1941 with
C.F. Edwards on shortwave echos arising from ionospheric scattering, and
a Bell Labs publication in 1948 on one aspect of a microwave repeater
system for television transmission from New York to Boston. Long after
the 1932-33 era, his merry-go-round was used as a mount for a wide
variety of antennas, including even microwave radar reflectors during
World War II. Jansky's own wartime work involved developing systems

which could locate and identify German U-boats through interception of their shortwave transmissions.

After the war Jansky of course followed with great interest the myriad discoveries in the field which came to be called radio astronomy. But his lifelong kidney ailment led over the years to very high blood pressure, in the end necessitating a diet of little more than rice and fruit. But even this was not enough and he died of a stroke in February 1950 at age 44. He never received any scientific award for being the first man to detect and study radio waves from outside the earth.[15]

WAS JANSKY STOPPED BY FRIIS FROM DOING MORE WORK ON THE STAR NOISE?

A controversy has arisen over whether Friis hindered Jansky from continuing his investigations on the star static. The evidence I have given above would seem to indicate that the answer is "Yes," but the situation deserves more discussion. The controversy began in 1956 with the publication of the first popular book on radio astronomy by a science writer. In *The Changing Universe* John Pfeiffer argued that "Rarely in the history of science has a pioneer [such as Jansky] stopped his work completely, at the very point where it was beginning to get exciting." He then painted a picure wherein Jansky wanted very much to continue, but was thwarted by his practically-minded superiors.[16] Jansky's brother C. Moreau, in a eulogistic article in 1958, also stated that his brother was "transferred to other activities" and would have preferred to continue his work in "radio astronomy." Friis replied to all this in a 1965 article in *Science:*

> In 1938, Karl dropped the study of star noise and, some 17 years later, I was criticized by people who thought that I had stopped him. This was not true. Karl was free to continue work on star noise if he had wanted to, but more than five years had passed since he had made his epochal discovery, and not a word of encouragement to continue his work had appeared from scientists or astronomers. They evidently did not understand its significance. Also, Karl would have needed a large steerable antenna to continue his work, and such antennas were unknown to us at that time. Radio astronomy, as such, did not then exist, and neither Karl nor I had the foresight to see it coming.

About 1965 the Bell Labs Publication Department conducted its own investigation into Jansky's work and the reasons why he discontinued his work on star noise (Kestenbaum 1965). Interviews with many of Jansky's former colleagues were conducted. More recently, I have myself conducted many such interviews with Jansky's relatives and former colleagues. The analysis below is based largely on both sets of interviews.

Harald T. Friis (1893–1976) was a Danish immigrant with a highly successful career in research and development in a wide variety of radio fields.[17] Over forty years at Bell Labs he acquired about thirty patents, several medals from engineering societies, and even a Danish knighthood. Besides his technical expertise, he was known for his techniques of management which contributed in large measure to the productivity of the Holmdel group. He had an aggressive personality and ran a tight ship; John R. Pierce called him the "baron of Holmdel."

On the other hand Jansky, although fiercely competitive in sports and games, did not have a forceful character when it came to dealing with conflicts in professional matters. This attitude probably was inculcated by his father, from whom he often sought and received advice:

I do not know what to make of Friis' attitude....Do not antagonize him. Keep on consulting him as formerly. He is your boss, and loyalty ultimately pays no matter whether it is deserved. [CJ:KJ, 9 May 1933]

Jansky was not one to rock the boat. His letters to his father, as well as the recollection below of his widow, give a fair picture, I think, of a typical interaction regarding the question of resuming work on the star static:

Harald says that Karl never expressed to him a desire to continue work on his star noise.[18] How incredible!... Periodically, over the years that Karl worked under Friis, he would come home and say, "Well, Friis and I had a conference today to discuss what my next project should be, and, as usual, Friis asked what I'd like to do, and, as usual, I said, 'You know, I'd like to work on my star noise,' and, as usual, Friis said, 'Yes, I know, and we must do that some day, but right now I think such-and-such is more important, don't you agree?'" [Alice Jansky:C. Moreau Jansky, 11 February 1958]

A poll of Jansky's colleagues at Holmdel and nearby reveals a large majority agreeing with Friis's viewpoints expressed above.[19] The majority testify to the practical aspects of their research in the 1930s and furthermore state that they never saw Jansky disgruntled or frustrated, which they reason certainly would have happened if he had been making repeated unsuccessful requests to continue on his star noise. But the lack of overt discontent is undoubtedly explained by Jansky's attitude of being a "team player," as discussed above. One final piece of evidence which has come to light is a letter from Lloyd Espenschied, a senior Bell Labs engineer in the 1930s, objecting to Friis's assertion that Jansky was free to continue his star noise research:

> I have a distinct remembrance of Ralph Bown [Friis's boss] having discussed with several, including myself, the question of the continuance of Karl's star-noise observations. Ralph could not see that the Bell System could ask its subscribers to pay for an investigation that had proven to be so definitely in the realm of Science. I was inclined to agree with him, but thought we were obligated to see to it that such work was continued by an appropriate [outside] agency....I had the impression that Karl's star work was called off so far as Bell was concerned; probably he was led to agree to it; too bad Karl isn't here to speak for himself.[20]

It thus seems that yes, Friis did not allow Jansky to continue his work in any major way, but also that Jansky did not press the point on those occasions when he was turned down. Given Friis's style, he may well have felt that something did not even qualify as a request unless detailed plans were forcibly advanced. Furthermore, from Friis's point of view they were all working for the Telephone Company in the middle of a Depression, and although Bell Labs was somewhat freewheeling, it was still a laboratory oriented towards a specific mission: applied research and development bearing on electrical communications. This probably influenced Jansky's stance in later years, as in his 1937 article, that the star static *was* of practical import. One of Friis's management doctrines was that the worthiness of any research job could be checked by asking: "If I owned the entire Bell System, would I pay for it?" (Friis 1971; also see Espenschied letter above). On such a criterion, investigation of the Milky Way does not stand much of a chance, especially with no urgent petitions from the astronomical community (see below). The Depression was strongly molding Jansky's attitude, too. He had chronic health problems and a young family to support and was not going to jeopardize his job -- he realized that there were few other places, if anywhere, where he could find such a first-rate, well-paying position.

And so for about four years Friis and Jansky went back and forth like partners in a strange dance, but there never was any doubt as to who was leading.

REACTION OF THE SCIENTIFIC COMMUNITY TO JANSKY'S WORK

Other contemporary investigations

Outside of Bell Labs, too, very few radio engineers or physicists looked upon Jansky's work as more than an interesting curiosity. One attempt by Jansky to stir up activity was a proposal to the 1934 U.R.S.I. General Assembly in London that the star noise be studied at a variety of stations scattered around the globe and at differing frequencies. Commission III on "Atmospherics," headed by E.V. Appleton, in fact did formally adopt such a resolution ("U.R.S.I." 1935), but it came to nought.

The only observations of any depth (before the Reber era[21]) were carried out in 1936 by Gennady W. Potapenko (1895-1979) and Donald F. Folland, a professor of physics and his student at the California Institute of Technology. Potapenko was a colorful Russian émigré who worked on a variety of problems in physics, their only common denominator being that they involved radio technology, the shorter the wavelength the better. He read Jansky's articles, decided he should try to confirm these results for himself, and got Rudolph M. Langer, a theorist in the same department, thinking about possible mechanisms of emission. Langer worked up a theory which he presented at the end of 1935 to a meeting of the American Physical Society at Berkeley, California (Langer 1936). He argued that dust grains in the interstellar medium become highly ionized by starlight, and could be considered to have quantum energy levels. When an electron recombined with such a grain, discrete spectral lines would be emitted; for substantial emission at Jansky's wavelength of ~15 m, the theory then required grains of size about one micron.

Potapenko assigned Folland the task of thinking about the extraterrestrial radiation and the best way to detect it. In February 1936 Folland presented a seminar in which he summarized Jansky's results and then worked out a rough estimate of the flux densities implied by Jansky's published data. Comparison with that expected for blackbody emission at a temperature of 10,000 K then revealed that the star noise was of much higher intensity and therefore undoubtedly not arising from thermal processes. Folland built a sensitive receiver operating at precisely Jansky's wavelength of 14.6 m, since if Langer were right, the radiation was narrow-band in nature. Potapenko and Folland's first antenna (Fig. 9) consisted of two one-meter diameter loops at a spacing of ~2.5 m on an equatorial mount. They were plagued by manmade interference at sites on the Caltech campus in Pasadena and on a nearby farm in Arcadia, but finally were successful in the spring of 1936 in a remote valley of the Mojave Desert. As they rotated the two loops about the sky, a weak peak in signal was found from the general direction of Sagittarius, a peak which moved in the expected manner as the night wore on. But after only one night of observations, a strong wind blew down the antenna and mount and they decided to switch to a less cumbersome and more portable setup. This was simply a 10 m long wire which was attached to the top of a 8 m mast and carried about in azimuth, May-pole fashion, by one man while the other took readings on the receiver (Fig. 10). With this rig, which also had the advantage of higher directivity, they got more definite results over several nights in the desert. During the summer Folland also repeated these experiments with the same setup in the mountains east of his native Salt Lake City, Utah.

These results were too rough to publish, however, and it was clear that a large antenna was needed to do the job properly. Potapenko therefore enlisted the expert help of Russell W. Porter, one of the key men then working on the design of the Mt. Palomar 200 inch telescope. Porter produced the concept shown in Figure 1 of Greenstein's article in this volume, a 180 by 90 ft rhombic antenna mounted on a rotating assembly very similar to Jansky's. Potapenko was all ready to launch into a full-scale project, but he was not able to convince Robert A. Millikan, head of the Physics Department and President of Caltech, that the required $1000 would be well spent.[22] Realizing he was stymied in his desire to look upwards, Potapenko switched his energies to a more lucrative venture -- using radio waves to probe for subterranean oil

Figure 9. The double-loop antenna with which Potapenko and Folland first detected galactic radiation in the spring of 1936. The pier, about 2 m high, provided an equatorial mount for the rotatable loops, each ~1 m in diameter. Photograph taken on the roof of the Norman Bridge Physics Laboratory at the California Institute of Technology. (Fig.'s 9 and 10 courtesy G.W. Potapenko)

Figure 10. The single-wire antenna (wire not visible), pictured here at Big Bear Lake in the San Bernardino Mountains, with which Potapenko and Folland carried out further measurements near Los Angeles. The wire was carried about the mast, May-pole fashion, while readings of the received intensity were taken inside the 1924 Chevrolet.

fields! Radio astronomy would not return to Caltech until twenty years later.[23]

There was one other effort, on the theoretical side, which paid attention to Jansky's strange hiss. Fred L. Whipple and Jesse L. Greenstein, respectively instructor and graduate student at the Harvard College Observatory and both of whom had been involved with amateur radio in the 1920s, first learned of Jansky's work in late 1936 through his article in *Popular Astronomy*. Greenstein was then in the last stages of his thesis work, under Bart Bok, involving calculations of the scattering of starlight by interstellar dust grains. Only in the previous decade had the enormous importance of the interstellar medium, both dust and gas, been brought home to astronomers and much work at Harvard was then being directed towards an understanding of interstellar astrophysics (see Greenstein's article in this volume for further discussion). I asked Greenstein why he became fascinated with Jansky's results:

> I guess it was probably because cosmic static was a rather romantic idea. [But even] for a sensible young astronomer - I was 27 - the idea of learning anything about the center of the Galaxy directly with cosmic static was exciting. My thesis work, for example, indicated that we would never see the center of our Galaxy....that nothing would come through. Interstellar absorption blinded the astronomer at optical wavelengths towards the center, and here was this challenge that the *only* thing Jansky saw *was* the center.

Whipple and Greenstein set out to explain Jansky's signal in terms of heated dust concentrated at the galactic center. They knew that in general interstellar dust had a temperature of only 3 K, so they supposed that fully one-tenth of the mass of the galactic center consisted of dust grains a good bit larger ($\sim$100 μ in diameter) than those in the solar neighborhood. In this way virtually all of the starlight in the galactic center was absorbed by grains while the grains were still large enough to radiate with some efficiency at the required wavelength of 14.6 m. The emitted intensity of the dust was then calculated using S. Chandrasekhar's new theory of spherically symmetric radiative transfer.

It was one thing to do the astrophysics, but quite another to interpret the signal levels which Jansky published. Despite seeking help from G.W. Pierce of the electrical engineering department, their first attempt at converting Jansky's (1932) value of "0.39 microvolts per meter for 1 kHz bandwidth" went far astray, essentially through forgetting the factor of 120π for the impedance of free space. Then, making very optimistic assumptions about the conditions in the galactic center, they were able to derive a temperature of ~800 K, which produced a radio intensity at the earth agreeing with their mistaken value for Jansky's intensity. Excited about this agreement, they even arranged for a press release to the local newspapers, but soon discovered the error in the conversion factor. In the end, their paper published in 1937 in the *Proc. of the Natl. Acad. of Sciences* states that the dust cannot be heated to more than 30 K, a factor ten thousand short of that needed.

Although this first detailed attempt to understand the origin of the Milky Way radiation ended in frustration, Whipple nevertheless also entertained thoughts about checking Jansky's results at other wavelengths:

I did talk Harlow Shapley, who was the Director, into giving me $50 to buy an acorn [tube] set. My thought was to put a rhombic antenna on top of the 61 inch telescope dome [at Harvard's Oak Ridge station]. Then one would have a rotating antenna that could look for sources. That was the idea...but it was just too hard to get going on it, and I didn't really make a serious attempt to bring in the electronics people at Harvard -- the physics department might have been enlisted, they might not have. [F.L. Whipple interview, 1979]

General reaction of astronomers

If Jansky's work on star noise did not excite physicists or fellow engineers, then perhaps it might have fallen to the astronomers of the 1930s to investigate this new phenomenon which opened up the electromagnetic spectrum by a factor of a million. But although at first there was some enthusiasm, it quickly waned. Jansky tried to communicate with the astronomical community through the 1933 *Popular Astronomy* article already mentioned. There was also some correspondence with leading astronomers such as Henry Norris Russell at Princeton University and Joel Stebbins of Wisconsin; Stebbins was enthused enough in one popular talk to call the discovery of the "Jansky center of the Milky Way" the greatest achievement since Lindbergh's solo flight across the Atlantic six years earlier. Harlan T. Stetson, an unconventional astronomer who was then Director of the Perkins Observatory in Ohio, gave Jansky's discovery prominent treatment in his 1934 popular book *Earth, Radio, and the Stars.* In 1933 he also suggested in a letter that his Observatory's new 69 inch reflector be outfitted with a microwave receiver to do a systematic search of the heavens!

In October 1933 Jansky shared the evening program at the American Museum of Natural History in New York City with Annie Jump Cannon of Harvard University -- he spoke on "Hearing Radio from the Stars" (once again with a direct hook-up to Holmdel for the hiss) and she on "Unravelling Stellar Secrets." A few months later he went to New York to hear a similar popular talk by Harlow Shapley, Director of the Harvard College Observatory, and took the occasion to introduce himself to Shapley after the talk. Only a few weeks before, Shapley had heard about Jansky's work in the form of a report, by one Mr. King in the Harvard physics department, which had "incited vigorous comments, pro and con."[24] Shapley asked Jansky about the cost of doing such radio work and was somewhat disheartened by the answer, which included several years of salaries and development effort. Jansky later had second thoughts and in a follow-up letter emphasized that he had overestimated the cost, outlined the minimum equipment needed to study the star static, and closed by saying:

> I would be very much interested in seeing my experiments tried by someone else and would be willing to help in any way possible.[25]

Shapley's reply was hopeful, but noncommittal:

> Possibly I can arouse interest with some of the local radio people. I may write you again about details if the local programs are not already too overcrowded.[25]

Apparently things went no further.

One other sign of interest was a letter which Jansky received circa 1935 from a radio engineer in Chicago asking if Bell Labs were planning to do any more work on the star noise. Jansky had to reply in the negative, but his correspondent, Grote Reber, did not let that dampen his enthusiasm for the subject (see the following article in this volume). Regarding the astronomers, Reber gave his opinion in interview many years later as to whether they were shortsighted:

> I wouldn't say the astronomers were shortsighted. You have to
> remember, in that day even the photoelectric tube was a
> mysterious black box; when it came to vacuum tubes and
> amplifiers, tube circuits and all the rest of it, they just
> didn't have any comprehension of these matters. And they didn't
> build radio sets -- they weren't even radio amateurs. If they
> needed a radio, they went out to a store and bought one. And
> consequently, well, from their point of view it would be foolish
> to embark on anything like this. The chances of them going
> wrong would be about a hundred to one....These kinds of
> electromagnetic waves just weren't part of their repertoire.

Astronomy in the 1930s was only just exploiting the full
power of the reflector over the traditional refractor telescope, of
photographic techniques over traditional visual observations. One of the
most central observational techniques was photographic spectroscopy of
stars, yielding a wealth of information which allowed the first confident
opinions on how the stars evolve. Electronics were not part of the
observatory and no observatory director would think of hiring a radio
engineer instead of a conventional astronomer. Besides, if Jansky could
specify the source of the radio noise no better than either "Milky Way"
or "non-Milky Way," what could an optical astronomer do with such gross
accuracy?[26]

The world of decibels and superheterodyne receivers was
simply too far removed from that of binary star orbits and
Hertzsprung-Russell diagrams. The supreme night-time quiet of the
observatory dome was the antithesis of the rumbles, clicks, crashes,
fluttering, grinds, grunts, grumbling, and *hiss* of radio communications.

WHY DID JANSKY SUCCEED?

While there can be no doubt that Jansky's discovery was
serendipitous, it was more than just lucky -- it also possessed another
aspect of any true instance of serendipity. Horace Walpole, inspired by
the ancient tale *The Three Princes of Serendip*, coined the word *serendipity*
in the eighteenth century and his original definition involved not only
an accidental felicitous happening, but one where sagacity also played a
key role. In the case of Jansky we have seen that such sagacity was
expressed in a tenacious inquisitiveness which led him to delve into an

incidental phenomenon which was of no importance for the problem at hand. His training as a physicist, rather than as an electrical engineer, undoubtedly strengthened this trait. Jansky himself commented along these lines in a letter written only five months before his death:

> As is quite obvious, the actual discovery, that is the first recording made of galactic radio noise, was purely accidental and no doubt would have been made sooner or later by others. If there is any credit due me, it is probably for a stubborn curiosity that demanded an explanation for the unknown interference and led to the long series of recordings necessary for the determination of the actual direction of arrival.[27]

There was of course also the direct contribution to the discovery from the advanced antennas and receivers available at Bell Labs. And this equipment in turn resulted from the strong staff of research engineers which A T & T had assembled to attack the technical problems of telephony.[28] The quality of the components in Jansky's set-up was therefore no matter of luck, although his particular combination fortuitously was precisely that needed to detect and recognize extraterrestrial radio waves: a steerable antenna with reasonable directivity, a stable and sensitive receiver, a recording apparatus and methodology ideally suited for the eventual discovery, and a frequency of operation which (we recognize today) yields about as strong a signal as possible from the Milky Way.[29]

The final ingredient was the pure happenstance that the 1931-33 period of Jansky's investigations occurred at the minimum phase in the 11 year cycle of solar activity. This meant two things. First, the sun itself was not emitting powerful radio bursts, bursts which at times of solar maximum are often much stronger than the Galaxy's radiation. Thus it was the Milky Way, and not the sun, which began radio astronomy. Second, and more importantly, ionospheric effects such as refraction and absorption were relatively small and stable. This allowed the extraterrestrial hiss component to be much more easily sorted out on the 1932 strip chart recordings; this point is particularly brought home by an inspection of Jansky's solar-maximum data published in 1937, where the situation is far more confused.

Jansky therefore succeeded through a mix of fortune and sagacity and the right environment. There can be no doubt that he is the father of radio astronomy -- but (shifting metaphors) only in the sense of finding and sowing the seed, not in raising the crop. Through a combination of circumstances, his discovery fell on stony ground and was not to yield fruit until the technical demands of a war created a new generation of men and equipment. If he had not made the discovery, then surely someone else would have done so within the following decade and the ensuing development of radio astronomy would not in the least have been different. But the tale of the beginning of a new era in astronomy might well have been not nearly so fascinating.

Figure 11. A modern reduction (Sullivan 1978) of Jansky's traces of 16 September 1932 (Fig. 4). The contour map is in galactic coordinates, in which 0° latitude corresponds to the plane of the Milky Way and 0° longitude is towards the galactic center. Contours are normalized to a peak value of 100, corresponding to ~100,000 K in brightness temperature at 20.5 MHz. The positions of the sun and Jupiter on that day, as well as several other radio sources discovered later, are indicated. This map is offset by about 5° because an incorrect value of 13 sec for the time constant of Jansky's receiver was used; the correct value is 30 sec.

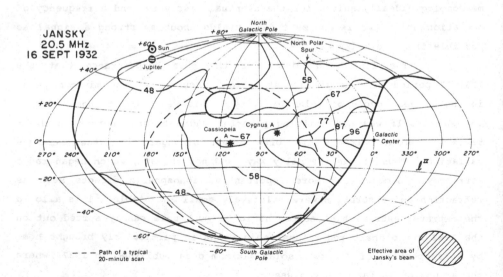

NOTES

1. Letter of 5 May 1933, Karl Jansky to his father, Cyril M. Jansky (KJ:CJ). All excerpts from letters in this article not otherwise attributed are from either the C.M. Jansky papers in the University of Wisconsin archives or from the private collections of Alice Jansky Knopp, widow of Karl, and David B. Jansky, his son. These collections together cover well the period 1928-37 except for a 22-month gap in 1928-30.

   Various other quotations in this article originate in interviews conducted by me over the past decade as research for a project on the history of radio astronomy.

2. The name honors Karl E. Guthe, a German-American physicist with whom his father trained at the University of Michigan at the turn of the century.

3. Despite his small size (5 ft 7 inches tall and 140 pounds weight), he played on the varsity ice hockey team at Wisconsin. Throughout his life he was a fierce competitor, excelling also in tennis and baseball as well as indoor games such as ping pong and bridge.

4. C.M. Jansky interview (1973). Karl was ten years younger than C. Moreau (1895-1975) and his early career closely paralleled his brother's: both studied physics at Wisconsin, took jobs at Bell Labs, and pursued careers in radio engineering. C. Moreau, however, only worked at the Labs for a couple years and then joined the electrical engineering faculty at the University of Minnesota. From 1930 onwards he headed the prominent consulting firm of Jansky & Bailey in Washington, D.C., specializing in technical and administrative aspects of radio communications and broadcasting.

5. Details of the early development of Bell Labs may be found in Fagen (1975) and Hoddeson (1980, 1981).

6. The term *T.U.* refers to a *transmission unit*, later known as a decibel. Given the state of education in electrical engineering at most universities in the 1920s, it is not at all clear that Jansky would have been better served by such a degree. Most electrical engineering training then concentrated on power engineering, in essence not worrying about any frequencies higher than 60 Hz. In fact a good argument can be made that an important component in Jansky's discovery was his training as a physicist rather than as an electrical engineer.

7. Information given by Jansky in a 1937 paper enables an estimate of 5 db to be made for the noise figure of the receiver, corresponding to a system temperature of ~1000 K.

8. Most of Jansky's laboratory notebooks and monthly work reports are available at the Bell Telephone Laboratories Archives in Murray Hill, New Jersey. Most quotations in this

chapter are from the original notebooks and work reports, but some are unfortunately only available secondarily through a 1956 article about Jansky's work by G.C. Southworth. In particular it is distressing that laboratory notebook No. 10136, entitled "Astronomy," which apparently covers the critical 1928–1937 period, is now missing from the Archives. Photocopies of ~40 of its pages were made in the 1960s, however, and these survive. They indicate that during the critical 1932–33 period Jansky did not keep a detailed notebook concerning data analysis or scientific reasoning, but rather used it for more mundane purposes such as calibrations and circuit details.

9. The antenna was designed for 14.5 m wavelength, but Jansky later found that in fact 14.6 m was the correct wavelength of the vacant band which he had found.

10. *Earth currents* are currents measured between two grounded terminals, separated by distances anywhere from 100 m to 100 km.

11. Skellett recalls that he made the initial suggestion regarding sidereal time (interview, 1977). In his formal publication, Jansky (1933b) acknowledges Skellett's help "in making some of the astronomical interpretations of the data."

12. Given the great degree of expertise in antennas at Holmdel, I do not understand why Jansky apparently never appreciated that the modest vertical directivity of his Bruce array could profitably be used to locate the source of emission. Perhaps it was the communications engineer's overriding concern with the *azimuthal* direction of arrival of long distance transmissions. In any case a recent analysis (Sullivan 1978) of the characteristics of Jansky's antenna indicates that its effective half-power beamwidth was 24° in azimuth (as he measured) and 36° in elevation, with maximum gain at an elevation of 24°. When this information is applied to the one day's worth of Jansky's data extant (Fig. 4), an interesting all-sky contour map emerges (Fig. 11, at end of article).

13. No record of such a study has been found.

14. The "big antenna system" was the Multiple Unit Steerable Antenna (MUSA), a 3/4 mile long array of six rhombics operating over a range of 5 to 25 MHz. MUSA was able to change its elevation angle of maximum response (through quickly and automatically adjustable relative phasing of its elements) and thus follow signals varying in arrival angle as a result of ionospheric fluctuations. In 1937 Friis and C.B. Feldman published a detailed description of this system, including even a few individual measurements in the autumn of 1935 of star static on 10 and 19 MHz.

15. In 1947 Jansky was made a Fellow of the I.R.E., a modest award. But contrast this with the case of Southworth, who in

1942 first detected microwaves from the sun, also at Holmdel, and who in 1946 received for this work the Louis Levy Medal of the Franklin Institute.

16. Pfeiffer asserts, among other things, that Jansky at one stage proposed to build a 100 ft diameter dish, but was turned down on grounds of cost. His source for this was Grote Reber, who recalls Jansky telling him this sometime before 1940 (letters, Reber:WTS3, 18 July 1983 and Reber:Pfeiffer, 31 January 1955 (Ohio State University Archives)).

17. Friis tells his own story in his autobiography *Seventy-five Years in an Exciting World* (1971).

18. Alice Jansky is here reacting to the following statement of Friis's made in a 23 January 1958 letter to C. Moreau Jansky: "Karl never expressed to me or any of his associates a desire to continue his work on star noise. He felt that astronomers should follow up his discovery and it disappointed him that they did not."

19. These colleagues included A.C. Beck, R. Bown, A.B. Crawford, D.H. Ring, J.C. Schelleng, and A.M. Skellett. Viewpoints on the other side were expressed by C.B. Feldman and G.C. Southworth.

20. L. Espenschied:H. Friis, 24 August 1965 (Friis papers, Library of Congress).

21. Reber constructed his antenna in 1937 and began observations in 1938. One other early, unsuccessful experiment was carried out in May 1935 by John H. DeWitt, a radio engineer working in Nashville, Tennessee. Using a hand-held yagi with reflector on a few nights, DeWitt was not able to detect any 300 MHz signals from the general direction of the Milky Way. He finally succeeded, however, in 1940 at 111 MHz (but never published any results) and later won fame for heading the U.S. Army team which in 1946 first bounced radar off the moon.

22. Fritz Zwicky (1969) wrote that he also was part of this group seeking funds for an antenna; he remembered the requested amount as $200, while Potapenko recalled $1000 (interview, 1975).

23. Potapenko and Folland were observing at a time of maximum in solar activity; their measurements may well thus have been strongly influenced by ionospheric variations.

24. H. Shapley:KJ, 16 January 1934 (Shapley papers, Harvard University).

25. KJ:Shapley, 13 March 1934; Shapley:KJ, 23 March 1934 (Shapley papers).

26. Perhaps if Jansky had indeed drawn the astronomers more of a

visual representation of his data, such as a contour map of
the sky as discussed in Note 12 and illustrated in Figure 11,
he might have elicited more interest.

27. KJ:E.V. Appleton, 28 September 1949. This was a letter of
     appreciation to Appleton for remarks made in his Presidential
     address to the 1948 U.R.S.I. General Assembly in Stockholm.
     Appleton devoted most of his talk to Jansky's discovery and
     the post-war rise of radio astronomy, and stated that
     "Jansky's work has all the characteristics of a fundamental
     discovery."

28. Note also the remarkable circumstance that A.M. Skellett, who
     conducted experiments involving the detection of the echoes
     of radio pulses off ionized meteor trails in 1931-32 at Bell
     Labs, did in effect the first *radar* astronomy in precisely the
     same milieu where Jansky began radio astronomy.

29. A study of atmospherics on 5 to 20 MHz published by
     R.K. Potter in 1931 came close to having these essential
     ingredients. The difference was that Potter employed
     antennas of little directivity, as well as a method whereby
     he only recorded the *peak* values of the "crash" atmospherics,
     thus largely ignoring any low-level signals. Nevertheless,
     Potter does make reference in his paper to "exceptional cases
     of atmospheric noise of a hissing character" and to
     occurrences of a "brief hiss" at times; whether these were
     extraterrestrial in origin is problematic. Jansky's
     colleague A.C. Beck (1983) further recalls that it was common
     knowledge at Holmdel that an excess was often measured when
     comparing the noise level from an antenna with that from a
     matched resistive load. In retrospect it seems certain that
     much of this was galactic radiation, but this could only have
     been established with the aid of a steerable antenna and
     long-term study.
          Grote Reber has also related another "near miss," told
     him by Gordon H. Stagner. In 1928 Stagner was an operator at
     an R.C.A. trans-Pacific shortwave communications station near
     Manila, The Phillipines. Stagner spent a good deal of time
     measuring total circuit noises on a variety of antennas at
     frequencies of 5 to 25 MHz. At one stage he noticed that the
     background hiss seemed to be varying with time of day and
     began an investigation, but was soon told to confine himself
     to his assigned duties.

30. I am deeply grateful to all of the people who over the years
     have shared with me their insights regarding the early years
     of radio astronomy. I especially thank the interviewees
     quoted herein and the members of the Jansky family, who have
     been extremely helpful. I thank the many archivists and
     librarians without whose diligence and assistance such
     research would be impossible. Finally, I acknowledge support
     from the History and Philosophy Section of the National
     Science Foundation for portions of this work.

REFERENCES

Beck, A.C. (1983). Personal recollections of Karl Jansky's work. *In* Serendipitous Discoveries in Radio Astronomy (ed. K. Kellermann). Green Bank: National Radio Astronomy Observatory.

Edwards, C.F. & Jansky, K.G. (1941). Measurements of the delay and direction of arrival of echoes from near-by short-wave transmitters. Proc. Inst. Radio Eng., 29, 322-9.

Fagen, M.D. (ed.) (1975). A History of Engineering and Science in the Bell System: The Early Years (1875-1925). Murray Hill, N.J.: Bell Telephone Laboratories.

Friis, H.T. & Feldman, C.B. (1937). A multiple unit steerable antenna for short-wave reception. Proc. Inst. Radio Eng., 25, 841-917.

Friis, H.T. (1965). Karl Jansky: his career at Bell Telephone Laboratories. Science, 149, 841-2.

Friis, H.T. (1971). Seventy-five Years in an Exciting World. San Francisco: San Francisco Press.

Hoddeson, L. (1980). The entry of the quantum theory of solids into the Bell Telephone Laboratories, 1925-40; a case-study of the industrial application of fundamental science. Minerva, 18, 422-47.

Hoddeson, L. (1981). The emergence of basic research in the Bell Telephone System, 1875-1915. Tech. and Cult., 22, 512-44.

Jansky, K.G. (1932). Directional studies of atmospherics at high frequencies. Proc. Inst. Radio Eng., 20, 1920-32.

Jansky, K.G. (1933a). Radio waves from outside the solar system. Nature, 132, 66.

Jansky, K.G. (1933b). Electrical disturbances apparently of extraterrestrial origin. Proc. Inst. Radio Eng., 21, 1387-98.

Jansky, K.G. (1933c). Electrical phenomena that apparently are of interstellar origin. Pop. Astron., 41, 548-55.

Jansky, K.G. (1935). A note on the source of interstellar interference. Proc. Inst. Radio Eng., 23, 1158-63.

Jansky, K.G. (1937). Minimum noise levels obtained on short-wave radio receiving systems. Proc. Inst. Radio Eng., 25, 1517-30 (erratum: 26, 400).

Jansky, K.G. (1939). An experimental investigation of the characteristics of certain types of noise. Proc. Inst. Radio Eng., 27, 763-8.

Jansky, K.G. (1948). I.F. amplifier [part of a series edited by H.T. Friis on "Microwave repeater research"]. Bell Tel. Sys. Tech. Pubs., Monograph B-1565, 44-9.

Jansky, C.M. (1958). The discovery and identification by Karl Guthe Jansky of electromagnetic radiation of extraterrestrial origin in the radio spectrum. Proc. Inst. Radio Eng., 46, 13-15.

Kestenbaum, R. (1965). Karl Jansky and radio astronomy. Murray Hill, N.J.: Bell Telephone Labs (unpublished report).

Langer, R.M. (1936). Radio noises from the galaxy. Phys. Rev., 49, 209-10.

Pfeiffer, J. (1956). The Changing Universe: The Story of the New Astronomy. New York: Random House.

Potter, R.K. (1931). High-frequency atmospheric noise. Proc. Inst. Radio Eng., 19, 1731-65.

Southworth, G.C. (1956).   Early history of radio astronomy.   Sci.
        Monthly, 82, 55-66.
Stetson, H.T. (1934).   Earth, Radio and the Stars.   New York: McGraw
        Hill.
Sullivan, W.T., III (1978).   A new look at Karl Jansky's original data.
        Sky and Tel., 56, 101-5.
"U.R.S.I." (1935).   Resolution No. 7 of Sub-commission 1 of Commission
        III. Proc. U.R.S.I. General Assembly, 4, 116.
Whipple, F.L. & Greenstein, J.L. (1937).   On the origin of interstellar
        radio distrubances. Proc. Nat. Acad. Sci., 23, 177-81.
Zwicky, F. (1969).   Discovery, Invention, Research (pp.90-1).   New York:
        Macmillan.

A 7.5 meter diameter Würzburg reflector used for observations near Paris circa 1950. These wartime German dishes were of great use to several early radio astronomy groups. In the early 1940s they were the closest rivals to the size and accuracy of Reber's dish.

EARLY RADIO ASTRONOMY AT WHEATON, ILLINOIS[†]

G. Reber
Bothwell, Tasmania, Australia

My interest in radio astronomy began after reading the original articles by Karl Jansky (1932, 1933). For some years previous I had been an ardent radio amateur and considerable of a DX [Distance Communication] addict, holding the call sign W9GFZ. After contacting over sixty countries and making WAC ["worked all continents"], there did not appear to be any more worlds to conquer.

It is interesting to see how the mystifying peculiarities of short-wave communications of 1930 gradually have been resolved into an orderly whole. The solar activity minimum of the early thirties must have brought with it abnormally low critical frequencies. Many a winter night was spent fishing for DX at 7 Mc/s when nothing could be heard between midnight and dawn. It is now clear that the MUF [maximum usable frequency] over all of North America was well below 7 Mc/s for several hours. An hour after sunset when the west coast stations disappeared 14 Mc/s went dead. These years would have been a very fine time for low-frequency radio astronomy. The now appreciated long quiesence of the sun during the latter half of the seventeenth century (Schove 1955) would have been even better!

One further recollection is that on these quiet nights it was always possible to make the receiver quieter by taking off the antenna. This receiver uses a regenerative detector and one RF [radio frequency] stage. The detector and RF stage are tuned separately so that the latter may be gradually tuned across the former. When this was done with the antenna off, no appreciable change could be heard in the sound of rushing

---

water. When the same thing was done with the antenna on, the rushing
sound increased several times in loudness at the resonant frequency of the
RF stage. Whether or not this difference was due to cosmic static or
merely to some vagary of the receiver is not readily resolved at present.
Near the next solar activity minimum it might be interesting to try the
old receiver out again with a similar antenna, which was an inverted "L"
40 feet high and 200 feet long. The above experiences were at 225 West
Wesley Street, about 70 yards from 212 West Seminary Avenue where the
rest of the observations were conducted. All this property now belongs
to the Illinois Bell Telephone Company.

ORGANIZED EXPERIMENTS

In my estimation it was obvious that Jansky had made a fundamental and very important discovery. Furthermore, he had exploited it to
the limit of his equipment facilities. If greater progress were to be
made, it would be necessary to construct new and different equipment
especially designed to measure the cosmic static. Two fundamental problems presented themselves. These were, are, and will continue to be: How
does the cosmic static, at any given frequency, change in intensity with
position in the sky; and how does cosmic static, at any given position in
the sky, change in intensity with frequency? To solve these problems
would require an antenna which could be tuned over a wide frequency range,
provide a narrow acceptance pattern in mutually perpendicular planes, and
be capable of pointing to all visible places in the sky.

About this time I had completed a modicum of academic learning.
Two features stood out as pertinent. First, geometrical optics demonstrates
that the angular resolving power of a device is proportional to its aperture in wavelengths. Consequently, for a given physical size aperture,
the angular resolving power is proportional to frequency. Obviously a
high frequency was indicated. Secondly, Planck's black body radiation law
shows that for radio frequencies at any probable temperature, the intensity per unit frequency bandwidth of radiant energy is proportional to the
square of the frequency. Again, a high frequency was indicated. Thus, on
the face of the prevailing ignorance, a very high frequency should be used
since much better resolution would be secured and very much more energy
would be available for measurement. Some frequency in the decimeter
region was indicated.

RADIO TELESCOPE

Consideration of the antenna problem showed that any type of wire network would be exceedingly complicated since several hundred minute dipoles would be needed. The only feasible antenna would be a parabolic reflector or mirror. By changing the simple focal device it would be possible to tune the mirror over a very wide frequency range. About this time Barrow (1936) published some experiments on circular waveguides. The patterns of these apertures seemed just the thing for looking into a mirror. Thus I conceived the idea of using a single dipole inside a short length of waveguide at the focus of the mirror. It turned out to be a very satisfactory, simple arrangement providing good shielding against radiation other than from the desired direction. The focal point of the mirror was placed in the aperture plane of the waveguide.

If optical practice were to be followed, the mirror should be on an equatorial mounting. This would have been very complicated and exceedingly expensive. Even an alt-azimuth mounting would be prohibitive in cost and could not have been used profitably in the confined location available. Therefore, a meridian transit mount was decided upon. The mirror was to be as large as possible consistent with available funds. One inquiry was made for a purchased framework of steel. The American Bridge Company offered a design which consisted of a circular billboard 50 feet in diameter and 10 feet thick with a horizontal axis through the center mounted upon two vertical pillars 30 feet high. The billboard was to have a four leg parapet 35 feet high on one side with counterweights and locking mechanism on the other side. Since the mirror skin and supports plus turning motors were extras, I adjudged their offering price of 7000 dollars, erected, to be excessive! This was in 1936. Not many years later I wished I had been more of a speculator. In any case, I decided to design the mirror and do the job myself.

The mounting had been decided upon. Only one more problem remained, distortion of the surface by bending. This could be reduced greatly by using a deep framework. Various materials were considered. From an academic point of view, the best material for preventing bending is the one with the highest ratio of modulus of elasticity to density and not the highest ratio of strength to weight. Of the common materials, steel and aluminum are about equal. However, although much inferior, wood was decided upon because of cheapness and ease of working. Even with wood, the bending on a mirror 30 feet in diameter could be made negligible by

using a deep framework. Most of the wooden members were 2 inches x 4 inches cross section and from 6 to 20 feet long. All joints were fastened by
steel gusset plates of various widths and 1/8 inch thick. The bolts were
1/2 inch in diameter. All pieces were given two coats of paint before
assembly and two coats afterwards. The framework was given added coats
every couple of years. When disassembled in 1947, all pieces were found to
be dry, solid, and as good as when put together. With reasonable care, it
could have lasted indefinitely. The skin was 26-gauge galvanized iron in
45 pieces of pie; nine on the inside and 36 on the outside. These were
supported on 72 radial wooden rafters cut to a parabolic curve. The sheet
metal was not pre-formed but was allowed to drape snugly over the rafters
and was fastened with a flat head brass wood screw at every lineal foot.
The joints were overlapped about 2 inches and fastened with bolts every
foot of lineal joint. All joints were spaced half way between rafters.
The over-all diameter of skin was 31 feet 5 inches. Since the focal length
was 20 feet, a relatively flat mirror was required. This, in conjunction
with the multiple joint skin, allowed a good smooth surface to be secured.
The major roughness was due to the bolt heads. The exact accuracy was unknown. However, the mirror platform was flat to plus or minus an eighth
of an inch in all positions as found by measurement, so the skin had a
commensurate parabolic accuracy. All the wooden pieces, including the
lattice parapet, were cut, drilled, and painted by me personally. Part time
assistance of two men was secured on the foundations, metal parts, and
erecting the structure, with the exception of the skin which I personally
put together piece by piece. The entire job was completed in four months;
from June to September, 1937 [Figures 1, 2, and 3]. The over-all weight
was less than two tons. This mirror usually emitted snapping, popping, and
banging sounds every morning and evening. The rising and setting sun caused
unequal expansion in the skin and the various pieces would slip over one
another until equilibrium was attained, no matter how tightly the joining
bolts were pulled up. When parked in a vertical position, great volumes
of water poured through the center hole during a rain storm. This caused
rumors among the local inhabitants that the machine was for collecting
water and for controlling the weather. The center hole, 2 feet in diameter,
was included just in case it would be desirable to use the mirror for centimeter waves with an elliptical secondary mirror at prime focus. A similar opening was built into the carriage, so that a very long focal length
could be secured if the receiver were placed below the bottom of the

Figure 1. The author at the time of these experiments (ca. 1937).

Figure 2. The 31 foot paraboloidal reflector with which all observations in Wheaton, Illinois were conducted. This photograph (ca. 1939) shows the 187 cm (160 Mc/s) resonating-cavity drum at the focus.

carriage. This Gregorian scheme was never used. The lower parts of the
machine provided an enticing structure for all the children of the neighborhood to climb upon. However, they were prevented from getting on top
by the overhanging nature of the skin.

When working at the focal point, the mirror was tipped far
south to an elevation of about 10°. The structure had a peculiar attrac-
tion for private flyers who would examine it from many directions and distances in their putt-putt airplanes. More than once, when one of these
flyers would approach down the beam from the south, I had the sensation
that a motorcycle was coming up out of the ground right through the center
back of the mirror. Obviously, the mirror also had good acoustical properties.

This radio telescope was acquired by the National Bureau of
Standards in 1947 and was erected on a turntable at their field station
near Sterling, Va. About 1952, it was disassembled and the parts sent to
Boulder, Colorado. Recently, the National Bureau of Standards (NBS) has
made the antenna available to the National Radio Astronomy Observatory on
an indefinite loan basis for exhibition and demonstration purposes. [The
antenna is in fact still in existence at NRAO, Green Bank, West Virginia.-
Ed.]

ELECTRONIC APPARATUS AND TESTS AT 9 CM

The shortest possible wavelength in the middle 1930s was on
the order of 10 cm. An RCA type 103A end plate magnetron was acquired for
general testing. This tube could be operated from 2500 to 5000 Mc/s, depending upon the electrode voltages and magnetic field applied. The frequency limits really were due to the resonance of the internal anode
structure which had a natural period of about 3300 Mc/s. At optimum, nearly 1/2 watt could be put into a small lamp bulb. Audio modulation was
applied in the end plate circuit.

Quite a variety of crystal detectors were experimented with.
All could be made to work, more or less. However, a clear amber piece of
zinc sulfide, known as sphalerite, was by far the best since it was sensitive all over, no matter where the cat whisker was placed. The rest of
the receiver consisted of a four-stage audio amplifier using 6F5 triodes
giving a gain of the order of 100 db. It was peculiarly free from micro-
phonics and was very reliable. Using this equipment, a variety of experiments were conducted on cavity resonators, etc. (Reber 1938, 1939)

Figure 3. The turning mechanism for changing declination on
the antenna, constructed around a differential gear for a Ford
Model A truck.

Figure 4. Electronic apparatus for 9 cm operation (ca. 1936).
Various style horns are being tested using a magnetron as signal generator (left) and a crystal detector in the cylinder on
the right.

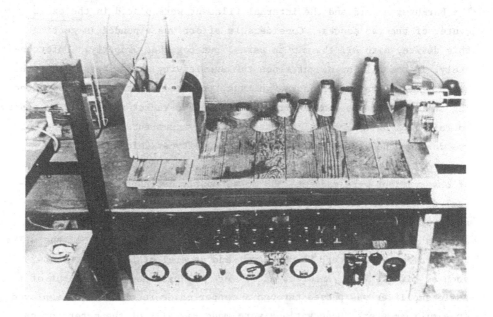

[Figure 4]. For close work, a 0-200 microammeter was used directly in the crystal circuit. The magnetron was tried as a detector in place of the crystal, but was found to be worthless because of very great shot noise voltage. Two other attempts were made to improve upon the crystal detector.

One was a special small diode entirely made of tungsten and pyrex. The spacing from anode to cathode was carefully adjusted to about 0.005 inch and the physical structure so arranged that the anode became part of a tunable line or cavity. Energy was fed into a cavity by a small dipole and hairpin loop. In spite of all efforts, this device was markedly poorer than the crystal. The significance of electron transit time was only beginning to be appreciated.

Next, an elaborate Barkhausen tube (Hollmann 1934) was constructed on the theory that the virtual cathode beyond the grid could be made to come as close as desired to the anode and thus cut down on the electron transit time loss. The tunable system consisted of lecher wires three half-waves long. Sliding shorting-bars at each end provided outside adjustment. The tube seals were at the voltage nodes at third points and thus reduced capacity effects. A glass bellows allowed for unequal expansion between the glass envelope and the tungsten lecher bars. The anodes were small semicylinders at the voltage peak in the center of the system. The Barkhausen grid and the internal filament were placed in the exact center of the two anodes. Considerable effort was expended in getting this device, with all the minute parts, put together properly. Unfortunately, it was all for naught since the sensitivity was no better than the diode. Considerable shot-noise voltage, produced by stray electrons finding their way to the anodes, created a terrific racket and it was necessary to fall back on the crystal detector. This vacuum tube construction work was done by the glass experts at the University of Chicago.

During the spring and summer of 1938, a considerable number of observations were made at 3300 Mc/s, mostly during the day. The crystal detector in its cavity, a short length of waveguide horn, and the audio amplifier were mounted just behind the mirror focal point. The antenna was parallel to the celestial equator. Various parts of the Milky Way, sun, moon, Jupiter, Venus, Mars, and several of the bright stars, such as Sirius, Vega, Antares, etc., were all examined. The output of the audio amplifier was passed through a copper-oxide rectifier and displayed on a microammeter. Observations were made visually of the meter indication and were tabulated, sometimes at minute intervals, sometimes at longer intervals, such as an hour. Some small irregular fluctuations were

encountered, but no repeatable results were secured which might be con-
strued to be of celestial origin. All this was rather dampening to the
enthusiasm. Admittedly, the sensitivity of the system was quite poor.
However, the frequency was 160 times as great as Jansky used and the pre-
sumed black body radiation intensity should be 26,000 times as great. If
anything could be deduced from these efforts, it seemed to be that the relation between celestial radiation intensity and frequency did not conform
to Planck's law.

ELECTRONIC APPARATUS AND TESTS AT 33 CM

Consideration showed that it would be best to lower frequency
a bit and build more conventional electronic apparatus using triode tubes
with the aim of greatly increasing the sensitivity. A couple of years
prior to this, RCA had brought out the type 955 acorn triode. These commercial tubes were of different internal construction and of considerably
inferior performance to the experimental tubes described by Thompson (1933)
and Rose. The early 955's also were better than the later versions. These
early type tubes had a smaller diameter cylinder and a hemispherical top,
compared to the flat top of the later tubes. A pair of tubes were connected in a push-pull arrangement with the grid pins soldered together, since
this circuit determined the highest operating frequency. The plate and
cathode leads were tuned by lecher systems. The former determined the
frequency and the latter the strength of oscillation. A pair of the early
tubes could be made to oscillate up to nearly 1000 Mc/s. In 1938, these
early tubes were still abundant and several oscillators were made for testing receivers, etc. Only about fifty volts were required on the anodes.

About this time, RCA brought out the type 953 acorn diode with
the anode lead coming out of the top. An adjustable cylindrical resonator
was constructed which could be used with either the 953 or the above described tungsten diode. These diodes and crystals were all tested for
sensitivity and were found to be about equal at 910 Mc/s and some two
orders of magnitude poorer than the regenerative detector. Also, it was
now observed that the tungsten diode had a faint rattle sound associated
with it which increased in strength as the filament temperature was increased. This rattle was not present in the 953 with its low-temperature
cathode. Apparently, the rattle was due to minute bits of tungsten boil-
ing off the white hot filament. This is typical of the kind of thing
which may be overlooked when no comparative observations are possible.

A new cavity resonator with an iris was constructed, for use
at the focal point of the mirror, out of a steel drum container for 100
pounds of white lead. The dimensions of this drum determined the operating
frequency of 910 Mc/s. A half-wave dipole was placed a quarter-wave length
from the back of the cavity. A half-wave lecher wire went from the center
of the dipole through a hole in the rear of the cavity and was coupled by
a small loop to the filament tuner of the push-pull detector. Thus, the
entire antenna system resonated in a three half-wave mode. The detector
and audio amplifier were placed in a small drum attached to the rear of the
cavity. The entire arrangement was supported in circular bands, so the
drum could be turned on its axis to change the plane of polarization received, since it was thought that there might be some variation with the
plane of polarization of the still-to-be-found celestial energy.

During the autumn of 1938 and during the following winter, a
variety of observations, both by day and by night and with various polari-
zations, were made at 910 Mc/s [Figure 5].

Figure 5. Control box for the receiver (ca. 1938), with 33 cm
(900 Mc/s) oscillator on top. Hanging on the wall is the power
supply.

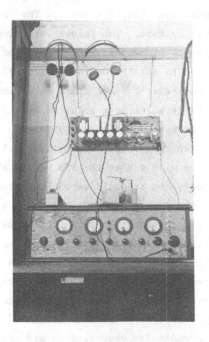

All the same objects were examined again without any positive results. In a measure, it was disappointing. However, since I am a rather stubborn Dutchman, this had the effect of whetting my appetite for more. Here was a circumstance where the frequency had dropped nearly two octaves and the sensitivity had improved by two orders of magnitude and still nothing resulted. Perhaps the actual relation between intensity of the celestial radiation and frequency was opposite from Planck's law.

All this old equipment is still in existence and it might be interesting to set it up again and measure just what absolute sensitivity was achieved, both at 3300 and at 910 Mc/s.

ELECTRONIC APPARATUS AND TESTS AT 187 CM

By then, autumn 1938, it was perfectly clear that a further great increase in sensitivity was necessary and that attempting to operate at exceedingly high frequencies was wrong. The resolution would have to be whatever it came out, for better or for worse. Concurrent with these experiences, I had been following in the literature various articles on the input resistance of triodes by Ferris, wide-band amplifiers by Percival, and trying to understand a rather deep book about random fluctuations by Moullin. Also I had been gaining experience in the radio receiver industry. It was clear that a big jump in sensitivity could be made by changing from a crystal to a regenerative detector. Another big jump would be to a superheterodyne receiver. Also, a "superhet" with an RF stage was much better and two RF stages still better. The tunable feature of a superhet seemed of little value. What were important were the RF stages. Also, it seemed that a wide bandwidth should be used, which again ruled against a superhet. Finally, I decided upon a multistage tuned radio-frequency amplifier with as wide a bandwidth and as high a frequency as feasible.

RCA brought out the 954 acorn pentode in 1935; the first ones were exceedingly poor, having many internal shorts. However, by purchasing new tubes from the autumn 1938 production, instead of dealer stock, some respectable samples were secured. About this time a small NBS bulletin (Dunmore 1936) was issued which described a multistage amplifier using 954 tubes with coaxial line resonators. It was tunable from about 100 to 300 Mc/s and over-all gain data were included. This bulletin was of much assistance in determining a suitable design. The wide range tunable feature seemed of no particular value and tended to introduce feedback.

Also, the gain dropped rapidly above 200 Mc/s. A frequency somewhere near 150 Mc/s appeared suitable since it already was clear that a low internal random-fluctuation voltage only could be secured if the first-stage gain was fairly high, such as eight or more.

The idea of a cavity resonator at the focus of a mirror still seemed good compared to an open dipole, and I decided to continue using a cavity. This relatively low frequency would require a large but light cavity, which should be cheap and with as little welding as possible. Inquiry disclosed that aluminum sheet 1/16 inch thick could be secured in pieces 6 feet wide and 12 feet long. One piece could be formed into a cylinder about 4 feet in diameter and 6 feet long. The Alcoa Company fabricated this resonator, including various size irises for the aperture, at a very nominal price. In essence, the operating frequency of 162 Mc/s was determined by the size of a drum which could easily be made from a standard size sheet of aluminum.

The NBS design was revised to provide capacity tuning over a range of 150 to 170 Mc/s (Reber & Conklin 1938), and the whole assembly of copper water pipe and copper plate was brazed into one piece by the local blacksmith [Figure 6]. After fitting the by-pass capacitors and connections, an over-all trial was made and the entire five stages worked immediately. Furthermore, the well-known (by then) thermal-fluctuation voltage easily could be found when the first tuned circuit was tuned through resonance; this was very encouraging. No signal generator was available to measure the sensitivity, but obviously it was good. This amplifier was then attached to the back of the large aluminum drum. The antenna was quite similar to that used at 910 Mc/s but with additional fine-tuning adjustments.

Before hoisting this cumbersome assembly atop the parapet of the mirror, some ground tests were made. These provided quite a shock, since all kinds of man-made electrical disturbances now could be heard which before were not known to exist. The main one was caused by automobile ignition sparking. However, this trouble mostly disappeared after 10:00 P.M.; this was reassuring. Thus on the first Saturday when help could be obtained, the whole assembly was placed atop the mirror and trials were started. This was in the early spring of 1939.

During the day, no worthwhile results could be secured because of the multitude of automobiles in continual operation. This disturbance leaked into the drum from the back around the edge of the mirror.

Cars at the front side and in front of the mirror could hardly be detected. Thus, the shielding action of the drum really was effective. The output of the receiver came down a coaxial microphone cable as a small dc voltage. About three-quarters of this voltage was bucked out by a battery and the remainder was fed into a dc amplifier. The display was a microammeter. About two hours were required for equipment warmup. After 10:00 P.M., disturbances quieted down and observations were made in earnest. Data were taken by manually recording the meter indication every minute. Continual aural monitoring was employed to delete those times when interference was present. These data were then plotted as meter reading versus time.

During the night, good smooth flat reproducible plots were secured, but nothing could be found which moved along in sidereal time. In early March, the plane of the galaxy to the south still crossed the meridian after sunrise. While the auto disturbances did not become really bad until about 10:00 A.M., the amplifier was subject to gain variations when the sun came up. This was found to be due to unequal thermal expansions which caused some of the resonators on the interstage couplers to become out of tune compared to the others. On cloudy mornings the trouble was absent and some plots were secured which seemed to show excess energy when the plane of the galaxy crossed the meridian. These occasions were few since the cloudy morning had to appear on a weekend when I was not at work in the city. By early April, the plane of the galaxy was crossing the meridian during hours of darkness and good reproducible plots were secured every night when observations were made. It was now apparent that cosmic static from the Milky Way had really been found and that it was of substantial strength, especially to the south [Figure 7]. As the months went by, the more northerly parts of the galaxy became available. However, the cosmic static became weaker and dropped nearly to the limit of the system sensitivity at a declination of 20° north. These results confirmed Jansky in a general way.

During the summer of 1939, a variety of celestial objects were examined but nothing convincing could be found, except from the Milky Way. Particular effort was made to find the sun but it was lost under compound equipment difficulties. Only thickly clouded days were possible for observation since thermal effects were induced by scattered clouds. Also, it developed that the 953 diode had another internal trouble. The velocity potential of the diode was in series with the signal voltage. This velocity potential is dependent upon cathode temperature and effective spacing

Figure 6. The five-stage 160 Mc/s tuned radio-frequency amplifier (right) and control box with which the "cosmic static"
from the Milky Way was first detected in the spring of 1939.

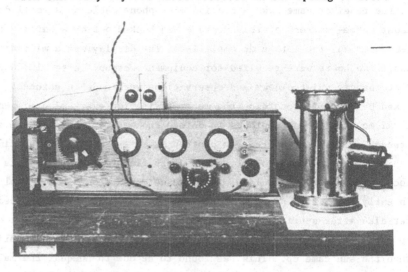

Figure 7. Ammeter readings, taken at one minute intervals on
11 May 1939, as the Milky Way passed through the 12 degree
beam of the antenna while set at a declination of -10°. Right
ascensions are labelled at hourly intervals and P marks the
time when the galactic plane was centered in the beam (at a
new galactic longitude of 10°). The 160 Mc/s received power
increases downwards and is proportional to the square of the
readings. The apparent rise in reading over the first 30 minutes is due to the equipment warming up.

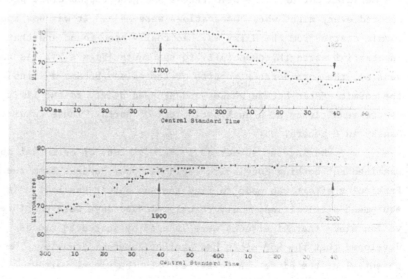

between anode and cathode. During quiet periods, the velocity potential
was stable. However, when a strong auto ignition disturbance was imposed,
there was a small internal rearrangement of active areas on the cathode.
Thus, the diode velocity potential would be different after, compared to
before, a strong burst of ignition sparking. Consequently, a shift in the
dc zero level occurred irregularly up and down after each objectionable
vehicle went by. Several 953's and 955's were tried but to no avail. A
200,000 ohm diode load resistor was used at the time. The solution seemed
to be to move everything out to the country, which then was not feasible.
These preliminary results were then published (Reber 1940a, 1940b). The
intensity was guessed at by noting the effect of resistance shunts across
the first tuned circuit; no calibrated signal generator was available then.
In any case, the observed intensity was far below that encountered by
Jansky. Obviously, the source of cosmic static was some new and unknown
phenomenon.

The above success further whetted my appetite on the basis of,
"If a little is good, more is better." A survey of the sky was contemplated; this would mean collecting a lot of data. Obviously, an automatic
recorder was a primary necessity. Thus, a General Radio dc amplifier and
Esterline Angus 5 ma recorder were purchased early in 1940. Also, new
power supplies with entire ac operation and automatic regulation were built
to replace the old dc system with its manual voltage adjustments. The new
steady power supply had a great effect on receiver stability and resulted
in effective long-term sensitivity. A few all night vigils were made with
this automatic apparatus to gain confidence in its operation and to gain
experience in how interference manifested itself.

In addition, a decent signal generator was imperative if any
quantitative measures were to be secured. Since one could not be purchased,
I built my own signal generator rather as a copy of a Hazeltine machine
designed for the old low-frequency television band. My design uses a
WE 316A tube, with a tuning range of 140 to 200 Mc/s, a reference output
level of 1 volt across 10 ohms, and an inductive attenuator to 120 db. No
detectable leakage could be found. The case is all brazed copper pipe and
plate by the same blacksmith mentioned above and the cover is sealed water
tight using lead foil gaskets.

Finally by 1941, things seemed to be in order for starting a
survey of the sky. Preliminary results were published the following year

(Reber 1942) with an analysis of the theoretical conditions governing receiver sensitivity. Later studies merely have confirmed these early investigations. Even today, explorers of the Milky Way at wavelengths less than
a few meters do not receive any net celestial energy, but in fact merely
measure changes in the net rate at which their equipment dissipates energy
into the sidereal universe via the antenna beam. Further efforts to detect
the sun failed, as mentioned above.

Returning now to the theory, it is worth mentioning that a
significant improvement in signal-to-ripple ratio of 3 db may be secured by
using a push-pull full-wave rectifier instead of the common half-wave type
which loses half of the random fluctuation peaks. This is not particularly
important on wide bandwidths, but I have found it to be worthwhile at low
frequencies, where only a few kilocycles per second of bandwidth may be
secured due to crowded channels.

IMPROVED APPARATUS AND TESTS AT 187 CM

Before much data had been accumulated, a greatly improved receiver was designed on the basis of the above and other equipment studies
(Reber 1944a). The type 954 tube was still the best tube available in
1941. Improvement in sensitivity only seemed possible by increasing the
bandwidth or lengthening the integration time. Since auto ignition disturbance was so severe, it was desirable that the integration time should remain short, in order that the auto disturbance might quickly clear when an
offending car had passed. Thus I decided to widen the bandwidth to the
maximum feasible. Wide-band couplers of the Y type were designed using
coaxial elements. The load resistance turned out to be about 7000 ohms for
an 8 Mc/s bandwidth, with some impedance step down from plate to grid.
Again the whole affair was brazed into one piece. Because of the circuit
complexity, considerable time was required to align everything. It would
have been nearly impossible without the signal generator. Finally, a five-
stage receiver emerged having a gain of about 90 db over the frequency band
of 156 to 164 Mc/s, compared to only 0.16 Mc/s bandwidth of the earlier
design. To maintain bandwidth, the diode load resistor was reduced to
15,000 ohms. This had the effect of reducing and stabilizing the velocity
potential. The wide bandwidth eliminated the detuning caused by the temperature effects and the resistor on each grid did wonders for holding the
stage gain constant. This receiver is nearly impervious to mechanical and

electrical shock. The traces became sharp and clear; sensitivity and stability were sufficient to read to 0.001 of total output voltage during a run
of several hours [Figures 8 and 9]. It was operated for several thousand
hours, merely by replacing a 954 from time to time. The antenna system
was broad-banded by removing the front iris, so that the drum had full
aperture toward the mirror, and by replacing the wire dipole inside with
two aluminum cones, also made by Alcoa, having a 15° angle of rotation.

        Now that really worthwhile electronic equipment was at hand,
a complete survey of the sky was undertaken early in 1943. Data also could
be secured during the day, except that the traces were quite rough due to
auto disturbances. For some reason, the sun was not tried until September.
A relatively high-gain setting was used, corresponding to the weaker parts
of the galaxy then under observation. On the very first try the quiet sun
put the pen hard against the pin at full scale for half an hour near meridian transit. Two further days' tries were required before proper on-scale
readings were obtained. The observations continued daily up to the middle
of 1944 before a complete coverage of the available Milky Way was secured.
During these years, the sun was at low activity and the solar traces were
all very much alike and uninteresting. The results were published (Reber
1944b) and included a polar diagram of the antenna pattern taken on the
strong source in Cassiopeia [Figure 10]. At the time, it was not realized
that at the center of these contour circles there was actually a minute
object of very high surface brightness. If the solar data reported were
changed into units of temperature, then the sun could be represented by a
disk one-half a degree in diameter at a temperature of slightly less than
a million degrees. This had no meaning at the time.

        The first man-made electronic interference appeared during this
survey. It was caused by badly adjusted IFF [Identification Friend or Foe]
transceivers in aeroplanes. The "squitter" could be heard for many miles
when the plane crossed the antenna acceptance pattern. The ignition systems in commercial planes were shielded sufficiently well so that no sparking could be detected. A few small private planes were heard, but these
rarely operated at night.

APPARATUS AND TESTS AT 62 1/2 CM

        When it became apparent, toward the end of 1943, that the situ-
ation was fairly well in hand at 160 Mc/s, I cast about to see what could
be done at higher frequencies to improve the resolution. An increase of

Figure 8.   Sample strip chart recordings at 160 Mc/s taken in
1943 as the Milky Way drifted through the telescope beam at
declinations of (top to bottom) -30°, -20°, -10°, and 0°. The
bottom recording is particularly afflicted with interference
from an electrical storm.

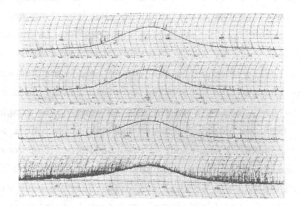

Figure 9.   The back-end rack for the second 160 Mc/s receiver
with which the survey shown in Figure 10 was taken in 1942-43.
From top to bottom, the main components shown are a 130 to 200
Mc/s signal generator, strip chart recorder and its DC amplifier, and power supplies.

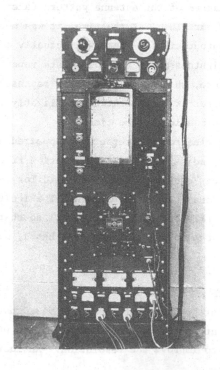

two to one would be the least acceptable and three to one would be more
interesting and significant. Cosmic static, clearly now, had an inverse
intensity versus frequency relation. Consequently, the effective sensitivity should be at least equal to, and preferably surpass, that achieved at
160 Mc. The use of a rather anemic commercial signal generator covering a
frequency range from 400 to 550 Mc also became available. Therefore, I
decided on a new operating frequency of 480 Mc/s.

The only tube offering any hope as an RF amplifier, to which
I had access, was the RCA orbital beam type A5588A. A four-stage amplifier
providing a gain of over 100 db on a bandwidth of 10 Mc/s was constructed
in 1944. These tubes certainly would amplify but that is all that can be
said for them. The secondary emitter had a life of only 50 hours or so and
the internal fluctuation voltage was exceedingly high.

A re-examination of the antenna focal apparatus was made to
improve its efficiency (Reber 1944c) and broad-band characteristics. The
drum had desirable properties of shielding which should be retained, but it
was cumbersome and tended to be rather sharply resonant. The concept of
its operation was that the field radiated from a virtual image point in
the aperture plane which was made coincident with the focal point of the
mirror. The drum acted more or less like an ellipse which transformed this
virtual image point back to the antenna. I decided it would be best to do
away with this image transformation and place the tips of a cone antenna
directly at the focal point of the mirror. Since the field from a cone
antenna is nominally radial, a hemispherical shield should be provided
over the back half of cones with a radius of about one-half wavelength.
Several models were built and tested (Reber 1947).

A couple of months' observations were made during the spring of
1945 using the four-stage amplifier and the new focal point apparatus. The
results were very poor since the Milky Way could not be detected, nor even
the quiet sun. On one day only was the sun detected, when it was radiating
in a steady but enhanced manner. The significance of this was not realized
at the time. Just how poor this equipment was could easily be guessed by
the fact that auto ignition disturbance rarely was detected. Since these
electron multiplier tubes cost thirty dollars each and could be seen to
rapidly die in a few hours, the cost of observation averaged several dollars an hour. In fact, the best part of the life of a new tube was used up
in getting the set aligned. Thus, not only were the results practically
nil, but operation was very uneconomical.

Figure 10. Contour map, as published in the *Astrophysical Journal* in 1944, of 160 Mc/s intensities over the entire northern sky in equatorial coordinates. One contour unit corresponds to 10^{-22} W cm^{-2} (circular degree)$^{-1}$ (Mc/s)$^{-1}$. [A recent analysis indicates that one contour unit in fact corresponds to 130 K in brightness temperature. - *Ed.*] The beamwidth is approximately 12°. Starting at the highest galactic longitudes in the north, we can now recognize the influences on the map of Cassiopeia A, Cygnus A and Cygnus X, the galactic center region in Sagittarius, and Vela X.

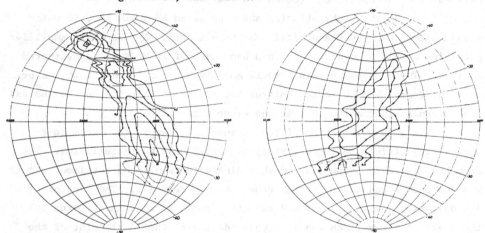

Figure 11. Similar contour map, as published in the *Proc. Inst. Radio Engineers* in 1948, of 480 Mc/s intensities. Contour units are the same as in Figure 10. [Recent analysis indicates a value of 25 K brightness temperature. - *Ed.*] The beamwidth is now about 4° and the Cygnus region at +40° declination is resolved into Cygnus A, to the west, and Cygnus X. Comparison with Figure 10 indicates the steep spectrum of the galactic radiation.

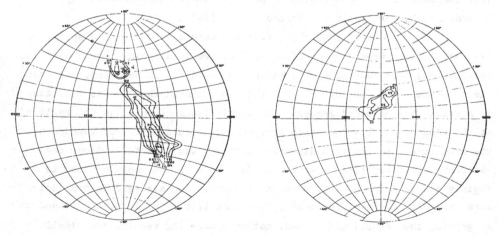

IMPROVED APPARATUS AND NEW TESTS AT 62 1/2 CM

Even before the four-stage amplifier was finished, it was clear that it would be rather ineffective. By using special dispensation, I secured some GE 446B lighthouse tubes in the summer of 1945. These had markedly less gain but were an immense improvement in life, stability, simplicity, and particularly low internal-fluctuation voltage. A new six-stage amplifier with a 6 Mc/s bandwidth was built using these high grade triodes [Figure 12 (at end of article)].

Observations commenced in the summer of 1946. Success was immediate on both the Milky Way and the sun. Also, automobile ignition sparking now was very objectionable during the day. The galactic radiation was markedly weaker than at 160 Mc/s, as was expected. However, the quiet sun was much stronger. Again the gain was set much higher than need be, so that the quiet sun held the pen at full scale for a quarter-hour on the initial try. There seems to be no way of guessing these matters in advance, especially with such an unknown phenomenon. In fact, it seemed that perhaps the sun was a black body radiator, but the temperature would have had to be nearer a million degrees than the optical value of six thousand degrees.

Most daytime observations were still unattended since I continued to work in the city. Examination of the solar records showed several cases where automobile ignition interference was abnormally strong and persistent during a half hour or so near solar transit. Why the cars should pick this time to be particularly objectionable seemed quite mysterious. In time, a similar circumstance occurred on a weekend when I was present. Listening observations and pointing the antenna beam away from and towards the sun quickly unravelled the situation. The sun was, in fact, emitting hiss-type transients which rose to great amplitude and fell again in a second or less. These occurred repeatedly and in an overlapping manner, so that for a couple of minutes at a time the pen would remain hard at full scale. Apparently there were many small independent sources of intense transient solar radio waves such as I had never encountered before (Reber 1946). The phenomenon became more frequent by the spring of 1947. Also the background intensity rose and fell with irregular periods of a week or more (Reber 1948a). Obviously, circumstances in the sun were far different now than in 1943 and 1944. Also, as expected, much more detail was found in the structure of the Milky Way (Reber 1948b) [Figure 11]. The

Cygnus region was divided into two loops. One of these turned out to be
the now well-known colliding galaxies source of minute dimensions [Cygnus
A], as was previously suspected (Reber & Greenstein 1947). The other is
known as Cygnus X and seems to be behind a dust cloud. One of the loops
in Taurus later was identified as the Crab Nebula. A long slender source
in Virgo now is suspected to be the resultant of local galactic clustering, that is, clustering of galaxies in the neighborhood of the Milky Way.
These were the last set of observations completed at Wheaton, Illinois.

EQUIPMENT FOR USE AT 21 CM

During the autumn of 1945, I met H. C. van de Hulst. He explained his theoretical work on neutral hydrogen and asked for my estimate
of the possibility of detecting it. Since I did not know anything about
the matter, I could not guess. Neither could he guess whether or not the
line would appear in emission or in absorption. At the time I had only the
unsatisfactory experience with A5588A tubes at 480 Mc/s. If the line were
to be observed in absorption, the matter would be practically hopeless; if
in emission, then there might be some possibility. In any case, before
trying more microwave experiments, it seemed best to have the lighthouse
tubes going at 480 Mc/s. By the autumn of 1946, this had been accomplished.

Consideration was then possible for work at 1420 Mc/s. First,
there was no test equipment. To remedy this, I built another signal generator covering the range of 1200 to 1600 Mc/s. The oscillator was modified
from the transmitter section of an ABJ type IFF equipment and it uses a
2C43 tube with an inductive attenuator. Experiment quickly showed that
none of the GE lighthouse tubes would be satisfactory amplifiers at 1420
Mc/s. About this time Sylvania brought out a better version known as the
5768 rocket tube, which was similar to some European tubes. A prototype
stage was tested and found satisfactory. The necessary movable selectivity
was to be obtained by an echo box bought on the surplus market; it has a
bandwidth of about 50 kc/s. It was intended to construct a multistage
amplifier using the Sylvania tubes and insert the echo box in the chain.
Provided that the amplifier was good enough to detect the continuum, it
seemed likely that a fairly strong absorption line might be detected. If
the line were to appear in emission, then so much the better. The entire
setup was never completed because operations were closed at Wheaton during
the early summer of 1947. In view of present day knowledge, it seems likely
that the experiment would have been successful.

ADDENDA

The above was written in Australia during August and September 1957 from memory and without recourse to log books, etc. located in Wheaton and Hawaii. Thus, there may be some slight error in minor facts, for example, the date when certain tubes became available, etc.

Two errors are worth correcting. First, the data (Reber 1940b) purporting to show results from The Andromeda Nebula were really chance drifts in the early equipment. Selected charts secured much later can be used to prove its existence. Equally good charts are available to deny it. The detectability of this object was on the edge of equipment capabilities as already explained (Reber 1948b). If all the data had been combined statistically, as is common practice today, the Andromeda Nebula certainly would have appeared. This process is really one of greatly lengthening the integration time.

Secondly, using fluctuation voltage it is possible to deduce the size of the elementary charges causing this voltage (Reber 1942). When the result was published, a value corresponding to three electrons was given. This was an error caused by an incorrect and much too low value of capacity. Afterwards, the mistake was found and the proper value of electron charge secured in agreement with other methods.

These old experiments at Wheaton were quite thrilling at the time. My present experiments (Reber 1958) at the other end of the spectrum in Tasmania using cosmic static at kilometer wavelengths fully equal the old in the realm of the unexpected.

Much remains to be done.

Figure 12. Seven-stage lighthouse-tube amplifier used for 480 Mc/s observations, as in Figure 11.

REFERENCES

Barrow, W. L. (1936). Transmission of electromagnetic waves in hollow tubes of metal. Proc. IRE, 24, 1298–1328.

Dunmore, F. W. J. (1936; 1935). A unicontrol radio receiver for ultra-high frequencies. Proc. IRE, 24, 837–849; J. Res. NBS, 15, Research Paper RP856.

Hollmann, H. E. (1934). The retarding field tube as a detector for any carrier frequency. Proc. IRE, 22, 630–656.

Jansky, K. G. (1932). Directional studies of atmospherics at high frequencies. Proc. IRE, 20, 1920–1932.

Jansky, K. G. (1933). Electrical disturbances apparently of extraterrestrial origin. Proc. IRE, 21, 1387–1398.

Reber, Grote (1938). Electric resonance chambers. Communications, 18, 5–25 (December).

Reber, Grote (1939). Electromagnetic horns. Communications, 19, 13–15 (February).

Reber, Grote (1940a). Cosmic static. Proc. IRE, 28, 68–71.

Reber, Grote (1940b). Cosmic static. Astrophys. J., 91, 621–632.

Reber, Grote (1942). Cosmic static. Proc. IRE, 30, 367–378.

Reber, Grote (1944a). Filter networks for UHF amplifiers. Electronic Indus., 3, 86–198 (April).

Reber, Grote (1944b). Cosmic static. Astrophys. J., 100, 279–287.

Reber, Grote (1944c). Reflector efficiency. Electronic Indus., 3, 101–216 (July).

Reber, Grote (1947). Antenna focal devices for parabolic mirrors. Proc. IRE, 35, 731–734.

Reber, Grote (1946). Solar radiation at 480 Mc. Nature, 158, 945.

Reber, Grote (1948a). Solar intensity at 480 Mc. Proc. IRE, 36, 88.

Reber, Grote (1948b). Cosmic static, Proc. IRE, 36, 1215–1218.

Reber, Grote (1958). Between the atmospherics. J. Geophys. Res., 63, 109–123.

Reber, Grote (W9GFZ) & Conklin, E. H. (1938; 1939). UHF receivers. Radio, no. 225, 112–161; no. 235, 17–177; no. 236, front cover.

Reber, Grote & Greenstein, J. L. (1947). Radio frequency investigations of astronomical interest. Observatory, 67, 15–26.

Schove, D. J. (1955). The sunspot cycle 649 BC to AD 2000. J. Geophys. Res., 60, 127–146.

Thompson, B. J. (1933). Vacuum tubes of small dimensions for use at extremely high frequencies. Proc. IRE, 21, 1707–1721.

OPTICAL AND RADIO ASTRONOMERS IN THE EARLY YEARS

Jesse L. Greenstein
California Institute of Technology, Pasadena, Cal., USA

Radio noise from space was detected by Karl Jansky in 1931, working at the Bell Telephone Laboratories (Jansky 1933). This revolutionary discovery broke the barrier confining astronomical knowledge to the information contained, and the relevant physics, within the narrow band of wavelengths accessible (an octave and a half), and to positions and motions under purely gravitational forces. Jansky's wavelength was ten million times longer than that of light. His signals were radiated from the galactic center, 10,000 parsecs distant. The long wavelengths he used resulted in low angular resolution. There was no radial velocity information, no sharp spectral features (the first line was found twenty years later). For such reasons, and perhaps because he was an electrical engineer, no astronomer beat a pathway to his door; in fact I have never met any astronomer who personally knew him. Public recognition came only as an article in the New York Times (May 5, 1933) and a radio interview. His relevant bibliography includes only seven entries over the years 1932 to 1939, and he died young (see the article by Sullivan in this volume for further information on Jansky). As a summer resident of New Jersey seashore resorts in the early 1930s, I wore golf knickers, possibly even a hip flask, and drove an open car with a rumble seat (oh nostalgia!) past the giant antennas of the transatlantic radio transmitters for which Jansky's studies of noise background were to find the best operating wavelengths. Although I felt no premonitory twinges, I met my wife there, soon became interested in Jansky's results, and my life became linked with that place and time.

In 1936, G.W. Potapenko and D.F. Folland of Caltech carried a receiver into the Mojave desert, and with a simple, rotatable antenna confirmed Jansky's results. Potapenko tried to persuade R.A. Millikan to fund an antenna the size of a boxcar, on a rotating wooden frame. But the cost, $2000, proved too expensive; only a pencil drawing exists

by Russell W. Porter (whose sketches of the 200-inch reflector were so
effective), showing the scale and the simplicity of the design (Fig. 1).

Figure 1. A proposed Caltech radio antenna, dated 1936.

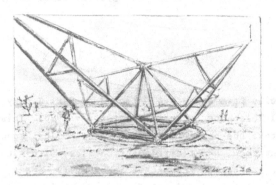

        Theorists paid a little attention; the first was R.M. Langer
(1936). At Harvard soon afterwards F.L. Whipple (then a young faculty
member) and I (a graduate student) attempted to explain Jansky's results
quantitatively and to suggest a source (Whipple & Greenstein 1937).
Our explanation was thermal radiation from heated dust; we at first did
not understand Jansky's engineering units and the effective collecting
area of his antenna. We could have visited Jansky at Bell Telephone
Laboratories but, unfortunately, we never did. At that time, dust and
calcium and sodium were the only known constituents of interstellar
material. The known gases could not radiate at Jansky's wavelength.
We failed to account for the strength of his signal by a factor near a
hundred thousand. The theoretical model for our radiative-transfer
computations assumed that dust was concentrated, with the stars, to the
center of our Galaxy. We found that the dust temperature might rise to
30 K, ten times hotter than near us in the Galaxy. But Jansky's low
frequency data required over 100,000 K. Our attempt was typical of ideas
of some astrophysicists for fifteen more years. As the radio observers
found more and more intense sources, astrophysics responded with hotter
thermal processes, e.g., in the ionized hydrogen gas in space, which
reached 10,000 K (but was optically thin). We now realize that most
radio noise from discrete sources is non-thermal in origin, a fact
only apparent after critical discoveries in the early 1950s.

World War II brought physicists and engineers into the
development of radar, high-frequency receivers and antennas of large
aperture and high gain. American astronomers first met such people at
the M.I.T. Radiation Laboratory; for instance, E.G. Bowen brought high-
frequency tubes from England and others came to know R. Hanbury Brown.
Many such Englishmen, like John Bolton, decided to work in Australia at
the end of the war in the Pacific. The Sun was detected in the radio by
J.S. Hey and given a high security classification. The first publications in America were by Grote Reber, the ultimate scientific amateur,
a radio engineer who built his own 30-foot tiltable paraboloid in his
back yard. Self-taught, ingenious, with a few thousand dollars he found
the galactic center at 160 Mc/s, later at 480 Mc/s. He published maps in
the Astrophysical Journal (1940; 1944; also see the article by Reber in
this volume). He saw features in his Galaxy map, e.g. a concentration
in Cygnus. His results showed that dust obscuration was negligible at
radio frequencies; he also developed an approximate free-free emission
theory. His articles faced somewhat difficult refereeing at the Yerkes
Observatory where O. Struve was Journal editor. Struve and G.P. Kuiper
first visited Reber's laboratory; later I also did, and became involved
with him, and other Yerkes theorists, interpreting the radio spectrum.

After the war, many others extended the observations of the
Sun and Galaxy with improved equipment, 60-200 Mc/s (Hey 1946; Hey,
Phillips & Parsons 1946; Pawsey 1946). The spectral energy distribution
of sources which emerged decreased with frequency — a power law which
was quite impossible to explain by a thermal source, where flux is proportional to frequency squared, if optically thick, and nearly flat, if
thin. Why did we pursue a thermal explanation? Probably because most
things radiated thermally, as did the surfaces of stars and planets, or
fluoresced, as did gaseous nebulae. Knowledge of the major components
of interstellar space was limited. Lasers and masers were not invented
and few believed high-energy physics had much role in the astronomical
Universe. For example, a leading cosmic-ray physicist in a lecture at
Yerkes said, in the late 1940s: "The only thing cosmic about cosmic
rays is their place of origin, which is unimportant." Such hindsight
makes for unpleasant breast-beating, and astronomers were not alone in
conservatism. Some points should be made in partial explanation:

(1) Early radio flux measurements had so poor an absolute cali-
    bration that the shape of the continuum was truly uncertain.
(2) Only post-war did we learn of the high brightness tempera-
    ture of solar noise storms which had blanketed British radar.
(3) The concept of a plasma containing relativistic particles
    was unfamiliar; magnetic fields were required for cosmic-ray
    isotropy, but their existence was established only in sunspots.
(4) Primary cosmic-ray electrons were rare or non-existent as
    observations then stood; they could be of secondary origin.

    The Yerkes group (Henyey & Keenan 1940; Greenstein, Henyey &
Keenan 1946) therefore pursued further the theory of thermal free-free
radiation by hot, ionized gases. Using the standard formulae, including
quantum-theory factors, of D.H. Menzel and C.L. Pekeris (1935), it is
possible to relate classical and quantum physics by the correspondence
principle. Thus the strengths of free-free, bound-free and bound-bound
hydrogenic transitions can be derived. The review by Reber and myself
(1947) mentions excited fine-structure transitions of hydrogen and the
high n-value recombination lines such as n = 340 near 5 meters! For us,
emission lines from ionized hydrogen were implicit in thermal theory
(and were indeed found much later). A thermal model worked for the baseline, quiet-sun, coronal emission (temperatures above 1,000,000 K), but
failed for the strong solar bursts or any power-law energy distribution.
It failed hopelessly at Reber's and Jansky's low frequencies. Had we
thought deeply, we might have predicted the existence of masers through
stimulated emission in dense, ionized gas (see van de Hulst's contribution in this volume). But attention was soon focussed on a new type
of observation that became part of the solution. In 1949, attempting to
test a stellar-atmosphere prediction that S. Chandrasekhar had made,
W.A. Hiltner and J. Hall independently found optical polarization caused
by interstellar dust. The only obvious explanation was that organized
magnetic fields were present in the Milky Way, containing enormous
total energy. E. Fermi (1949) had tried to understand the acceleration
of protons to cosmic-ray energies by moving, magnetized clouds. The
energy density of cosmic-ray protons proved comparable to that in the
required magnetic fields. If the cosmic ray protons detected on Earth
had been accompanied by high-energy electrons, a framework for understanding radio-astronomical sources would have existed using ideas in
the wind before 1948. But cosmic-ray electrons are rare because they

lose energy so rapidly; it is only in the non-equilibrium regions of
our Galaxy (and strong radio sources) or shocked gas near exploding
stars that they live their brief and glorious span. The explanation of
the polarization in space by aligned dust grains (Davis & Greenstein
1951) did in fact require higher magnetic fields than were thought
reasonable, matching those found only in the 1970s in dense molecular
clouds. The fields in strong radio sources had to be amplified by the
same types of phenomena producing the relativistic electrons. Fermi
(1954) discussed some of the problems of production, leakage and decay
of cosmic rays, and of the interstellar magnetic fields.

      Reber and I (1947) wrote the first post-war resumé of the
exciting new discoveries in radio astronomy. Writing that review of a
continuously changing experimental science was an educational experience
for me, as an astrophysicist. Regions of emission other than the
galactic center were found numerous. Surveys had been made over a wide
range of frequencies. A rapidly fluctuating signal was found from a
small region in Cygnus, difficult to explain as refraction in an
electron cloud (now interpreted as scintillation from a "point" source).
Theories reviewed were found generally unsatisfactory, because of the
high brightness temperatures of the sources. The solar phenomena then
known included intense noise storms, circular polarization and possible
gyromagnetic radiation from sunspots. An addendum I wrote four months
later mentions flare-associated impulsive bursts, the detection of the
"quiet" sun, the optical thickness of the corona at low frequencies
and the desirability of a search for hyperfine emission of hydrogen.
Data had far outrun interpretation, and little we reported had involved
any cooperation with astronomers. The former radar wizards did their
own interpretation, starting careers in England and Australia as a new
breed, radiophysicists. In the United States radio telescope building
began five years later at sites near, or in connection with, optical
astronomers, e.g., Harvard, Michigan and Cornell. At first the natural
academic links had been with electrical engineering groups such as the
Naval Research Laboratory, Ohio State, and Cornell, often supported
from military funds. A major impetus came from the Netherlands, where
prediction of the 21 cm line by H.C. van de Hulst (1945) combined with
the strong interest of J.H. Oort in galactic structure and rotation.
(On my oral thesis examination in 1937, Harlow Shapley had asked me how
neutral interstellar hydrogen might be detected. The importance of H I

for galactic structure and star formation was obvious, but I had no
ready answer, knowing only how to find ionized hydrogen.) It was natural
for solar radio astronomy to have strong connections with those who
studied solar and flare activity (e.g., the High Altitude Observatory).
International organization started with URSI, as a meeting place for
electrical engineers. Later the IAU provided a home and a Commission,
and astronomical journals actively solicited radio astronomy papers.

Given the European dominance of pre-war astrophysics, it is now
interesting to re-read a fascinating paper by a leading theorist, A.
Unsöld (1949). While the conclusions are often wrong, Unsöld paid
careful attention to what the experimenters had found and tried to
explain their results, scaling from the radio phenomena found in the
Sun, as an example. He introduced many of the leading concepts of the
high-energy Universe in this analysis:

(1) Solar radio noise cannot be thermal; only at its lowest
level is the quiet-sun flux explicable by thermal, free-free
emission from the chromosphere and corona.
(2) Solar cosmic rays are produced by moving magnetic fields in
surges and prominences; particles up to 10 GeV are produced, and
locked onto sunspot lines of force in gyromagnetic motion.
(3) Extrapolating from the Sun to galactic cosmic-ray particles
a magnetic field in space near a microgauss is required if the
field is to produce the isotropy in direction of arrival.
(4) The magnetic energy density is nearly the same as that of
cosmic rays, with both greater than the random kinetic energy
of gas clouds, stars, or the energy density of starlight.

But even such good, pioneering astrophysics may be premature. In his
attempts to account for the discrete radio sources, then known to be
common but not yet identified, Unsöld invoked the existence of a class
of numerous, highly active, faint M-dwarf flare stars, with very strong
magnetic fields and noise storms of supersolar strength. He explained
the supposed radio variability of Cygnus A, known then as a point
source, as super-flares on a nearby M dwarf. He was wrong, but he had
been misled by the radio astronomers; the Cyg A source was not itself
variable, but twinkled like a star in the turbulent ionospheric plasma.

While Unsöld was wrong on the discrete radio sources, we are

not much better off today, since we still posit many "magical" devices.
Thus, interstellar magnetic fields originate from initial conditions of
galaxy formation; relativistic electrons may come from supernova explo-
sions. We require a family of phenomena in small, intense radio sources —
supersonic gas motion, magnetic shocks, violent accelerations — which
we hope may be indirect consequences of energy released during collapse
under gravity. Very-long-baseline interferometry reveals in the heart
of a quasar motions apparently faster than light; optical and near-
ultraviolet spectroscopy show very hot surrounding gas and jets of
invisible plasma. Displacing the origin of radio noise into a deep
potential well, perhaps that near a black hole, seems plausible, but is
still worth some concern. While solar cosmic rays are soft, a few
hundred MeV, the electrons in intense radio sources reach a few
hundred GeV, and must be continuously replaced.

The change from the old model based on scaling up the active
Sun to the present mysterious central engine came about through the
identification of strong extragalactic radio sources such as Virgo A
(M87) and Cygnus A, the realization of their enormous distances, based
on the cosmological interpretation of their redshifts, and the further
identification of other faint, disturbed galaxies with large redshifts.
Equally important were the identifications of two supernova remnants,
Cassiopeia A and Taurus A (the Crab nebula). (As new wavelength regions
opened, the latter has since been involved in other major discoveries,
such as the optical, X-ray and gamma-ray pulsar.) Central to this major
advance were two remarkable astronomers of the Mount Wilson and Palomar
Observatories staff. Unfortunately, they published little, and their
worldwide, often handwritten correspondence has not been assembled.
Walter Baade was a classical astronomer, while Rudolph Minkowski had a
little more training in physics. Both were superb observers, with the
patience, skill and observing time on the Mount Wilson 100-inch and on
Caltech's then new Palomar 200-inch needed to study the faint objects
which were identified with radio sources of small angular size. They
were not young (58 and 56) in the critical years 1950-51 when they
responded to the suggestions of radio astronomers to search for optical
counterparts to the strongest radio sources. Their two classical papers
(Baade and Minkowski 1954a,b) end with the acknowledgment: "We are
greatly indebted to the members of the radio astronomy groups in Sydney,
Cambridge and Manchester for their generous communications of information in advance of publication." The accuracy of radio positions

had gradually increased, from a degree down to one or two minutes of
arc. The radio astronomers, consulting catalogs of all types of
known nebulae, sometimes found positions roughly coincident with a
radio source, but had no way to measure distance. A radio source
roughly coincident with a bright,large galaxy usually proved to be
intrinsically weak (as is our own Galaxy). The two strongest, Cas A
and Cyg A, had no easily seen optical counterparts. Radio resolution
and pointing accuracy were 100 times inferior to the optical results.
Since both disciplines needed each other, the time was ripe. For
almost two years, letters and discoveries were exchanged freely (for
example, see the article by Smith in this volume). In 1949-52, Baade
and Minkowski used both the 48-inch Schmidt, the new, wide-field
mapping telescope, and the 200-inch to pursue investigations stimulated
by radio astronomers. For us in Pasadena, this stimulation led
to the eventual founding of our radio astronomy group, under John
Bolton, in 1955, and to the Owens Valley Observatory, operating by
1959. This was first staffed, or visited, by many with whom Baade
and Minkowski had corresponded — Bolton, G.J. Stanley, F.G. Smith,
B.Y. Mills, P. Scheuer and K. Westfold.

Figure 2. (Left) Walter Baade (1893-1960)(photo: early 1950s);
(right) Rudolph Minkowski(1895-1976)(photo: early 1940s).

Those who know Baade and Minkowski only as names have missed the rare experience of two extraordinary men. I am indebted to the photo-library of the Mount Wilson and Las Campanas Observatories for the two portraits from their historical collection shown in Figure 2. Walter Baade had come to Pasadena in 1931, from Hamburg. He was born in Westphalia; he was an intense person, with excellent taste in wine, humor, food and conversation. He had even better taste in astronomical puzzle-solving, aided by an enormous memory for astronomical facts. He was committed to a program of studying the distances and stellar populations of nearby galaxies, the cosmic distance scale, and the expansion of supernova shells. Yet he could always find time to browse, as in his critical 1952 discovery of the two types of Cepheids, the first major step in the great enlargement of the distance scale of the expanding Universe. Rudolph Minkowski, born in Strasbourg, came to Pasadena in 1935 as a refugee from Germany. He was a large, gentle person, also fond of astronomical puzzles, and more familiar than Baade with atomic physics and spectroscopy. He studied the physics of, and the expansion in the Crab, and the nature and spectra of supernovae. Lunches at Caltech with these two, at least weekly, made the 1950s a precious decade for many of us. These two classical astronomers had remarkable freedom in their speculations, combined with the best observing talent at a time when photography was still the dominant technology. They were always re-educating themselves. One might say that these best practitioners of a mature, well-instrumented, experimental science were ready with open minds to look at the novelties revealed by the pioneering instruments in the new, radio-wavelength region.

Taurus A, after an early radio position by Bolton(1948), had been identified with the Crab nebula, a supernova remnant, by Bolton, Stanley & Slee (1949). Smith (1951) then found a radio position good to a few minutes of arc, and Mills (1952) to a minute of arc; furthermore, the radio size and shape resembled that of the Crab nebula. This information could not have fallen into better hands, given Minkowski's study (1942) of the spectrum of the filamentary expanding cloud, and of the amorphous inner continuum whose spectrum he did not understand. Baade (1942) had also studied the expanding filaments and identified the Crab as the remnant of the supernova of A.D. 1054. Both worried about the possible stellar remnant, and noted the featureless spectrum of one of the close central pair of faint stars. Minkowski established

that it was hot and dense and did not radiate like a blackbody. There
were many false starts, including errors in the computation of continua
of hot plasmas. When Minkowski and I (1953) undertook a full theoretical
re-analysis of the spectrum, we found that no thermal source could
explain the central star, or the optical and radio energy distribution
of the amorphous nebular mass with its nearly flat continuum lacking
photo-ionization jumps. We noted, with alarm, that the Crab emits
radio waves twelve orders of magnitude stronger than the Sun. Bolton had
written us that the Crab's emission dropped only slowly at high radio
frequencies, which was not explained. Nor could we explain how ultraviolet radiation from the central star, no matter how hot, could
maintain the luminosity of the nebula for 900 years. We estimated that
the shock-heating resulting from turbulence would produce temperatures
above a million degrees. I mentioned nuclear-energy sources, suggesting
possible abundance anomalies produced in the explosion of the collapsed
remnant of a massive star, since hydrogen was deficient relative to
helium, and carbon or lithium might have been produced from the decay of
radioactive beryllium. Our first mention of the synchrotron process was
not until 1954 (Minkowski and Greenstein 1954) quoting "newly-arrived"
papers by Shklovsky (1953a,b). While our quantitative study had produced
little real progress, we felt that the origins of both radio and optical
radiation were linked to an unknown, energetic process.

        In a certain sense, the "new physics" required explains the
failure of these early interpretations; we had not specifically
invoked cosmic ray electrons. But such theories cannot have much
slowed the progress of the radio observations. The radio astronomers
improved their techniques. The Universe seemed full of the unknown.
When an important new technology becomes available, such "accidental"
discoveries are normal, and most theories prove, in retrospect, to
have been conservative. An argument raged in England as to whether the
majority of the unidentified sources were stars or galaxies, since
faint M dwarfs and faint galaxies both are isotropically distributed,
as were the radio sources. Radio observers seemed reluctant to ascribe
to galaxies the powerful energy sources needed if the radio sources were
distant objects. They reasoned this way since most identified galaxies
were weak emitters, i.e., normal galaxies. With hindsight the theorists
come off better if one reads today the (unpublished) "Proceedings of
the Conference on Dynamics of Ionized Media" (April, 1951) held under

the chairmanship of Professor H.S.W. Massey at University College, London. I am grateful to Tommy Gold for a copy. In a paper there, Gold has gems such as: "The most favorable conditions [for radio sources] would be expected in the neighborhood of collapsed, dense stars. Their magnetic field must be stronger than before collapse...." This suggests pulsars sixteen years before their discovery. Also: "If for example one supposes cosmic rays to be as intense in the whole Galaxy as they are here, then it would suffice if one part in a million of the power they dissipate by collisions appeared in the form of radio noise." While high-energy electrons rather than protons are needed, the suggestion is close. If one combines these two ideas -- collapsed, but massive objects, and radiative loss by high-energy electrons -- one has a good contemporary answer. What useful conclusions can we reach from such American and British struggles for understanding? Theory may often delay understanding of new phenomena observed with new technology unless theorists are quite open-minded as to what types of physical laws may need to be applied; conservatism is unsafe. Likewise, poor data often throws plausible theories off track, since theorists may trust current "discoveries" based on incorrect data. In astrophysics, historically, theories have only seldom had predictive usefulness as guides to experimenters. But as an observer, I believe that good new observations may shed a brilliant light, as was the case with the use of the improving radio data by Baade and Minkowski.

Before 1950 identification of radio sources fared poorly, in that only 7 of the 67 known sources had been identified. Why? Now, two-color photographs of the sky of the Palomar-National Geographic Society Sky Survey, using the 48-inch Schmidt, make it possible for anyone to make a first search for, and to identify radio, infrared or X-ray sources. In the early 1950s, however, this northern sky map was only beginning to become available. It had high positional accuracy, reached beyond 19th magnitude, had resolution of 1 or 2 arcseconds, and finally covered the northern two-thirds of the sky. Minkowski was charged with the inspection and acceptance of plates as taken, as well as preparing the copies for distribution, but no copies yet existed. In addition, Baade and Minkowski had a large share of the dark-sky observing with the new 200-inch telescope and could exploit its enormous power and fast focal ratio. The work on the two classical papers on radio sources

by Baade and Minkowski (1954a,b) was initiated in 1952. They first show
in beautiful photographs the peculiar structure of extragalactic sources
and the unusual emission-line spectra, observed with a high-resolution,
faint-object spectrograph designed by I.S. Bowen. The supernova remnants
had extraordinary filamentary structure, in giant bubbles of gas, with
hypersonic velocity differences between filaments (Cas A). Some of the
extragalactic radio sources had unusual structure on the best photographs, e.g., Cyg A, Vir A = M87, Per A = NGC 1275, Cen A = NGC 5128.
They also had strong emission lines, a fact which later proved a common
and useful feature for identifications and velocities of radio galaxies.
Cyg A was interpreted as two galaxies in collision. M87 had a one-sided
polarized jet (Baade 1956a). NGC 1275 had strong emission, as did the
Seyfert galaxies, but was apparently undergoing collision, because two
systems of velocities co-existed. Further, independent of distance,
Cyg A radiated more energy over the radio frequencies than in the
optical region, a truly remarkable fact. The collision hypothesis was
an unfortunate trap for them (and myself), since it apparently provided
sufficient energy from gas-cloud collisions at high relative velocity.
Many intrinsically weak radio sources proved to be nearby "normal"
galaxies, with eight listed for which Cambridge or Manchester positions
lay near apparently large galaxies. Thinking about identifying sources
as stars, they say: "A slim chance may exist of obtaining positions of
required accuracy from occultations of sources by the moon" — the
method later used by C. Hazard, M.B. Mackey and A.J. Shimmins (1963) to
locate 3C 273 accurately, permitting the first identification of a
quasar with high redshift by M. Schmidt (1963). The most striking
single result by Baade and Minkowski was that one of the most intense
radio sources, Cyg A, was an 18th magnitude galaxy at a redshift of
$z = 0.056$, or 16,830 km/s. With their value of the Hubble constant
this implied a distance of about 30 million parsecs (now revised to
250 Mpc). Its luminosity was extremely high, and was concentrated in
the radio frequencies. Strong radio galaxies were indeeed different!

        Soon afterwards, other classical astronomers solved the very
non-classical problem of the Crab nebula and its synchrotron continuum
(Oort and Walraven 1956). Magnetic fields and high-energy electrons to
100 GeV, constantly renewed, were needed. Baade´s (1956b) observations
of the continuum through polaroid showed nearly complete polarization.
He had already established the variability of hazy features near the

central star. The high-energy Universe was being forced on us by nature.

The beginning of the high-resolution radio maps came soon. R.C. Jennison and M.K. Das Gupta (1953) used an interferometer to image Cyg A in detail, and found it was double-lobed, with radio emission outside the optical galaxy. By 1960 when interferometers revealed the double-lobed structure to be common, simple positional coincidence no longer proved identification. Soon, on the theoretical side, the upwards revision of the distance scale meant that the volumes of the magnetized plasma were so large that constraints on the total energy were serious -- both in magnetic fields and relativistic electrons. Real doubts arose whether even nuclear energy could suffice, so that the invention of a radical new driving engine became a necessity.

As a result of such excellent early cooperation, observations by radio astronomy techniques clearly became essential, and especially so for cosmology. The pressure for one's own radio observatory was irresistible by 1952, when Baade, Minkowski and I began to press strongly for Caltech's entrance into the field. We were fortunate in our leadership, since President Lee A. DuBridge had run the wartime M.I.T. Radiation Laboratory, and my Division Chairman, R.F. Bacher, was familiar both with high-energy physics and our goals in optical astronomy at Palomar. Evidence accumulated that the radio sky was at least as informative as the optical. The new window contained emission and absorption lines. The high-energy Universe had led to a theoretical explanation, the synchrotron process, and an enigma, the energy source. A first major attempt to review the field and to produce a synthesis of knowledge occurred during a conference held in Washington, D.C. in January, 1954, sponsored by the National Science Foundation, the Carnegie Institution of Washington, and Caltech; I acted as chairman for the organizing committee. Abstracts published (Greenstein 1954) came from Australia, Manchester, Cambridge, the Netherlands, the Naval Research Laboratory, Ohio State, Canada, Cornell, Michigan and the High Altitude Observatory. Antenna-design topics included the long fought-over question of "big dish versus interferometer array". New lines, other than hydrogen, were predicted by C.H. Townes in a prescient paper. B.J. Bok discussed results on galactic structure using the 21-cm line. One practical effect was pressure for a cooperative radio observatory, which with NSF support, eventually became the National Radio Astronomy Observatory. There was clear realization that radio astronomy had an indefinitely long and happy future, and was deserving of broad support by physicists and astronomers. The "early years" had come to an end.

REFERENCES

Baade, W. (1942). The Crab nebula. Astrophys. J., 96, 199.
Baade, W. (1956a). Polarization in the jet of M87. Astrophy. J., 123, 550.
Baade, W. (1956b). The polarization of the Crab nebula on plates taken
       with the 200-inch telescope. Bull. Astr. Inst. Neth., 12, 312.
Baade, W. & Minkowski, R. (1954a). Identification of the radio sources
       in Cassiopeia, Cygnus A, and Puppis A. Astrophys. J., 119, 206.
Baade, W. & Minkowski, R. (1954b). On the identification of radio sources.
       Astrophys. J., 119, 215.
Bolton, J.G. (1948). Discrete sources of galactic radio-frequency noise.
       Nature, 162, 141.
Bolton, J.G., Stanley, G.J., & Slee, O.B. (1949). Positions of three
       discrete sources of galactic radiofrequency radiation.
       Nature, 164, 101.
Davis, L., Jr. & Greenstein, J.L. (1951). The polarization of starlight
       by aligned dust grains. Astrophys. J., 114, 206.
Fermi, E. (1949). On the origin of cosmic radiation. Phys. Rev., 75, 1169.
Fermi, E. (1954). Galactic magnetic fields and the origin of cosmic rays.
       Astrophys. J., 119, 1.
Greenstein, J.L. (1954)(ed.). Washington conference on radio astronomy.
       J. Geophys. Rsch., 59, 149.
Greenstein, J.L., Henyey, L.G., & Keenan, P.C. (1946). Interstellar origin
       of cosmic radiation at radio-frequencies. Nature, 157, 805.
Hazard, C., Mackey, M.B., & Shimmins, A.J. (1963). Investigation of the
       radio source 3C 273 by the method of lunar occultation.
       Nature, 197, 1037.
Henyey, L.G. & Keenan, P.C. (1940). Interstellar radiation from free
       electrons and hydrogen atoms. Astrophys. J., 91, 625.
Hey, J.S. (1946). Solar radiations in the 4-6 metre radio wavelength
       band. Nature, 157, 47.
Hey, J.S., Phillips, J.W., & Parsons, S.J. (1946). Cosmic radiations at
       5 metres wavelength. Nature, 157, 296.
Hulst, van de, H.C. (1945). The origin of radio waves from space. Ned.
       Tijd. Natuurkunde, 11, 210 (in Dutch).
Jansky, K.G. (1933). Electrical disturbances apparently of extra-
       terrestrial origin. Proc. Inst. Radio Engr., 21, 1387.

Jennison, R.C. & Das Gupta, M.K. (1953). Fine structure of the extra-
        terrestrial radio source Cygnus I. Nature, 172, 996.
Langer, R.M. (1936). Radio noise from the Galaxy. Phys. Rev., 39, 209 (abstr).
Menzel, D.H. & Pekeris, C.L. (1935) Absorption coefficients and hydrogen
        line intensities. Mon. Not. Roy. Astr. Soc., 96, 77.
Mills, B.Y. (1952). The positions of 6 discrete radio sources. Austral.
        J. Phys., A5, 456.
Minkowski, R. (1942). The Crab nebula. Astrophys. J., 96, 229.
Minkowski, R. & Greenstein, J.L. (1953). The Crab nebula as a radio
        source. Astrophys. J., 118, 1.
Minkowski, R. & Greenstein, J.L. (1954). The power radiated by some
        discrete sources of radio noise. Astrophys. J., 119, 238.
Oort, J.H. & Walraven, T. (1956). Polarization and composition of the
        Crab nebula. Bull. Astr. Inst. Neth., 12, 285.
Pawsey, J.L. (1946). Observations of million degree thermal radiation
        at a wavelength of 1.5 meters. Nature, 158, 633.
Reber, G. (1940). Cosmic static. Astrophys. J., 91, 621.
Reber, G. (1944). Cosmic static. Astrophys. J., 100, 279.
Reber, G. & Greenstein, J.L. (1947). Radiofrequency investigations of
        astronomical interest. Observatory, 67, 15.
Schmidt, M. (1963). A star-like object with large red-shift. Nature,
        197, 1040.
Shklovsky, I.S. (1953a). On the nature of the radiation of the Crab
        nebula. Dokl. Akad. Nauk. SSSR, 90, 983 (in Russian).
Shklovsky, I.S. (1953b). The problem of cosmic radio waves. Astr. Zh.,
        30, 15 (in Russian).
Smith, F.G. (1951). An accurate determination of the positions of four
        radio stars. Nature, 168, 555.
Unsöld, A. (1949). On the origin of the radio-frequency and shortwave
        radiation in the Milky Way. Zs. Astrophs., 26, 176 (in German).
Whipple, F.L. & Greenstein, J.L. (1937). On the origin of interstellar
        radio disturbances. Proc. Natl. Acad. Sci.(USA), 23, 177.

(l. to r.) John Bolton, Gordon Stanley, and Joe Pawsey in the early 1950s (courtesy CSIRO Radiophysics Division)

SECTION TWO: *Australia*

In the decade following World War II radio astronomy evolved from a minor curiosity to a strong scientific discipline, from small groups of equipment-oriented radio physicists and electrical engineers to major laboratories whose tenor was as much astronomy as radio techniques. This evolution took place in a similar fashion in many countries, but there were two nations where radio astronomy especially flourished. These were England and Australia. While it is not surprising to find England at the forefront of a scientific field in the middle of the twentieth century, Australia's presence requires more explanation.

The main elements in the Australian success story emerge from the articles in this section, by five of the key participants. First, the Radiophysics Laboratory in Sydney had in fact very close ties to the mother country and her strong tradition in radio science -- many of the staff members were originally British or trained in Britain. Second, the Laboratory had been at the forefront of radar development during World War II and, when the war ended, was *not* dissolved. Rather, the strong team was kept intact under CSIRO aegis while new recruits and directions for peacetime radio research were sought. Third, dynamic and wise leadership was provided by "Taffy" Bowen and Joe Pawsey -- two men whose contrasting personalities and styles of science led to just the right mix for exploring and exploiting the most profitable avenues into the radio sky.

A lightweight, air-warning radar system, typical of those developed during World War II at the CSIRO Radiophysics Division in Sydney (courtesy CSIRO)

THE ORIGINS OF RADIO ASTRONOMY IN AUSTRALIA

E.G. Bowen
Chief, CSIRO Division of Radiophysics, 1946-71.
Sydney, Australia.

1 THE IMMEDIATE POST-WAR ACTIVITIES OF THE DIVISION OF RADIOPHYSICS

Radio astronomy began in Australia at the Division of Radiophysics of CSIR (Council for Scientific and Industrial Research, later renamed CSIRO, Commonwealth Scientific and Industrial Research Organization) as a direct outcome of its involvement in radar research and development during World War II. The Division was established in 1939 with responsibility for developing radar for the Australian Army, Navy and Air Force - and later for the American Forces in the Southwest Pacific. What were the ingredients which led in 1946 to the development of radio astronomy?

The first and by far the most important of these was the decision by the Chairman of CSIR, Sir David Rivett, that at the conclusion of World War II, CSIR would be devoted only to peace-time research, and that defence research would be carried out by other agencies. This meant that a highly developed laboratory with a superlative staff became available for a wide range of researches and practical developments in a peace-time environment.

It would be easy to underestimate the importance of this decision. In later years we were to be reminded of its wisdom by the fate of some overseas laboratories which were not so fortunate. For many years they carried joint responsibility for civilian and military research. When a new research proposal came up, it often led to a bureaucratic argument as to whether it was predominantly civilian or military - and what priority it should be given. This argument would run right down through the staff structure, to the detriment of the job itself. We were spared this problem, and Australian science has reason to be grateful to Sir David Rivett for the clarity and firmness of his decision.

The next ingredient was that the staff, about two hundred strong, was already highly skilled in electronic research and development. They ranged from professors of physics to practical engineers from industry.

Many of them had spent months, if not years, at the best overseas laboratories and were saturated with the most recent electronic techniques. In view of later events, it is also rather remarkable that there was not a single astronomer on the staff, nor, for that matter, anyone who had done a university course in astronomy. It would be interesting to speculate how events might have developed had there been an astronomer on the staff of the Division in 1946.

Next in importance to the people was the store of special components of all kinds which had accumulated during the war years - magnetrons, klystrons, pulse-forming networks, pulse-counting circuits - the whole paraphernalia of a new electronic era. This was augmented by an event which occurred within a few weeks of the end of the Pacific war, which turned out to be particularly fortunate for us. A large part of the Pacific Fleet had assembled in Sydney before returning to the USA; also in Sydney were gigantic stores of radar and communications equipment. It was impracticable to return much of this to the USA and orders were given to destroy the surplus. So, huge quantities of technical equipment, including whole aircraft, were loaded on the deck of aircraft carriers, taken a few miles out of Sydney Harbour and bulldozed into the Pacific.

Our friends in both the Australian and the US Services were saddened to see this and allowed us to collect all we could lay our hands on. After a frantic few weeks loading our own trucks on the dockside, we ended up with a cornucopia of invaluable equipment, often brand new and in the original crates. I seem to remember two huge warehouses full of these good things near Botany Bay, which we were to draw on for many years to come.

In retrospect, it is clear that another important factor was morale. Some scientists, particularly nuclear scientists, came out of the war with a heavy guilt complex. They were worried by the fact that science had been used for unprecedented destruction, often of civilians, and they were horrified by it. We, on the other hand, had no such sense of guilt. Radar had been used predominantly to detect enemy aircraft and enemy submarines and ultimately to destroy them. In other words, it was a machine to destroy the engines of destruction.

So here was a well-equipped laboratory with a highly talented staff, permanent buildings, excellent workshop facilities, a command of modern technology, a bountiful supply of equipment and a burning desire to apply these to the welfare of mankind. What did we do and how did we react?

Some of the staff had unbreakable commitments to return to
positions in universities or in industry which they had vacated at the start
of the war. Others, like Sir Frederick White, who was then Chief of the
Division, were to move up in the hierarchy of CSIR. But a hard core of
perhaps seventy per cent remained, and to most of us the prospect was
entrancing. I myself had been involved from the beginning of the British
radar effort in 1935 and had completed a total of eleven years' unbroken
commitment to the war effort. This included thousands of hours in night-
fighter and sea-search aircraft, seldom under the safest or most salubrious
conditions. I was heartily tired of it and looked forward to less demanding
pursuits.

There were any number of avenues we could follow, several
examples of which are given in Section 6 below. Our policy was to try anything that gave promise of useful scientific results or practical application;
if successful, we poured in manpower and resources. Radio astronomy was to
become one of the most productive of these.

2 EARLY CONCEPTS OF RADIO NOISE

My first exposure to the concept of radio "noise" from outer
space came in 1935 when I joined the small team which Sir Robert Watson-Watt
assembled to build the first air-warning radars in Britain. We were concerned about the noise background which would appear in our receivers when
a directive or semi-directive antenna was pointed at an object in the sky,
in this case an incoming enemy aircraft. The radar equation had not been
formulated in explicit terms, but the factors involved were beginning to
form in our minds. My colleague A.F. Wilkins had the clearest ideas about
this, but I also remember a senior engineer of the Plessey Company, by the
name of Bailey, - I regret that I cannot remember his initials - giving a
lucid account of the type of coupling which would exist between an antenna
system and outer space, both in the transmitting and the receiving modes.
These ideas were not in the literature of that time, but they were very real
to us and of immediate practical significance. We were well acquainted with
Jansky's results and recognized that, after the many noise sources to which
we were exposed had been reduced or eliminated, we were ultimately exposed
to Jansky's cosmic noise. We called it just that - Jansky noise - by
analogy with Johnson noise and Schott noise.

It was some time before these concepts were expounded in a
formal way, and in that hectic period I hesitate to put a date on when they

crystallized out. The matching of an antenna to free space became formalized in terms of the characteristic *impedance of free space*, having the well-known value of 377 ohms. I first heard this expounded in a factual way by members of van Atta's antenna group at the Radiation Laboratory in the early 1940's, but the idea may have existed earlier.

The concept of an antenna system as a radiometer must have crystallized about the same time. It was clearly described in Southworth's pioneering measurement of solar temperature in 1942-43 (but not published until 1945). He quoted a solar temperature of 20,000 K at a wavelength of 10 cm and 10,000 K at 3 cm. At the time there was mild interest in why these measurements should differ from the 6000 K of optical astronomy, but the discrepancy was put down to instrumental factors.

Following this same precept, Dicke with his ultra-sensitive radiometer gave a surface temperature of the Full Moon of 290 K at a wavelength of 1.25 cm. This result was not published until 1946 but the measurement had been made a few years earlier at the Radiation Laboratory of MIT. As a member of the Radiation Laboratory at that time and a frequent visitor to the Holmdel Laboratory of the Bell Telephone Company, I was quite familiar with these results.

With these ideas as background and Hey's observations of radio noise from the Sun in 1942, there is little wonder that it was radar people who pioneered the re-birth of radio astronomy immediately after the war.

3 THE START OF SOLAR RADIO ASTRONOMY AT THE DIVISION OF RADIOPHYSICS

At the Division of Radiophysics it was the great J.L. Pawsey who was fascinated by the concept of an antenna as a radiometer. He was an antenna man from way back and, as an engineer employed by the EMI Company in England in the 1930's, he had been involved in antenna design for the first BBC television transmitting station at Alexandra Palace. He was a master of antenna theory and had many original design concepts to his credit.

Noise from the "Quiet" Sun

He began two experiments. One was to measure the temperature of the Sun at 200 Mc/s - a standard radar frequency much lower than that used by Southworth. His second experiment was to measure the temperature of the ionosphere at a wavelength short enough to be completely absorbed by the ionosphere, i.e. at a wavelength such that the ionosphere acted as a perfect black body.

For the solar experiment he used existing Air Force radar
antennas at Collaroy, a few miles north of Sydney, and another at Dover
Heights, within the city limits. In a very short time he produced a
dramatic result - that the temperature of the quiet, or undisturbed, Sun
was a million degrees Kelvin (1). This caused some rumblings of disbelief
at the time, but the explanation of the temperature gradient from 6000 K
at optical wavelengths to Southworth's 10,000 K at 3 cm, to 20,000 K at
10 cm, to a million degrees K at 1½ metres, was soon well established and
well understood. Pawsey's paper on this subject is now regarded as one
of the classics of radio astronomy.

The measurement of the temperature of the ionosphere was made
shortly afterwards (2). This Pawsey did, picnic fashion, in the Burragorang
Valley west of Sydney, an extremely deep gorge with nearly vertical sides
and therefore well protected from noise from terrestrial sources. He built
an antenna system pointing vertically at a frequency of 2 Mc/s and measured
temperatures which varied from 240 K to 290 K, in good agreement with those
which had previously been deduced from theoretical considerations.

Noise from the "Disturbed" Sun

Following Hey's suggestion of enhanced noise from sunspots,
Pawsey also noticed that the noise level from the quiet Sun increased many
times in intensity when sunspots were present. Within a few weeks, by a
delightful technique which became known as the sea interferometer, he
accurately located the source of this enhanced noise on the face of the
Sun and showed unequivocally that it came from the vicinity of sunspots
(3).

I have a vivid recollection of describing these results, prior
to their being published, at a lecture I gave at the Cavendish Laboratory
in Cambridge on September 20th 1946. For good measure I also mentioned
D.F. Martyn's observation of circularly polarized radiation from sunspots,
similarly unpublished. About thirty or forty members of the post-war
Cavendish team were there, including Martin Ryle. At the end of the
lecture, Ryle rose quickly to his feet and assured the audience that on
two counts I was dead wrong: The solar temperature could not possibly be
a million degrees, and there was something very wrong about Martyn's
observation of circular polarization. It was some time before they were
to change their minds!

There was a very different response in the USA a few months later, when, on December 3rd, I delivered a similar lecture at the newly built RCA Research Laboratory in Princeton, NJ. I knew many of the senior staff from the war years, but sitting in the front row was an individual whom I had never seen before. He was virtually jumping up and down with excitement at what he heard. After the lecture, he introduced himself as an optical astronomer who had been interested in radio astronomy since he was a graduate student at Harvard. His name was Jesse Greenstein! I like to think that he was the first of the optical astronomers to be converted to radio astronomy.

Fig.1 100 Mc/s Broadside Array at Dover Heights used as a sea interferometer, circa 1947.

4 THE BEGINNING OF GALACTIC AND EXTRAGALACTIC RADIO ASTRONOMY

John Bolton was the first to turn his attention to galactic radiation. He had been trained at Cambridge University, had taken a commission in the British Navy and finished up as Radar Officer on the aircraft carrier "UNICORN" in the Pacific Fleet. At the end of the war he elected to be paid off in Sydney and, in answer to one of our regular advertisements, was appointed an Assistant Research Officer in 1946. He was in fact the first of the recruits to be appointed to the Division post-war.

He was an extremely talented researcher, with a yen for working on his own with a minimum of assistance. He applied the technique of
accurate position-finding pioneered by Pawsey, and focused his interest on
the broad peaks in the radiation reported by both Reber and Hey. With
further refinement of the direction-finding technique, Bolton showed that
these peaks were very sharp indeed and was able to put a limit on their
size of 8 minutes of arc (4).

To us, this for the first time suggested that they were indeed
"point sources" and, assisted by Gordon Stanley, Bolton was soon able to
identify three of them with known optical objects (5). In all cases they
turned out to be astronomical freaks - the Crab Nebula, M87 and NGC 5128.

Fig.2 The 16 ft diameter, equatorially-mounted paraboloid
at Dover Heights

There has been controversy over who first discovered discrete
sources. At different times this has been attributed to Reber, to Hey,
and even to much later workers. To me the real discoverer was John Bolton.
A careful reading of Reber's original papers shows that he only conceived

of a maximum in galactic radiation, which was generally understood at that
time to be due to free-free transitions in interstellar space. The free-
free transition concept was one which I personally could not swallow. When
Bolton came along with his three identified sources, this to me opened up
a whole new concept of galactic radio astronomy - one which is very much
like that which we know today.

Fig.3 Bolton's 80 ft "hole-in-the-ground" antenna
at Dover Heights.

These were the first successes of the Radiophysics Division's
radio astronomy programme - Pawsey's observations of noise from the quiet
and the disturbed Sun, and Bolton's identification of discrete sources -
and they were tremendously exciting. These were the events which persuaded us to pour considerable resources and manpower into the programme,
and in this we were fully supported by the far-seeing Executive of the
CSIR. (Photographs of some of the aerials used in these experiments are
given in Figures 1 - 3).

If this sounds too simple and uncomplicated, let me hasten to
add that our plans were not without opposition, particularly from quarters
which were not hard to identify. We were told in no uncertain terms that

the "properly constituted centre for astronomical research was the Commonwealth Solar Observatory at Mount Stromlo." Someone had, apparently,
forgotten about the State Observatories in Sydney, Melbourne, Perth and
Adelaide, the first of which had been established as early as 1858! The
Commonwealth Astronomer (soon to become the Astronomer Royal) at a lecture
he delivered on "The Future of Astronomy in Australia" about this time
managed to fill a whole hour without once mentioning radio astronomy.
When asked afterwards "Where do you think radio astronomy will be in ten
years' time?", he replied "It will be forgotten." With that kind of advice
being delivered in the nation's capital, our task was not always easy.

5   OBSERVATIONS AND DEVELOPMENTS OF THE 1950's

By 1950, the Division's commitment to radio astronomy had advanced well beyond the exploratory state and was beginning to consolidate.
For the next thirty years it was to continue as the main part of the
Division's programme and to command the greater part of its budget. The
programme itself soon diversified to cover a complete range of solar,
lunar, planetary, galactic and extragalactic studies, and was to involve
a whole series of original instrumental developments, the variety of which
would be difficult to find in any other establishment.

Within a year of Pawsey's original observations on sunspots,
Ruby Payne-Scott, Yabsley and Bolton, observing independently at Dover
Heights and Potts Hill (another Radiophysics field station) noticed that
"bursts" of noise coming from sunspots arrived at quite different times
when observed at different frequencies (6). It happened that during
the war the Division - as part of a secret within a secret - had been
involved in the construction of receivers for surveillance of enemy radio
and radar transmissions. The basic method of carrying this out was to
scan rapidly over a 2:1 frequency range and to cover the whole band of
usable frequencies in a series of 2:1 steps. Within the Division there
was a substantial store of such receivers. These were ready-made for
spectral analysis of the noise from the Sun and they were quickly pressed
into service. In the inventive hands of Lindsay McCready and Paul Wild, a
"radio spectrometer" evolved which was to dominate the field of solar
studies for the next twenty years (7). This was another instance of a
device, designed in the first place for military use, repaying a massive
debt to fundamental science. An early example of the antennas for such a
spectrometer at the Division's field station at Dapto is shown in Figure 4
(for further information see Christiansen's article elsewhere in this volume).

Fig.4   Rhombic antennas for the radio spectrometer
at Dapto, New South Wales, of Wild and colleagues (1956).

In 1948 Piddington and Minnett refined Dicke's original observations on the Moon and gave accurate surface temperatures for both the
bright and the dark side of the Moon at a wavelength of 1.25 cm.   These
showed an intriguing phase difference of 45° between surface temperature
and actual phase of the Moon (8).   From a detailed consideration of the
thermal conductivity, the electrical conductivity and dielectric constant
of different models of the lunar surface, they were able to deduce that the
Moon was probably covered with a layer of non-conducting dust about 2 cm
deep.   A photograph of the aerial used in these measurements is shown in
Figure 5.
About the same time, Frank Kerr, following earlier work by the
US Signal Corps and by Bay in Hungary, independently received echoes from
the Moon on frequencies of 17.84 and 21.54 Mc/s (9).   With the low resolution then available it was difficult to extract good physical information
from these observations, but some information complementary to that of
Piddington and Minnett was gleaned about surface roughness.

Fig.5 Piddington and Minnett's 1.25 cm tracking antenna for
measurement of the surface temperature of the Moon, 1949.

A significant "miss" about this time was a failure to follow
through on a hint about line radiation. Well before 1949 I had been urged
by Rabi and Zacharias of Columbia University to look for line radiation in
the Galaxy. Unfortunately, we were to focus our attention on the lines
known to be most easily excited in the laboratory, like those from caesium
and related elements. We simply did not know about van de Hulst's wartime
suggestion of hydrogen as the most likely candidate. However, in 1951 Ewen
and Purcell were kind enough to let us know of their discovery of the H line
prior to publication and we were able to verify it three weeks later (10).
Frank Kerr was to become our principal worker in this field and, in a fruitful collaboration with Oort and his co-workers at Leiden, gave an early
account of the spiral structure of our own Galaxy (11). Further details
are given in the contribution to this volume by Kerr. The aerial at Potts
Hill which he used in these observations is shown in Figure 6.

Fig.6 The 36 ft diameter transit instrument at Potts
Hill used principally for H-line work.

     Many of the successes of the Division in the immediate post-
war period were made using ex-wartime equipment after minor modification
to fit it to a particular task. During the 1950's we realized that more
and more sophisticated instruments would be necessary. The days of hasty
improvization were over - more and more planning would be required and
much more money would be needed for capital expenditure. Little did we
realize how large those sums would become.

     For galactic and extragalactic research, two principal types
of instrument were possible - the large steerable paraboloid and the multi-
element interferometer, of which many different varieties were proposed.
At the technical level, argument about the relative merits of the two
types was long and often vehement. Fortunately, from a management point
of view, our choice was easy: We simply decided to support both to the
best of our ability. In this way, two principal types of instrument were
to emerge from the Radiophysics Division during the 1950's - the large
steerable paraboloid and the cross type of antenna invented by B.Y. Mills.

Cross antennas

Mills made the first suggestion for a cross-type antenna in 1952. It was a highly intriguing proposal but, if my memory is correct, not even Mills was completely convinced that it would work in the first instance. Obviously a trial was called for, and it was typical of the way we worked that an experimental model, with arms 120 feet long and operating on a frequency of 97 Mc/s, was quickly constructed at our Potts Hill field station. With this instrument Mills demonstrated that the principle of the cross antenna was sound. It worked first time and, although small in size, successfully gave a profile of the Sun and several radio sources (12).

This success encouraged us to increase the size and therefore the resolving power, and during the next few years the Division was to build no fewer than three large crosses at a larger field station located at Fleurs about forty kilometres west of Sydney. These were the Mills Cross (85 Mc/s), completed in 1954 (13), the Shain Cross (19.7 Mc/s) completed in 1956 (14) and the Chris-Cross (1410 Mc/s) in 1957 (15). They were among the great successes of the 1950's and were responsible for a large part of the Division's research output over that period. Photographs of the Mills Cross and Chris-Cross aerials at Fleurs are given in Figures

Fig.7 The 85 Mc/s Mills Cross at Fleurs Field Station with arms 1500 ft long.

7 and 8. Further information will be found in the contributions by Mills and Christiansen to this volume.

Fig.8   The 1410 Mc/s Chris-Cross at Fleurs Field Station.

The Parkes 64 m steerable telescope

As in optical astronomy, steerable parabolic antennas are a basic part of the instrumentation for radio astronomy; they played a prominent part in early galactic research, particularly in investigations of line radiation. As in other establishments, there was an urge to increase the aperture of such instruments to the largest possible dimensions.

Among the first options to be explored was a collaborative effort with our friends in the RAAF, with whom we had maintained a post-war connection. As early as 1949, we discussed with them the possibility of building a really large air-warning antenna, with linear dimensions of several hundred feet. Several designs were roughed out and costed, and at one stage there even seemed to be a possibility of going to a horizontal dimension of 500 feet. Our interest in the project was based on the real hope that, if built for defence purposes, we would have the use of the instrument for radio astronomy.

Unfortunately, this expectation was not realized and it became clear that the RAAF would be even more squeezed for funds than the CSIRO. Slowly but surely the prospect evaporated and some way had to be found for funding the project from within the Division's own budget.

So in 1952 a proposal was made for a cylindrical antenna lying on its back, 1000 feet long by 200 feet wide, constructed of five adjoining elements each 200 feet square. The cylinder would lie along an east-west line and would be steered by an arrangement of cables and winches in the north-south direction. An important characteristic of the plan was that the units would cost about £25,000 each (a decidedly large sum in those days) and that one segment would be built each year for five successive years at a total cost of £125,000. Had such an antenna been built, we would soon have realized the advantage to be gained by spreading the segments over a longer baseline. Sad to relate, this proposal was not proceeded with for lack of assurance that the money would be available.

Then a new factor appeared. During my frequent visits to the United States I used to discuss this and related problems with my old friends and colleagues from the war years. These were people with whom I had been involved in the days of microwave radar - *Vannevar Bush*, an ex-President of MIT, Chairman of OSRD (Office of Scientific Research and Development) during the war and at that time President of the Carnegie Corporation in Washington; *Alfred Loomis*, Chairman of the OSRD Microwave Committee during the war, a multi-millionaire, a Trustee of both the Carnegie Corporation and the Rockefeller Foundation and a genuine scientist in his own right; *Karl Compton*, Deputy Chairman of OSRD and at that time President of MIT; *Lee DuBridge*, ex-Director of the Radiation Laboratory of MIT and then President of Caltech - men of enormous stature in American science. They were all concerned by the fact that the USA seemed to be falling behind in an important branch of science and they asked my advice on what should be done about it. My advice was very simple - to set up an establishment with radio astronomy as its main objective and a large steerable telescope as its principal instrument. The proposal fell on receptive ears. It was these same people - the Carnegie Corporation in collaboration with Caltech - who founded and ran the largest optical telescope in the world, the 200 inch instrument at Palomar. The world's largest radio telescope would fall nicely into place alongside it.

We settled on a diameter of 300 feet and, in response to a request from Lee DuBridge, I wrote a detailed specification for such an

instrument in May 1952, which included approximate costs, together with
a programme of scientific activities which the telescope might be engaged
upon (16). This plan was not proceeded with in quite that form. At a
later date (in January 1955) I arranged for Bolton and Stanley to be
seconded to Caltech. This was to prove the starting point for radio
astronomy in California.

Back in Australia fortune turned our way once more. For reasons
which I have never completely understood, Van Bush and Alfred Loomis rather
suddenly took the view that this telescope could just as well be built in
Australia, with financial backing from the USA. Perhaps it was the feeling
that America already had too many of the world's good things, and that it
was time to come to the help of other parts of the world. Certainly, the
large Foundations like Ford, Rockefeller and Carnegie itself were soon to
change their emphasis from support for science to assisting social changes
in various parts of the world. A contributory factor was that, since the
start of the war, the Carnegie Corporation had accumulated $250,000 which
they were required to dispose of within the British Commonwealth. This was
made up of several smaller sums, payment of which had been suspended pending
the cessation of hostilities. In April 1954 the Trustees of the Carnegie
Corporation announced that the full sum of $250,000 would be granted towards
the construction of a giant radio telescope in Australia. These were the
very fortunate circumstances under which the Parkes radio telescope finally
got under way.

This was followed, not entirely by chance, by a further quarter
of a million dollars from the Rockefeller Foundation in December 1954. The
person mostly responsible was Warren Weaver, Director of Physical Sciences
for the Foundation. He again was an associate from the war years - a
highly talented mathematician with responsibilities for anti-aircraft and
naval gunnery predictors at OSRD. But there is no doubt that in the background Alfred Loomis also played a prominent part, as he had done for many
years in support of many different branches of science.

When this grant was made, Warren Weaver made a most important
stipulation. It was that the $250,000 from Rockefeller would only be paid
if the Australian Government matched it pound for pound; not only the
Rockefeller grant but all other grants from outside bodies. On hearing
this, the response was immediate. Sir Ian Clunies Ross, the Chairman of
CSIRO, went straight to the then Prime Minister, Robert Menzies, and within
a week or ten days the answer came that the Government would indeed match

the grants. In the event, the Australian Government did even better because
they over-matched, providing not only their share of the capital costs but
the running costs as well. The $250,000 from the Rockefeller Foundation
was followed by a generous addition of $130,000 about a year later when it
became clear that costs were escalating beyond the original estimates. We
had already cut the size of the telescope from 300 to 230 feet in an
effort to contain the costs and, in the end, a further cut to 210 feet was
necessary. Australian scientists will always be grateful for the extraordinarily generous grants from our American colleagues which made this
project possible.

On receipt of the grant from the Carnegie Corporation the
project became active and we engaged in preliminary design studies. The
basic concept was quite clear - a parabolic antenna of the largest possible
aperture, steerable over a large part of the sky - but the options were
wide open. We looked at everything - alt-azimuth and equatorial mounts,
holes in the ground; virtually nothing was excluded at that stage. We
sought advice from the engineering fraternity in Australia, the USA and
Britain and several things became very clear. In those days there was
simply no one in Australia who could handle the design work. Professor
Roderick of the School of Engineering at Sydney University and H.A. Wills
of the Aeronautical Research Laboratory were extremely helpful and both
advised that we should go to Britain for the design work. We knew that
the USA could certainly do it, but the costs were likely to be beyond our
means. So to Britain we went. There it was the great Sir Henry Tizard,
who played such an important role in wartime radar, who advised us first
of all to talk to Barnes Wallis. Incidentally, we also asked Tizard
about the possibility of raising extra funds from philanthropic bodies in
England like the Nuffield Foundation. This turned out to be unproductive
and he advised that our only hope lay in America - what about the Carnegie
Corporation? He did not realize that we already had the grant, or were
about to get it.

So Barnes Wallis was the first man we talked to, a man with a
tremendous reputation as the "dambuster", the designer of the Wellington
bomber and the builder of Britain's only successful airship, the R100. He
was immediately enthusiastic and full of ideas about the structural design
of a giant steerable telescope. These discussions started towards the end
of 1954 and in September 1955 his ideas had formed sufficiently for him to
write a report, which looks as impressive today as it did when we first

received it (17). To illustrate how advanced his thoughts were, Figure 9
shows a sketch taken direct from his report and placed alongside a schematic of the Parkes telescope as actually built. Many details are the same
or closely related, including the geodetic structure supporting the mesh
and details of the azimuthal and elevation drive system. Characteristically, his original drawing did not give a diameter, but he was quite certain
that such an instrument could be built up to a thousand feet in diameter
and he encouraged us to go for it and hang the cost! He refused to see any
problems. Deflection? He had two ways of solving this: At that time he
held patents for incompressible columns, that is, columns of steel or
aluminium which incorporated a servo compensator to adjust for any change
of dimensions under stress. One could break a structure made of such
columns, but one could not bend it! His other solution was one which has
been advocated by any number of people, namely automatic compensation of
the parabolic shape. As is well known, we did eventually build a form of
compensation into the Parkes telescope, the parabola-of-best fit concept,
but this came from Freeman Fox and Partners, not from Barnes Wallis.

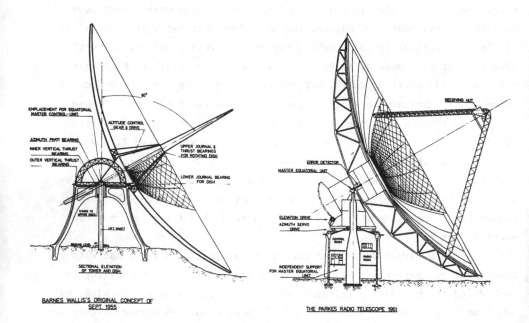

Fig.9 Barnes Wallis's concept (left) of a steerable telescope
in 1955 compared with (right) a schematic of the final design
of the Parkes radio telescope in 1961.

Not only did Barnes Wallis lay down important guidelines on the structural design, but virtually out of thin air, he invented the master equatorial drive system. He did this over lunch at the Athenaeum Club in London, a haven where distinguished scientists are apt to congregate to discuss their problems with kindred souls. Among others, Tizard is reputed to have solved many a wartime problem within those very walls. About half way through lunch, Barnes Wallis asked "How do you propose to drive this machine?" I gave him the conventional answer, which was that we would do it, analogue fashion, from a small equatorial mount built alongside the main telescope. It took him about thirty seconds - certainly not more than a minute. He pointed a finger at me and said "You're wrong. The place to put the equatorial unit is at the intersection of the two axes of rotation. You derive an error signal and servo this back to the main telescope. It's perfectly obvious that's the way to do it." He went away and patented it that same afternoon and from that moment it was part and parcel of the design. It was an enormously successful idea which was to be incorporated in many other giant telescopes.

It was one thing to have concepts from Barnes Wallis, but something else to have a feasible engineering design. We were still to face the problem of who would do the detailed engineering and carry out the calculations necessary to establish the integrity of the structure. Again we focused on Great Britain and, after further search, settled on Freeman Fox and Partners of London in 1956. We sent Harry Minnett to act as our liaison man and to participate in the design of the drive and control system, on which Freeman Fox did not claim to have particular expertise. There followed a three-year period - the engineering design phase - and although it bothered us at the time, it was three years extremely well spent. In the event, the telescope took three years to design and only two and a half years to build. There is an important message there for builders of exotic devices. There is no substitute for extremely careful study at the design stage; this is the way to save endless headaches later on.

The Freeman Fox design contract was completed in 1959, after which the MAN Company (Maschinenfabrik Augsberg Nürnberg A.G.) of West Germany was selected as prime contractor. We were extremely well served by both organizations and the telescope was completed during 1961 (18, 19, 20). It was commissioned at an impressive Opening Ceremony performed by the Governor-General of Australia, Lord de Lisle, on October 31, 1961. To us this was a very satisfying and extraordinarily rapid culmination of what was initially a very long drawn-out process.

At the time of writing, the telescope has completed twenty-one years of sterling service as a research instrument, during which a total of nearly a thousand research papers has been published. It seems destined to remain an important component in the Australian research scene until at least the turn of the century. Photographs of the telescope during construction and after completion are given in Figures 10 and 11.

The construction of the 64 m telescope at Parkes meant that more and more of the Division's resources were steered in that direction and it was inevitable that less attention could be given to the three cross antennas at Fleurs. Reluctantly, it was decided that the Division could no longer provide support for them and in 1962 they were handed over as a going concern to the Electrical Engineering Department of Sydney University. There they continued to produce good research results under the guidance of Christiansen, who had become a Professor at that Department. They continued to give a good account of themselves until converted to a synthesis instrument several years later.

Fig.10 The Parkes 64 m telescope in course of construction, 1959.

Fig.11 A night view of the Parkes radio telescope.

6 OTHER ACTIVITIES OF THE RADIOPHYSICS DIVISION

As mentioned at the outset, there was a host of research activities waiting to be exploited at the end of the war and the Division was not slow to explore them. Some of these are described below. For further information, reference should be made to *A Textbook of Radar*, Chapters XVIII and XIX, ed. E.G. Bowen, Cambridge University Press, 1954.

The cavity resonator

An early experiment was to exploit the high frequency of operation and the exceedingly high power generated by the resonant magnetron. It was simply to excite a 10 cm resonant cavity with the output of a 10 cm magnetron, by which approximately a million volts was developed across the cavity walls. A diagram of the original device is given in Figure 12 (21).

The cavity measured about four inches in each direction, yet in this extraordinarily small space electrons were accelerated to one million

volts. This made an admirable, compact X-ray machine and it was used as
such for a number of years.

An obvious development was to put many such cavities in series
and make a linear accelerator. We seriously considered proceeding in this

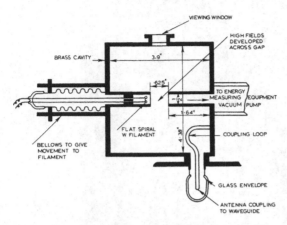

direction but were some-
what frightened by the
prospect. Such a step
would take us into
nuclear physics and put
us into direct compet-
ition with the gigantic
machines already being
planned in Europe and
the USA. Reluctantly,
therefore, we dropped
this option and left the
field to other, more
expert people.

Fig.12 A single-cavity resonator capable
of generating million-volt X-rays.

Electronic Computers

By the end of the war the revolutionary development of all
aspects of pulse generation and the attendant electronic gadgetry had led
to a variety of pulse-forming, pulse-counting and memory circuits. This
sparked the possibility of constructing an all-electronic computer instead
of the mechanical or semi-mechanical devices of the previous era. Microsecond pulses were commonplace at the end of the war and this alone provided the prospect of a machine capable of a million operations a second.

Any mathematical operation can be simulated by an electronic
circuit, but until pulsed radar came along many of the requisite circuits
had simply not been invented. At the end of the war the tools were available and the race began. Many centres became active in this field, notably
MIT and Princeton in the USA and Cambridge and Manchester Universities in
Great Britain.

The Division of Radiophysics was a little slow getting started,
but under the guidance of Pearcey and Beard a creditable machine was built
and was performing mathematical calculations by 1952 (22). This became
known as CSIRAC and was used for routine calculations for a good many years.

It is now in honorable retirement at the Victorian Museum of Applied Science
in Melbourne.

It was not among the first of the electronic machines, but was
contemporary with second-generation machines such as ILLIAC at the University of Illinois and EDSAC at Cambridge, and well up to them in performance.

Radar navigation

A great success of the war was the way in which radar was used
to navigate aircraft operating far from their home bases over land and sea.
The equipment used was intended for military use and was not immediately
adaptable to civil aviation, but there was enormous potential for development. Many navigation systems were proposed and worked upon around the
world, including several at the Radiophysics Division. The main requirement was to re-design military equipment in a compact form for civilian
use and to make it automatic. An early example was the Multiple-Track
Range (MTR) and Distance-Measuring Equipment (DME), which were built in
the Division and installed by Radiophysics personnel on commercial airlines
at an early date. Figure 13 gives the coverage of this system from Sydney

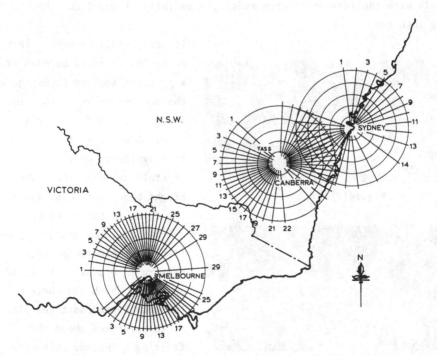

Fig.13 Coverage of an experimental track-guidance and distance-
measuring system on the air route from Sydney to Melbourne, 1947.

to Melbourne in 1947. The Distance-Measuring (DME) part of this system was
adopted by the Australian Department of Civil Aviation in 1950, and over
one hundred ground stations were installed over the continent within the
next few years. By 1954 the equipment became mandatory in all passenger-
carrying aircraft in Austraia. It was not until about ten years later that
a similar system came into operation in the USA and later still on the rest
of the world's airlines (23, 24).

        The DME system has continued in operation in Australia to this
day and is not due for upgrade until 1990. At the time of its introduction
it was every bit as innovative and successful as the INTERSCAN Landing
System, which was also to come from the same Division of CSIRO in 1980.

Cloud and Rain Physics

        The Cloud and Rain Physics work of the Division sprang from the
observation of rain and cloud echoes by both ground and airborne radars
during the latter part of the war. It was found that raindrops and ice
particles above a certain size were good reflectors of microwaves. During
the war this was regarded as a nuisance, and in some cases it interfered
severely with the very short-wave radars, especially around 1 cm, then
coming into use.

Fig.14(a)

It was obvious, however, that
radar itself could be used as
a powerful tool to investigate
the mechanisms by which rain
forms. Figure 14(a) shows a
vertical cross section of a
rain system in which ice
crystals are involved. The
bright band just below freezing level is the point at
which snowflakes melt. Figure
14(b) is the corresponding
pattern in a convective cloud
system in which the whole of
the growth process takes place
at temperatures above the
freezing point and only water
droplets are involved.

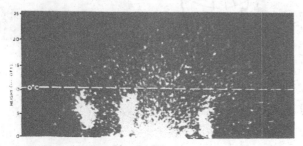

Fig.14(b)

Right at the end of the war the Division began an intensive programme of investigation of the physics of rain formation using both ground and airborne radars, and made practical application of this knowledge to weather modification. The work of this group deserves a short history of its own, but at this point let me simply quote from the words of Horace Byers, Professor of Meteorology at the University of Chicago and past-President of the American Meteorological Society, who said: "It is probably correct to say that no other organization contributed more to practical cloud physics during the period approximately from 1950 to 1961." (Horace R. Byers in *Climate and Weather Modification* (ed. W.N. Hess), Chapter 1, p.23, John Wiley, New York, 1974.)

If the last few pages have departed a long way from radio astronomy, I have done so to make an important point. We have often been asked how it was that in the years 1946 to 1960 a relatively unsophisticated country like Australia could support studies so new and fundamental as to amount to a revolution in astronomical thinking. One answer is simple. For every piece of basic science which was done in radio astronomy at the Division, there was an equivalent piece of applied research - in navigation, on computer applications and in weather modification.

For those years it can truthfully be said that our radio astronomers lived off the work of their applied colleagues. There are many institutes and laboratories around the world where, very properly, fundamental studies are carried out in their own right. But this was not true of the Radiophysics Division in the immediate post-war period, nor is it true today.

7   CONCLUSION

I am sometimes asked what was the philosophy of the Division of Radiophysics during the initial phases of radio astronomy, and what motivated us. I cannot say that we had an articulate philosophy as such, and will leave the professional philosophers to work this out. But our motivation was obvious.

It became clear to many of us during the war that there were all kinds of scientific avenues to be explored using the technology which had been developed for wartime use. The urgency of wartime work prevented this being done, but at the end of the war these ideas were crying out to be exploited. It was not only highly exciting to start these investigations,

but many of us felt an imperative urge to apply them to the extension of
knowledge and in ways which would be of benefit to society. If we had
a philosophy at all, it was to explore every possible avenue - to give any
promising proposal a try and, if it worked or seemed to have scientific
merit, to exploit it to the utmost. In this we received tremendous support
from Sir Frederick White and successive Chairmen of CSIRO. Much of the
camaraderie and sheer joy of living that went into those days came down
to the freedom we enjoyed and the confidence placed in us by successive
Executives of CSIRO.

Above all, we owe a tremendous debt of gratitude to Sir David
Rivett, whose philosophy was very simple - to appoint the very best men
available, give them the facilities, point them in the right direction and
let them go for their lives. This was the philosophy we followed within
the Division and it paid handsome dividends. We lived in fortunate days.

8 REFERENCES

As of 1982, the list of scientific papers on radio astronomy
from the Division of Radiophysics totals over seventeen hundred. The
following is a highly selective list of some of the early papers actually
referred to in the text.

1. Pawsey, J.L. (1946). Observation of million degree thermal radiation
    from the Sun at a wavelength of 1.5 metres. Nature, 158,
    633-4.

2. Pawsey, J.L., McCready, L.L. & Gardner, F.F. (1951). Ionospheric
    thermal radiation at radio frequencies. J.Atmos.Terr.Phys.,
    1, 261-77.

3. McCready, L.L., Pawsey, J.L. & Payne-Scott, Ruby (1947). Solar
    radiation at radio frequencies and its relation to sunspots.
    Proc.Roy.Soc. A, 190, 357-75.

4. Bolton, J.G. & Stanley, G.J. (1948). Variable source of radio
    frequency radiation in the constellation of Cygnus. Nature,
    161, 312-3.

5. Bolton, J.G., Stanley, G.J. & Slee, O.B. (1949). Positions of three
    discrete sources of galactic radio-frequency radiation.
    Nature, 164, 101-2.

6. Payne-Scott, Ruby, Yabsley, D.E. & Bolton, J.G. (1947). Relative
    times of arrival of bursts of solar noise on different radio
    frequencies. Nature, 160, 256-7.

7. Wild, J.P. & McCready, L.L. (1950). Observations of the spectrum of
    high-intensity solar radiation at metre wavelengths.
    I - The apparatus and spectral types of solar burst observed.
    Aust.J.Sci.Res. A, 3, 387-398.

8.   Piddington, J.H. & Minnett, H.C. (1949). Microwave thermal radiation
     from the Moon. Aust.J.Sci.Res. A, 2, 63-77.

9.   Kerr, F.J., Shain, C.A. & Higgins, C.S. (1949). Moon echoes and
     penetration of the ionosphere. Nature, 163, 310-3.

10. Pawsey, J.L. (1951). (Cable advising that Christiansen and Hindman
     had confirmed Ewen and Purcell's discovery of hyperfine
     structure of the hydrogen line in galactic radio spectrum).
     Nature, 160, 358.

11. Kerr, F.J., Hindman, J.V. & Stahr Carpenter, Martha (1957). The
     large-scale structure of the Galaxy. Nature, 180, 677-9.

12. Mills, B.Y. & Little, A.G. (1953). A high-resolution aerial system
     of a new type. Aust.J.Phys., 6, 272-8.

13. Mills, B.Y. (1955). The observation and interpretation of radio
     emission from some bright galaxies. Aust.J.Phys., 8, 368-89.

14. Shain, C.A. (1957). Galactic absorption of 19.7 Mc/s radiation.
     Aust.J.Phys., 10, 195-203.

15. Christiansen, W.N., Mathewson, D.S. & Pawsey, J.L. (1957). Radio
     pictures of the Sun. Nature, 180, 944-6.

16. Bowen, E.G. (1952). Draft Programme for a Radio Observatory, May
     1952. Unpublished report, CSIRO Division of Radiophysics,
     file A1/3/11/1, 11 pp.

17. Wallis, B.N. (1955). Giant Radio Telescopes, September 1955.
     Unpublished report, CSIRO Division of Radiophysics, file
     A1/3/11/32, 13 pp + 1 fig.

18. Bowen, E.G. (1963). Radio Astronomy and Giant Telescopes. In
     Clerk Maxwell and Modern Science, ed. C. Domb, Chapter V,
     pp. 89-101, Athlone Press, London.

19. Minnett, H.C. (1962). The Australian 210-foot radio telescope.
     Sky and Telescope, XXIV, 184-9.

20. Kerr, F.J. (1962). 210-foot radio telescope's first results.
     Sky and Telescope, XXIV, 254-260.

21. Bowen, E.G., Pulley, O.O. & Gooden, J.S. (1946). Application of
     pulse technique to the acceleration of elementary particles.
     Nature, 157, 840.

22. Beard, M. & Pearcey, T. (1952). An electronic computer.
     J.Sci.Instrum., 29, 305-11.

23. Beard, M. (1949). The Multiple-Track Range. Proc.Instn.Elect.
     Engrs., 96, Pt III, 245-51.

24. Burgmann, V.D. (1949). Distance-Measuring Equipment for aircraft
     navigation. Proc.Instn.Elect.Engrs. 96, Pt III, 395-402.

Ruby Payne-Scott, Alec Little, and "Chris" Christiansen circa 1949 at the Potts Hill field station of the Radiophysics Division (courtesy Christiansen)

THE FIRST DECADE OF SOLAR RADIO ASTRONOMY IN AUSTRALIA

W. N. Christiansen
Mount Stromlo and Siding Spring Observatories
Private Bag, Woden P.O. ACT 2606 Australia.

The first radio emissions detected from outside the earth's atmosphere were those from the Galaxy. It may seem strange, therefore, that so much of the early activity of radio astronomical groups was centred on Solar observations. This early focussing of interest on the Sun followed from war-time studies of radio interference to radar observations; most of the early radio astronomers were radar engineers and were familiar with this work or had taken part in it. Furthermore, they were not astronomers and the Sun as an astronomical object held less fears of the unknown than did the general field of astronomy; this must have had its effect. In Australia the largest of two groups, that in the Radiophysics Laboratory of CSIRO, was composed of physicists and engineers who at the end of the war had been switched from radar to radio astronomy. To them had been added a few physicists from other organizations. A second smaller group was at the Commonwealth Solar Observatory at Mount Stromlo. It included one professional astronomer, C. W. Allen, and an ionospheric physicist, D. F. Martyn.

The Sydney group directed by E. G. Bowen had as its scientific leader J. L. Pawsey, a physicist/engineer of infectious enthusiasm with the interest and ability to contribute in detail to each of the individual research projects being undertaken. His enthusiasm, combined with the youthfulness of the group and the interest inherent in exploring unchartered territory, produced an air of excitement that few scientific groups seem lucky enough to experience. With the eagerness of the young, the scientists shared new ideas with their colleagues as soon as the ideas were formed. This continual interchange produced a group strength that was much greater than the sum of the strengths of its members. The field work had a pioneering appearance. Each morning people set off in open trucks to the field stations where their equipment, mainly salvaged and modified from radar installations, had been installed in ex-army and navy

THE FIRST DECADE OF SOLAR RADIO ASTRONOMY IN AUSTRALIA

Fig. 1   The 4.8 m x 4.3 m ex-radar antenna used by Lehany and
Yabsley for solar observations at 150, 50 and 25 cm wavelength
during 1946-48.   The Yagi antennas of Payne-Scott and Little
used at metre wavelengths are in the background.

Fig. 2   The 10 cm wavelength radiometer of Piddington and
Minnett.   The "quarter-wave plate" used to convert linearly
polarised waves to circularly polarised is seen in the aperture.

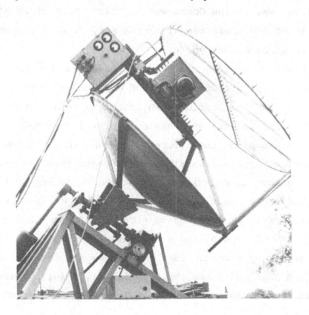

huts. At the field stations the atmosphere was completely informal and
egalitarian, with dirty jobs shared by all. Thermionic valves were in
frequent need of replacement and old and well-used co-axial connectors
were a constant source of trouble. All receivers suffered from drifts
in gain and "system-noise" of hundreds or thousands of degrees
represented the state of the art. During this period there was no place
for observers who were incapable of repairing and maintaining the
equipment. One constantly expected trouble.

One peculiarity of the Sydney group was the number of field
stations. This was partly the result of the taking over of a number of
former radar sites, but it continued because maintenance work and
observations at the same site by different groups produced mutual
electrical interference. Each group sought isolation for this reason
but also because it suited the style of work of most members of the
groups.

The earliest radio observations of the Sun consisted simply of
measurements of the intensity of the Solar radio emissions. Such
measurements were being carried out in several countries and required
techniques not used before to obtain the accuracy required for useful
scientific information. What was in need of investigation was how the
intensity of the solar emissions changed with the wavelengths of the
radiation and also how the emissions varied from day to day. It was
realised very early that there were several components of the radiation
and these had to be considered separately. The components could be
divided into two broad groups, one being quasi-constant in intensity and
the other variable; the first was more prominent at centimetre wavelengths and the second was very marked at metre wavelengths. As the
techniques used to investigate these markedly differ, we shall treat them
separately.

THE QUIET SUN AND THE SLOWLY-VARYING COMPONENT

In Australia daily measurements of the quasi-constant solar
intensity were made from 1946 for several years at wavelengths of 1.5
metres by Allen, at 150, 50 and 25 cm by F. J. Lehany and D. E. Yabsley
(see Fig. 1) and at 10 and 3 cm (Fig. 2) by J. H. Piddington and
H. G. Minnett (1951). A slow variation of the intensity from day to day
occurred at all of the four shortest wavelengths and, as was shown by
A. E. Covington in Canada (see his contribution elsewhere), this

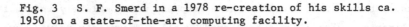

Fig. 3   S. F. Smerd in a 1978 re-creation of his skills ca.
1950 on a state-of-the-art computing facility.

variation was proportional to the area of visible sunspots. The lower
level of radiation, corresponding to zero sunspot area, was called the
"Quiet Sun" level. At a wavelength of 1.5 m Pawsey and Yabsley in 1946
found that this level appeared to correspond to a black-body temperature
of 10^6 K. Martyn pointed out that this corresponded to a coronal
temperature of 10^6 K, for which some optical evidence had already been
found (Martyn 1946). The equivalent temperature of the quiet Sun at
shorter wavelengths, however, was observed to diminish with wavelength,
corresponding to a decreasing optical depth in the corona. As the
wavelength decreased, the source of radiation came more and more from the
cooler inner regions of the solar atmosphere. S. F. Smerd and K. C.
Westfold (joined later by J. A. Roberts), who formed a theoretical group
in the Radiophysics Laboratory, worked out the theory of the quiet Sun
radiation by adapting the equation of transfer, used to investigate
stellar interiors, to the ray paths that they determined for the ionised
atmosphere of the Sun   (Fig. 3). This work (Smerd & Westfold 1949)

remains a standard reference to the present time. One of the predictions from this work (Smerd 1950) was that radio limb-brightening of the solar disk should be found at certain wavelengths.

At the same time as the quiet Sun was being investigated, the nature of the slowly-varying component was being studied.

The origin of this component near sunspot areas was confirmed at an eclipse in 1947 when A. E. Covington showed that a large change in radio emission occurred during the covering and uncovering of a large sunspot group by the Moon's disk (see his article). In the next year there was a solar eclipse visible in Australia. Yabsley, B. Y. Mills and I (Christiansen et al. 1949) established three widely separated observation posts for measuring the solar radiation at λ = 50 cm during the eclipse. This was Mills' introduction to radioastronomy — he had been working on a linear accelerator after his radar investigations ceased. It was also his last, as well as his first, investigation in solar astronomy.

We located eight regions of enhanced radio brightness, six at positions occupied or previously occupied by sunspots. The emitting regions were found to have an average diameter of 3 arcmin and a brightness temperature of about 5×10^5 K.

In order to carry the investigations further I felt it necessary to devise some method of viewing the Sun much more frequently than was possible with eclipse observations. This of course meant devising some antenna system of very great directivity. One path to this had been pointed out by L. L. McCready, J. L. Pawsey & Ruby Payne-Scott (1947). In order to locate a source of radiation on the Sun, they devised a sea-reflection interferometer in which an antenna placed on a cliff above the sea received waves directly from the Sun and also waves reflected from the sea. The interference fringes produced by the two sets of waves allowed them to find the position of the source with great accuracy.

This type of interferometer was only useful for locating a single point source, as was the two-antenna interferometer used at about the same time at Cambridge. However, in their 1947 paper McCready et al. noted that had they been able to vary the distance between the antenna and its image they could have obtained, by Fourier synthesis, the distribution of radio emission over the whole Sun. The idea was not followed up in Australia, but Fourier synthesis was soon developed

rapidly at Cambridge as is described by Scheuer elsewhere. At Sydney
the path followed to obtain high resolution was different and was aimed
at sources on the Sun that were variable and therefore not suited to any
method of observation that involved hours or days to complete. Our
solution followed closely on my previous work on short wave communication —
this time not on an array of rhombic antennae but on a linear array of
paraboloids. The idea occurred while reading a description of Bernard
Lyot's optical filter in which narrow frequency passbands are produced
at widely different frequencies. This may seem peculiarly indirect when
the analogy which is more obvious is the optical diffraction grating, but
to me as an antenna designer the cos n.cos 2n.cos 4n series of the Lyot
filter immediately suggested an antenna array and an array of arrays.
The feature of a spaced array of paraboloids was that the individual
elements could be made steerable to follow the Sun over many hours while
the array factor provided narrow multiple responses which could be spaced
so that only one at a time should fall on the Sun. These fringes would
scan the Sun in succession as it moved across the sky. The main
difficulty was cost, because in 1950 an antenna in which the materials
cost £100 was considered expensive. Since the antenna was to have a
one-dimensional resolving power of 1/20 degree, better than any antenna
in existence, Pawsey was enthusiastic and I was given permission to build
a 32 element array if it could be built for a material cost of £500. Our
mechanical engineer, K. McAlister, an Australian-born Scot who looked at
every bawbee at least twice, came up with the answer that he would produce
32 6-ft diameter paraboloids on mounts for £10 each. I had to rival this
with a cheap transmission line and amidst sarcastic remarks from my radar
minded colleagues I devised a pine-tree network of balanced twin-wire
lines to carry the currents from each antenna over equal lengths of line
to the receiver. Apart from cheapness and low ohmic loss the open-wire
lines would make phase adjustments extremely easy, I claimed, and Pawsey
agreed. He contributed a very simple device from his early days in
English television to suppress the unbalanced mode in the line.

         The array was built in 1951 along the east-west wall of the
Potts Hill Reservoir near Sydney (see Fig. 4). This site, which was
already being used by Payne-Scott & A. G. Little for their metre wavelength solar observations, had a flat path along each wall of sufficient
size to allow the installation of a 1000 wavelength array to operate at
21 cm. Each of the 32 paraboloids was mounted on a very simple polar
axis placed on a wooden pole (Fig. 5) and could be turned by hand to

Fig. 4   The first grating array at Potts Hill Reservoir, during construction in 1951.

Fig. 5   E. V. Appleton (center) and B. van der Pol (right) inspecting an antenna of the Potts Hill grating array, while attending the 1952 General Assembly of URSI (author in back).

follow the movement of the Sun. At quarter hourly intervals and in the
short time between successive scans of the Sun, this involved a 200 metre
run to push forward each of the 32 antennas by one "stop". Observations
showed the contribution of individual regions of enhanced radio brightness on the solar disk. They gave the size and brightness temperatures,
as well as the height of the emitting regions in the lower corona. The
conclusion from this investigation was that the regions corresponded
closely in size with areas of chromospheric plage and were located above
them. They were named "radio plage". Since the apparent temperature of
the regions did not appear to exceed markedly that of the corona, it was
concluded that they were simply regions of enhanced coronal density where
the high-temperature corona became optically thick. It is in these
regions also that non-thermal radiations originate.

In Fig. 6 a number of daily observations that J. A. Warburton
and I made during 1952 and 1953, a low sunspot period, are shown superimposed. The inner envelope of these measurements was taken to represent
the "quiet Sun" level and its shape suggested that the radio brightness
distribution of the Sun was far from circularly symmetrical. A second
array, this time along the north-south wall of the dam, was built in
order to scan the Sun from pole to pole (Fig. 7). These scans showed
immediately that the emission near the poles was considerably less than

Fig. 6   One-dimensional scans (resolution 3 arcmin) of the
Sun during a minimum sunspot period, 1952–3, taken near noon
each day.

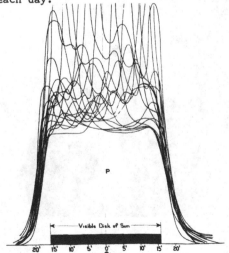

Fig. 7   The two Potts Hill gratings (1953), one east-west
and the other north-south, used for earth-rotational synthesis
observations of the Sun.

that near the Solar equator. At this stage it became obvious that the
two-dimensional brightness distribution over the quiet Sun should be
determined. Meanwhile at Cambridge the technique of observing radio
sources by means of aperture synthesis had been developing by the use of
two-antenna interferometers which had to be moved around for different
orientations and distances to obtain two-dimensional maps. The technique
however had not reached the stage where the phase angles of currents
could be determined with any certainty. Hence the maps derived were not
complete. It occurred to me that the two arrays at Potts Hill were
ideally suited to make a two-dimensional synthesis map of the Sun, since
with calibrated fixed arrays there was no problem of unknown phase angles.
The   steerable paraboloids could be turned to follow the Sun from dawn
to dusk and by this means most of the required range of scanning angles
of the Sun could be covered. Thus the Potts Hill arrays were used for
purely earth-rotational synthesis for the first time. The two-dimensional

map of the quiet Sun showed enhanced brightening near the limbs in the equatorial region (Fig. 8). Since this map was made, linear arrays of steerable paraboloids have become the normal tools for high resolution radio observations. This measurement at Potts Hill however, was the last use of aperture synthesis in the Radiophysics Laboratory for many years. The fact that the Fourier processes had to be done by hand, so that the map shown in Fig. 8 took half a year to compute probably deterred further work, but another factor was that those interested in high resolution observations in Sydney were dealing with sources that were variable or else were thought to vary because of ionospheric changes. Hence observations that required a long time to complete were treated with suspicion.

The two arrays placed at right angles at Potts Hill suggested a new type of antenna more suited to variable sources. While visiting Potts Hill one morning in 1953 Mills asked me why we did not couple the two arrays to produce high resolving power in two dimensions. During the ensuing discussion it was agreed that for this to be effective the centres of the two arrays must not be separated (as they were in the Potts Hill antenna), and also that some means had to be devised to multiply the outputs of the arrays. By the next morning Mills had devised the Cross Antenna consisting of a pair of thin orthogonal antennas with their outputs multiplied to give a single narrow response.

Fig. 8   The quiet Sun radio brightness distribution at
λ = 21 cm from observations with the two Potts Hill gratings
in 1952-3.

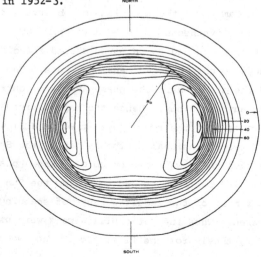

Fig. 9   Grating Cross antenna at Fleurs field station in 1957. Two 320 m arrays, each of 32 six-metre diameter paraboloids, operated as a Cross antenna at λ = 21 cm.

Although the idea of rearranging our two gratings to produce responses narrow in both dimensions was not taken up at the time, it was brought up again a year or two later by R. N. Bracewell. But again it was rejected on the grounds that the two individual orthogonal gratings were giving all the required scientific information on the bright solar regions, and that the only two-dimensional Solar map of any significance was of the quiet Sun and the two gratings had already supplied this by means of Fourier synthesis. Later, however, it was decided to build a crossed grating antenna at Fleurs, another Radiophysics field site, in order to produce daily maps of the Sun. This was done although the scientific prospects were not clear, and the cost, because of the necessary increase in collecting area, would be at least ten times that of the original pair of gratings. This crossed grating interferometer (Fig. 9) commenced operating at a wavelength of 21 cm in 1957 (Christiansen, Mathewson & Pawsey 1957) with a resolving power of 3 arc-min. At the same time Bracewell (by now at Stanford) was building a

Fig. 10   Some early radio "pictures" of the Sun at λ = 21 cm
(September 1957).

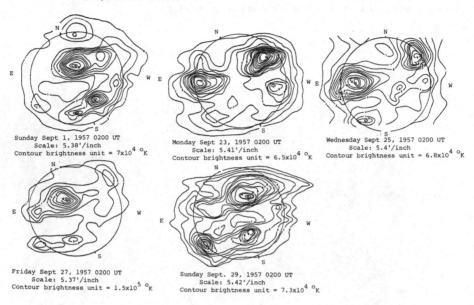

Sunday Sept 1, 1957 0200 UT
Scale: 5.38'/inch
Contour brightness unit = 7x10^4 ^{o}K

Monday Sept 23, 1957 0200 UT
Scale: 5.41'/inch
Contour brightness unit = 6.5x10^4 ^{o}K

Wednesday Sept 25, 1957 0200 UT
Scale: 5.4'/inch
Contour brightness unit = 6.8x10^4 ^{o}K

Friday Sept 27, 1957 0200 UT
Scale: 5.37'/inch
Contour brightness unit = 1.5x10^5 ^{o}K

Sunday Sept. 29, 1957 0200 UT
Scale: 5.42'/inch
Contour brightness unit = 7.3x10^4 ^{o}K

T-type grating to produce similar Solar radio maps at a wavelength of
10 cm. Both instruments produced Solar maps over a period of many years
(Fig. 10).

The old gratings at Potts Hill were eventually removed, one
being sent to India since Swarup & Parthasarathy (1958) had already used
this grating at Potts Hill for Solar studies, after modifying it to work
at a wavelength of 60 cm.

THE RAPIDLY VARYING RADIO EMISSIONS

The solar investigations described above were found to be
concerned with classical thermal emission. They led to a better understanding of the Sun's outer atmosphere, but to no spectacular discoveries.
Their real excitement was in the development of new instruments superior
in resolving power to any in existence. It was in the studies of the
rapidly varying radio emissions from the Sun that the most unexpected
and therefore interesting scientific results were obtained. In the
earliest work in Sydney these variable components were the ones that
aroused greatest interest, both because of their enormous intensity

Fig. 11   The time sequence of phenomena occurring with a radio "outburst" (Payne-Scott, Yabsley & Bolton 1947).

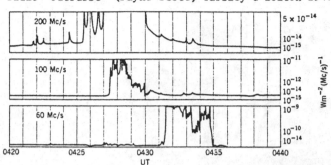

compared with the quiet Sun emission, indicating non-thermal origin, and because of their associated geophysical effects. Because these emissions took a number of different forms, an attempt was made to divide them into categories. Allen distinguished between two types of emission - the "outburst" associated with solar flares and the "noise-storm", a longer-lived phenomenon associated with large sunspot groups. Pawsey added a third type, the "isolated burst". In 1946 Martyn discovered that in a noise-storm the radiation was circularly polarized, indicating the presence at the source of an intense magnetic field.

In the following year two very important discoveries were made. With the sea interferometer described earlier, McCready, Payne-Scott & Pawsey (1947) located a source of intense radiation on the Sun above a large sunspot. The second discovery, by Payne-Scott, Yabsley & Bolton, was that during an outburst there was a delay between the measured start of the increase in intensity at different wavelengths - the shortest came first and the longest last. They suggested that this was due to a disturbance being ejected through the solar atmosphere and calculated that its velocity was about 500 to 700 km s^{-1}, of the same order as the velocity of particles reaching the earth after a solar flare.

Following this discovery, Pawsey immediately decided that isolated frequency observations were inadequate to study the signal delays, and what was needed was a swept-frequency receiver that could obtain dynamic spectra of bursts. He gave the task of building the receiver to McCready, a solid and trusted engineer, and to a young physicist not long out of the British Navy, J. P. Wild. At the same time, Payne-Scott and Little developed a means of rapid scanning of the

Fig. 12   J. D. Murray with the three-band radio spectrometer
at Dapto in 1952.

Sun to detect the location and movements of the source of radiation
during an outburst or a noise-storm (see below).

Wild and McCready's (1950) first radio spectrograph, completed
in 1949, was tunable in 1/10 second over the frequency range 70 to 130
MHz and was connected to a rhombic antenna usable over this range
without adjustment.   The rhombic antenna was constructed over a cross-
shaped frame and was pointed at the Sun by adjusting guy ropes attached
either to a vertical pole or to a point on the ground.   From the start
this equipment produced exciting results, and Wild, Murray & Rowe (1954)
rapidly went ahead to extend the frequency range by building three
separate receivers, each connected to its own rhombic antenna   (Fig. 12).
This system went into use in 1952 at the new location of Dapto, south of
Sydney, covering the frequency range 40 to 240 MHz.   Furthermore, by
building each antenna as a pair of rhombics in orthogonal planes,
polarisation measurements could be made   (Fig. 13).

The early measurements of the dynamic spectra of outbursts
(Wild 1950a) clearly confirmed the delay between the arrival of the
higher frequency and the lower frequency spectral features   (Fig. 14).
A delay rate of 4.5 seconds per MHz was typically found.   The velocity
of a disturbance producing radiation of different frequencies as it

Fig. 13   J. P. Wild adjusting the transmission line of one of
the three dual-polarization rhombic antennas of the Dapto
spectrometer (1952).

travelled outwards through the solar atmosphere was calculated to be
about 500 km s^{-1}, confirming the earlier spot-frequency measurements
which had tentatively led to a connection between the source and auroral
particles.

Simultaneously, from 1949 until 1953, the investigation of
Payne-Scott and Little (1951) was being carried out by means of a two-
antenna interferometer (Fig. 15) in which one channel was phase-rotated
so that the fringes were swept over the solar disk. They (1952) located
the source of noise storms at 97 MHz high in the corona above sunspots.
During a solar outburst the source was seen to move outwards from the
flare region at velocities between 600 and 3000 km s^{-1}, higher than the
velocities found by Wild, but of the right order of magnitude. At this
time the old terms for the bursts ceased to be used and were replaced by
the starker terms, Type I (the old storm burst), Type II (the old

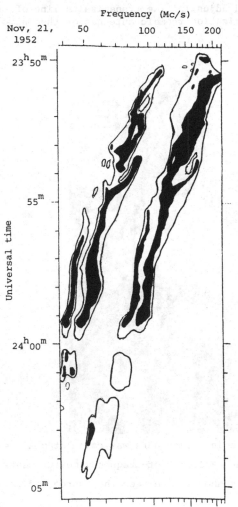

Fig. 14   Dynamic spectrum
of a Type II burst (Wild,
Murray & Rowe 1953).

outburst), Type III (the old isolated burst), Type IV (the longer lived
and circularly polarised second stage of some outbursts), and Type V (no
old name).

The big surprise in the spectroscopic work came from the
dynamic spectra of isolated bursts. The delay drifts for these were
1/20 second per MHz, i.e., nearly 100 times more rapid than for outbursts
(Fig. 16). If the same velocity calculation were made for the two types,
the velocity of the source in the case of isolated bursts came out in the
range 1/10 to 1/3 of the velocity of light (Wild 1950b). This long

Fig. 15   The rotating-fringe interferometer at Potts Hill
used by Payne-Scott and Little over the years 1949 to 1953
to locate sources of noise storms in the Solar corona.

Fig. 16   Dynamic spectrum of a Type III burst (Wild 1950b).

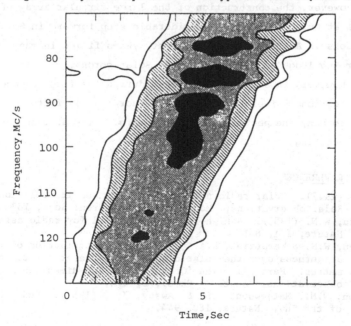

remained an intriguing mystery, as there were no known movements in the
Solar atmosphere corresponding to these velocities. The spectroscopic
results, including the observation of harmonic components of Solar
bursts, had confirmed the hypothesis that the origin of Type II and
Type III bursts was plasma oscillation produced by either shock waves or
streams of ionised particles travelling outwards through the solar
atmosphere. A further confirmation however was needed and this was the
location of different frequency components of a burst at different radial
distances from the Sun. This had to wait until Wild, Sheridan and
Neylan (1959) built a swept-frequency interferometer. The results of
this, combined with spectroscopic observations, gave the strongest
confirmation of the hypothesis. The exact mechanism of the production of
the solar bursts has provided a centre of activity for theoretical plasma
physicists to the present day.

CONCLUSION

It may be said that during the first decade of radio astronomy,
Solar work occupied the most prominent place in Australia. After that
time and with the building of the Parkes telescope the interests of many
radio astronomers shifted outwards from the Solar system to more distant
objects. However, the construction of the large circular array of
paraboloids at Culgoora marked a considerable step forward in Solar
investigations in Australia and over twenty years filled in many of the
gaps in our knowledge of processes in the Solar corona.

However, those of us who worked during the first decade of
Solar radio astronomy feel that we were extraordinarily fortunate to be
present at so many unexpected scientific events. Life at that period
was indeed exciting.

REFERENCES

Allen, C.W. (1947). Solar radio noise of 200 Mc/s and its relation to
         Solar observations. Mon. Not. Roy. Astron. Soc., 107, 386.
Christiansen, W.N. (1953). A high-resolution aerial for radio astronomy.
         Nature, 171, 831.
Christiansen, W.N. & Warburton, J.A. (1955). The distribution of radio
         brightness over the Solar disk at a wavelength of 21 centi-
         metres. Part III. The 'Quiet' Sun. Two-dimensional
         observations. Aust. J. Phys., 8, 474.
Christiansen, W.N., Mathewson, D.S. & Pawsey, J.L. (1957). Radio pictures
         of the Sun. Nature, 180, 944.

Lehany, F.J. & Yabsley, D.E. (1949). Solar radiation at 1200 Mc/s, 600 Mc/s, and 200 Mc/s. Aust. J. Sci. Res., A2, 48.

McCready, L.L., Pawsey, J.L. & Payne-Scott, R. (1947). Solar radiation at radio-frequencies and its relation to sunspots. Proc. Roy. Soc. (Lond), A190, 357.

Martyn, D.F. (1946). Polarization of Solar radio-frequency emissions. Nature, 158, 308.

Payne-Scott, R., Yabsley, D.E. & Bolton, J.G. (1947). Relative times of arrival of bursts of Solar noise on different radio-frequencies. Nature, 160, 256.

Payne-Scott, R. & Little, A.G. (1951). The position and movement on the Solar disk of sources of radiation at a frequency of 97 Mc/s. II. Noise storms. Aust. J. Sci. Res., A4, 508.

Payne-Scott, R. & Little, A.G. (1952). The position and movement on the Solar disk of sources of radiation at a frequency of 97 Mc/s. Part 3 - Outbursts. Aust. J. Sci. Res., A5, 32.

Piddington, J.H. & Minnett, H.C. (1951). Solar radio-frequency emission from localized regions at very high temperatures. Aust. J. Sci. Res., A4, 131.

Smerd, S.F. & Westfold, K.C. (1949). The characteristics of radio-frequency radiation in an ionized gas, with applications to the transfer of radiation in the Solar atmosphere. Phil. Mag., 40, 831.

Smerd, S.F. (1950). Radio-frequency radiation from the quiet Sun. Aust. J. Sci. Res., A3, 34.

Swarup, G. & Parthasarathy, R. (1958). Solar brightness distribution at a wavelength of 60 centimetres. Part II - Localized radio bright regions. Aust. J. Phys., 11, 338.

Wild, J.P. & McCready, L.L. (1950). Observations of the spectrum of high-intensity Solar radiation at metre wavelengths. I - The apparatus and spectral types of solar burst observed. Aust. J. Sci. Res., A3, 387.

Wild, J.P. (1950a). Observations of the spectrum of high-intensity Solar radiation at metre wavelengths. II - Outbursts. Aust. J. Sci. Res., A3, 399.

Wild, J.P. (1950b). Observations of the spectrum of high-intensity Solar radiation at metre-wavelengths. III - Isolated bursts. Aust. J. Sci. Res., A3, 541.

Wild, J.P., Murray, J.D. & Rowe, W.C. (1954). Harmonics in the spectra of solar radio disturbances. Aust. J. Phys., 7, 439.

Wild, J.P., Sheridan, K.V. & Neylan, A.A. (1959). An investigation of the speed of the Solar disturbances responsible for Type III radio bursts. Aust. J. Phys., 12, 369.

Joe Pawsey and Frank Kerr in the late 1940s (courtesy CSIRO Radiophysics Division)

EARLY DAYS IN RADIO AND RADAR ASTRONOMY IN AUSTRALIA

F. J. Kerr
University of Maryland

As in other countries, the first discoveries of Jansky and
Reber in radio astronomy made little impression in Australia. Rather,
interest in the subject was first aroused by news of the wartime discoveries of radio emission from the Sun by J. S. Hey in England and G. C.
Southworth in the United States. As World War II wound down in 1945, the
CSIR (later CSIRO) Radiophysics Laboratory in Sydney was ending its work
on the development of radar equipment for the South West Pacific theatre
of war. A wise decision was made to keep this highly experienced and
imaginative group of people together and to look for new scientific and
technical challenges. With the strong administrative encouragement of
E. G. (Taffy) Bowen and the scientific foresight of J. L. Pawsey, the
group turned its curiosity to radio astronomy, and in this way Australia
got an early start in the opening up of this new branch of science (see
the contribution by Bowen elsewhere in this volume). Those were exciting
days and new discoveries came rapidly as existing equipment could be
turned over to the exploration of the almost unknown radio sky.

"COSMIC NOISE"

Some experiments had been made in 1944 by Pawsey and Ruby
Payne-Scott from the roof of the Radiophysics Laboratory building in the
grounds of Sydney University, but the first significant observations were
made by pointing antennas of Air Force and Army radar stations near
Sydney at the Sun. The first years of solar work in Australia are
described by W. N. Christiansen elsewhere in this volume.

The first "cosmic" radio astronomy work followed soon afterwards when J. G. Bolton and G. J. Stanley used the sea interferometer at
Dover Heights to detect and identify the first "radio stars," as they
were called at that time. As in the solar case, J. S. Hey made the

pioneering discovery. He and his colleagues reported in 1946 that signals
from the constellation of Cygnus showed fluctuations resembling those
shown by the radio waves from the disturbed Sun. They drew the unexpected
conclusion that there must be a source of small angular size in this
region of the sky. Bolton and Stanley (1948) pointed the sea interferometer in this direction and located a source of angular size less than
their resolution (5 minutes of arc), which they called Cygnus A. There
was no outstanding optical object in this direction, and it took another
five years before Baade and Minkowski at Palomar identified this radio
source with a very distant galaxy (see the article by Greenstein elsewhere
in this volume).

The sea interferometer principle had been developed during
the war years, as a method which increased sensitivity in detecting and
measuring the height of low-flying enemy aircraft. The two "antennas" of
the interferometer are the physical antenna itself and its image below
sea level. Bolton, Stanley and Slee (1949) soon discovered several more
discrete sources of radiation with this system, and went on to identify
three of them with optical objects. The first such identification was
of the radio source Taurus A with the Crab nebula, the remnant of the
supernova explosion of A.D. 1054. The other two were Centaurus A and
Virgo A, the galaxies NGC 5128 and M 87.

The next major development in the study of so-called "cosmic
noise" was the survey by Bolton and K. C. Westfold (1950) of the accessible parts of the radio sky at a frequency of 100 Mc/s (to use the
then-current name for the unit of frequency). This survey was carried
out on the cliff site at Dover Heights, using a 3 x 3 array of yagi
antennas, with a beamwidth of 17°. It gave a plot of the distribution
of galactic radiation, and an early demonstration that radio methods
could be applied to the study of galactic structure.

Then followed the work of B. Y. Mills (1952), using a two-
element interferometer looking upwards or at points on the meridian (at
a frequency near 100 Mc/s). He discovered many more discrete sources,
and showed that there were two distinct classes, one of which was
concentrated to the Milky Way, while the other was distributed more
or less uniformly over the rest of the sky--in other words, galactic
and (what eventually came to be recognized as) extragalactic discrete
sources (see the contribution by Mills elsewhere in this volume for
further discussion). By this time, the development of the subject was
well on its way.

Figure 1   Joseph Lade Pawsey

The inspiration and driving force of all the early Australian work was Joseph Lade Pawsey (Figure 1). He was a brilliant research leader, who always knew the right questions to ask, usually simple ones. He actively led each part of the research program, but had the ability to develop the less experienced members of his group into independent thinkers who then became leading contributors in their respective fields. His untimely death in 1962 at the early age of 54 was a great blow to the development of radio astronomy, not only in Australia but throughout the world. Shortly before his death, he had been named to take over the Directorship of the National Radio Astronomy Observatory at Green Bank, West Virginia.

In his own account of the early years of Australian radio astronomy, Pawsey (1961) pointed out the way that simple equipment led to the development of more and more complex equipment in a step-by-step manner as each stage produced new phenomena that needed to be elucidated. He then went on to say:

"It should be noted that (this process) can only be followed effectively in a well organized scientific organization in which the scientific direction can very quickly make decisions and supply facilities for the really promising developments. In all too many cases elsewhere the energies of scientists are taken up in advertising the potentialities of their prospective investigations in order to obtain any support at all. The result is a neglect of the unspectacular preliminary probing investigations which are often such a vital ingredient in success."

RADAR ASTRONOMY

My own first work in the field was in radar astronomy. I
came into it many months after the end of the war, because I had been
finishing off an elaborate study of the correlation between the performance of Australian Air Force radar stations and meteorological conditions
(Kerr 1948).

In 1946, C. A. Shain and I began exploring the possibilities
of radar in astronomy. We started with a short period of work on echoes
from meteors, following the earlier experience of Lovell and others in
England, but we soon turned our attention to the Moon. In 1946 a group
from the U.S. Army Signal Corps (De Witt and Stodola 1949) were the first
to receive moon echoes, in "Project Diana " at Fort Monmouth on the coast
of New Jersey. They worked at 100 Mc/s, where the Earth's atmosphere is
effectively transparent; in contrast we carried out our experiments at
two frequencies near 20 Mc/s, where we could expect to be able to study
the ionosphere as well as the Moon (Kerr and Shain 1951). Another very
pragmatic reason for choosing this frequency range was that we could work
there without having to build our own transmitter, because we were able
to use transmitters of the short-wave broadcasting station, Radio
Australia. We set up a rhombic antenna in a well-shielded valley at
Hornsby, near Sydney. This antenna pointed at a low elevation in the
same direction as that of the North American service of Radio Australia,
400 miles away. We soon learned all about the motions of the Moon,
because our observations were limited to the 15-20 days of the year when
the Moon rose into our beam in the small hours of the morning; only then
were the transmitters not being used for broadcasting because the
ionosphere was not a good reflector, precisely the conditions we needed.

We controlled the transmitter over a landline from Hornsby,
and for most of the time we sent out continuous-wave pulses which were
just over two seconds in length. (The round trip to the Moon and back
takes about 2½ seconds.) In this way we were able to study the short-
term fluctuations of the signal as well as the longer-term fading.
Sample echoes are shown in Figure 2. We showed that the shorter-term

Figure 2 Three successive moon echoes obtained with 2.2-second signals, illustrating the rapid fading (Kerr & Shain 1951).

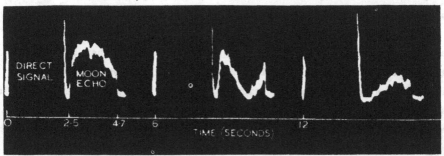

variations were due to the libration of the Moon, the slow rotation on its own axis which results in us seeing the Moon in a slightly different aspect from hour to hour or day to day. The libration resulted in varying interference between the very numerous components of the echo that came back from different parts of the Moon's rough surface. The longer-term fading (of the order of minutes) was shown to be due to ionsopheric effects, but since we could not measure polarization, we were not able to demonstrate the relation of this fading to Faraday rotation in the ionosphere; this point was established some years later by J. V. Evans at Jodrell Bank.

After the lunar radar work, I carried out an extensive and pioneering study of the likely characteristics of echoes from the Sun and the planets, and what equipment would be needed to study them (Kerr 1952). We decided, however, not to undertake the construction of the large transmitter that would be required, but to concentrate all our attention on the reception branch of radio astronomy.

21-cm HYDROGEN STUDIES

I spent a year at the Harvard College Observatory in 1950-51, in order to broaden my knowledge of astronomy. I happened to be there on the famous day (March 25, 1951) when H. I. ("Doc") Ewen and E. M. Purcell first detected the 21-cm line from neutral hydrogen in interstellar material in the Galaxy. H. C. van de Hulst, one of the pioneering Dutch radio astronomers, was also by chance at Harvard at that time. Purcell called us all together on the morning after that overnight discovery. Cautious physicist that he was, he wanted to see confirmation of the

detection before publishing the result. He suggested that van de Hulst
and I should cable our respective institutions in Holland and Australia
to report the discovery and ask whether early confirmation would be
possible.

        The Dutch group were already preparing for a 21-cm experiment
themselves, following van de Hulst's prediction in 1944 that the line
might be detectable astronomically, and C. A. Muller and J. H. Oort were
able to announce their own detection of the line very soon afterwards. In
Australia, no attempt to look for the line was being undertaken at the
time. However, on receipt of my cable to Pawsey, W. N. Christiansen and
J. V. Hindman started on a crash program and were able to assemble the
necessary equipment and make a detection after a short period of six weeks.
The first papers from the Harvard and Leiden groups were published together
in _Nature_ in 1951, with a short cable from the Australian group appearing
as an addendum. This whole episode was a fine example of international
cooperation, which has always been the hallmark of most of the relationships in radio astronomy.

        Figure 3   36-ft paraboloid at Potts Hill, near Sydney,
        c. 1953. The telescope is mounted as a transit instrument.
        Brian Robinson is seen working on the receiver.

Christiansen and Hindman carried out a preliminary 21-cm survey and in a 1952 article showed the general distribution of the peak intensity in the line profile over most of the sky. They further found a doubling of the line in many directions, "possibly delineating two spiral arms of the galaxy."

On returning to Australia in 1951, I became involved in 21-cm work. Hindman and I supervised the building of a 36-foot dish, which was large for its day and whose construction was made possible by restricting its motion to only along the meridian, as in the Green Bank 300-foot telescope today (Figure 3). We also built the world's first "multi-channel" receiver, which had all of four 40 kc/s channels! To cover a whole profile, we would do a series of drift curves day by day, stepping our four channels progressively across the desired frequency range, with a one-channel overlap. Full coverage of a broad profile in the Galaxy required 8-10 observing days.

We began with a survey of the Magellanic Clouds, the first extragalactic objects studied at 21 cm (Kerr, Hindman and Robinson 1954). We showed that each Cloud contained more gas than expected from the amount of dust it was believed to contain (Figure 4), and also that each Cloud is rotating. B. J. Robinson joined in this work as a young graduate

Figure 4 First map of hydrogen in the Magellanic Clouds (Kerr, Hindman & Robinson 1954). The integrated brightness contour unit is 10^{-16} W m^{-2}sterad^{-1}.

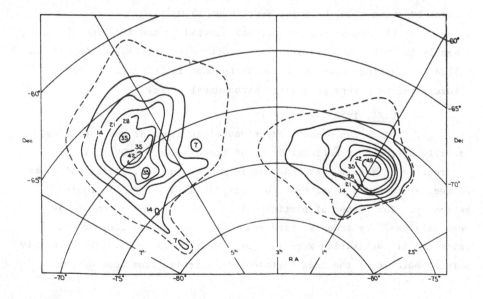

student from the University of Sydney. Next we surveyed the Southern
Milky Way (Kerr et al. 1957), finding significant differences from the
picture already obtained for the North by the Dutch group, especially
in the rotation curves derived from the 21-cm data. In 1957, I went to
Leiden for several months and Gart Westerhout and I joined together our
respective maps to derive the first overall look at the hydrogen in the
Galaxy (Oort, Kerr and Westerhout 1960).

One of the interesting events in this period was the agreement of the International Astronomical Union to define a new system of
galactic coordinates primarily in terms of the 21-cm observations of the
galactic plane region. Many conventional astronomers had been reluctant
to regard radio astronomy as a respectable branch of the subject, and
this was one of the key ways in which radio astronomy came of age.

The first system of galactic coordinates had been defined
in 1932 from optical observations of the Milky Way, and therefore
referred to the mean plane of nearby material. In the 1950s it was
becoming apparent from radio observations (both line and continuum) that
the conventional plane deviated from the plane of distant material by
$1\frac{1}{2}^{\circ}$. Also it was considered desirable to move the zero of longitude to
the vicinity of the galactic center. IAU Commission 33 set up a
committee consisting of A. Blaauw, C. S. Gum, and J. L. Pawsey to advise
on a new system. They made use of all available radio and optical
observations, but depended mainly on the "principal plane" of the
neutral hydrogen (Gum, Kerr and Westerhout 1960).

The recommended system was adopted by the IAU General
Assembly in 1958. Shortly after this work was completed, Colin Gum was
killed in a skiing accident on the Matterhorn in Switzerland, pre-
maturely ending a very promising astronomical career.

URSI, 1952

Since the present paper was first given at the 1981 General
Assembly of the International Union of Radio Science (URSI), it is
interesting to recall that URSI met in Sydney in 1952. This choice can
be seen as mainly a compliment to Australia's early work in radio
astronomy. The number of participants in that General Assembly was
certainly small by today's standards. This was partly because the
Union and its activities were less prominent in those days, but probably
more so because of the long distance of Australia from most of the

participating countries. The meeting brought many URSI people into
contact with the Radiophysics Laboratory for the first time, even though
the group photograph for the meeting (Figure 5) contains only 174 people,
and a substantial number of these were of course local. One of the other
interesting photographs from that meeting (Figure 6) contains almost all
the people connected with 21-cm work in 1952.

Figure 5 Participants in the URSI General Assembly
at the University of Sydney in 1952.

OTHER SPECTRAL LINES

Although none of the early line discoveries were made in
Australia, there were two discussions of the possibility of detecting
particular lines. Well before the discovery of the 21-cm line, Bowen
was urged by I. I. Rabi, and later by J. R. Zacharias, to look for
nuclear resonances in the Galaxy. This followed the original work of
the Columbia group on nuclear resonance, with elements like cesium high
on the list of materials known to resonate at radio frequencies. Discussions were held, but as far as I know no observations were carried
out.

Figure 6 The 21-cm group at the 1952 URSI General Assembly.
From left to right: F. J. Kerr, J. P. Wild, J. V. Hindman,
H. I. Ewen, C. A. Muller, W. N. Christiansen.

   J. P. Wild (1952) considered the spectrum of hydrogen at
radio frequencies, including the fine-structure lines. He found that the
Sun might give a detectable absorption line at about 10,000 Mc/s,
corresponding to the $2^2S\frac{1}{2}$ - $2^2P3/2$ fine structure transition, but that
other solar lines were not likely to be observed. (This line has in
fact still not been detected, and it is possible that the broadening is
too great for the line to be observable in the Sun.)

   In 1956, Yardley Beers, who was then at New York University,
spent a year in Sydney, based at the National Standards Laboratory, which
shared a building with the Radiophysics Laboratory. He carried out a
detailed study of lines (especially from molecules) which might be
detectable in the interstellar gas. But we radio astronomers were
unfortunately all too busy with our own projects to spare time for line
searches which could have changed the history of radio astronomy.

THE PARKES TELESCOPE

The commissioning in 1961 of the 210-foot telescope at Parkes, 200 miles northwest of Sydney, provided a new focus for Australian radio astronomy. Planning for a "giant radio telescope" (or GRT) began in the early fifties, with Taffy Bowen providing the leadership for this new endeavor. Many conceptual designs were considered before a straight-forward parabolic dish design was accepted. The design, produced by Freeman, Fox and Partner of London, was well in the forefront at the time. One of the radical features was the "master equatorial" in which an analog device with an optical link controlled the pointing of the telescope, carrying out the conversion from altitude-azimuth to equatorial coordinates. The structure was built by the M.A.N. Company of West Germany, who added a substantial German touch to the population of the town of Parkes for a year or so. An early historical photograph is shown in Figure 7.

Figure 7   J. L. Pawsey marking the spot for the center of the Parkes 210-ft telescope, October 1958.

During the design phases, H. C. Minnett acted as the Radio-physics Lab's liaison with the engineers in London, an important task. In 1960, J. G. Bolton returned to Sydney after spending several years heading the radio astronomy group at the California Institute of Technology, in order to look after the construction phase of the tele-scope. In that period, I was involved with some of the design details as chairman of the users' liaison committee, which was concerned with seeing that the telescope, and especially its control system, were as suitable as possible for the astronomical observers.

For example, this is believed to be the first telescope which used a digital display system for all the coordinates, a suggestion I was responsible for.

The impressive structure was finally ready in 1961, and was commissioned on October 31 of that year. In spite of the fact that the Parkes site was chosen in large part because of the reputed low average wind speed, the opening day was notable for 50 m.p.h. winds and a great deal of dust.

Radio astronomy in Australia had begun in 1946 with leftover wartime equipment and an experimental approach. With the Parkes telescope in operation, observations could be carried out on a stable high-performance system, and they were now generally directed to clear astronomical objectives.

REFERENCES

Bolton, J. G. & Stanley, G. J. (1948). Variable source of radio-frequency radiation in the constellation of Cygnus. Nature, 161, 312-3.

Bolton, J. G., Stanely, G. J. & Slee, O. B. (1949). Positions of three discrete sources of galactic radio-frequency radiation. Nature, 164, 101-2.

Bolton, J. G. & Westfold, K. C. (1950). Galactic radiation at radio frequencies. I. 100 Mc/s survey. Aust. J. Sci. Res., A3, 19-33.

Christiansen, W. N. & Hindman, J. V. (1952). A preliminary survey of 1420 Mc/s line emission from galactic hydrogen. Aust. J. Sci. Res., A5, 437-55.

De Witt, J. H., Jr. & Stodola, E. K. (1949). Detection of radio signals reflected from the moon. Proc. IRE, 37, 229-42.

Ewen, H. I. & Purcell, E. M. (1951). Radiation from galactic hydrogen at 1,420 Mc./sec. Nature, 168, 356-7.

Gum, C. S., Kerr, F. J. & Westerhout, G. (1960). A 21-cm determination of the principal plane of the Galaxy. Mon. Not. Roy. Astron. Soc., 121, 132-49.

Kerr, F. J. (1948). Radio superrefraction in the coastal regions of Australia. Aust. J. Sci. Res., A1, 443-63.

Kerr, F. J. (1952). On the possibility of obtaining radar echoes from the Sun and planets. Proc. IRE, 40, 660-6.

Kerr, F. J., Hindman, J. V. & Robinson, B. J. (1954). Observations of the
        21-cm line from the Magellanic Clouds. Aust. J. Phys., 7,
        297-314.

Kerr, F. J. & Shain, C. A. (1951). Moon echoes and transmission through
        the ionosphere. Proc. IRE, 39, 230-42.

Mills, B. Y. (1952). The distribution of the discrete sources of cosmic
        radio radiation. Aust. J. Sci. Res., A5, 266-87.

Muller, C. A. & Oort, J. H. (1951). The interstellar hydrogen line at
        1,420 Mc./sec., and an estimate of galactic rotation.
        Nature, 168, 357-8.

Oort, J. H., Kerr, F. J. & Westerhout G. (1958). The Galactic System
        as a spiral nebula. Mon. Not. Roy. Astron. Soc., 118, 379-89.

Pawsey, J. L. (1961). Australian radio astronomy. Aust Sci., 1, 181-6.

Wild, J. P. (1952). The radio-frequency line spectrum of atomic hydrogen
        and its applications in astronomy. Astrophys. J., 115,
        206-21.

The 1953 cricket club of the CSIRO Radiophysics Division: radio astronomers include (back row, first from left) John Bolton, (second) Dick McGee, (third) J.A. Warburton; (front, second) Paul Wild, (fourth) Kevin Sheridan (photo courtesy McGee)

Radio astronomy the hard way at Fleurs field station near Sydney (1956): (l. to r.) Flo (the lorry), Bruce Slee, Alec Little, a technician, and Kevin Sheridan (courtesy CSIRO Radiophysics Division)

RADIO SOURCES AND THE LOG N - LOG S CONTROVERSY

B.Y. Mills
School of Physics, University of Sydney, N.S.W. 2006, Australia.

The Sydney-Cambridge controversies of the 1950s concerning the
nature of radio sources and their role in cosmology remain of interest
because the roots of the controversies appear to lie in differing philoso-
phies of scientific research as much as in observational discrepancies.
They also clearly show the influence of instrumental limitations of the
time on the beliefs of astronomers and thus contain a message for the new
generation.

The controversies have been discussed at length by Edge &
Mulkay (1976) but there remains a great deal which can be said. Edge &
Mulkay aimed to give an objective account but did not have direct access
to Australian material or contact with those of us involved in the controversies. Here I try to present my personal viewpoint of the time, as much,
that is, as I am now able to reconstruct. My uncertain memory of some of
these ancient events has been bolstered by access to the correspondence
files of the Division of Radiophysics, with the kind permission of
Dr. R.H. Frater, Chief of the Division.

EARLY YEARS

Nowadays it is difficult to appreciate the abysmal ignorance of
the nature of radio sources prevalent around 1950. We were struggling with
inadequate instruments to make physical sense of some completely new and
unexpected phenomena. There were but two types of instrument, interferometers comprising pairs of low directivity antennas (or the equivalent
single antenna on a cliff-top overlooking the sea) and single small-size
antennas operating at low frequencies with poor resolution. Accordingly,
the sky appeared to be populated by randomly distributed 'point sources',
which produced interferometer patterns, and a broad band of rather featureless emission concentrated to the plane of the Milky Way. This purely
instrumental result seemed to correspond to the general appearance of the

visible sky and led to models comprising dark "radio stars" distributed
through the Galaxy much like the visible stars. The Cambridge group, under
the leadership of Martin Ryle, strongly espoused this view.

My own realization that it was likely to be incorrect came when,
in early 1950, I put into operation a three antenna interferometer with two
spacings of 60 m (20 λ) and 270 m (90 λ). This was my first major project
in radio astronomy. Earlier I had made some observations of the source
Cygnus A in collaboration with A.B. Thomas (Mills & Thomas 1951). We had
used the solar interferometer of Little & Payne-Scott (1951) after they
had finished their daily monitoring of solar bursts. These observations
had stimulated my interest and I decided to push ahead, replacing the small
Yagi antennas, suitable for observing the Sun, with much larger broadside
arrays. A photograph of one of the broadside arrays is reproduced in
Figure 1. At the same time it was clear that a new field station was

Figure 1. One of the broadside arrays of the three antenna
interferometer at Badgery's Creek (1950-1952). The frequency
was 101 MHz and the dimensions of the antenna approximately
9 m x 6 m   (Mills 1952a, b).

required, free from the electrical interference which plagued the solar
station located near Sydney. I found an excellent site at Badgery's Creek,
a C.S.I.R.O. cattle research station located some 50 km from the city. In
all these activities I was strongly supported by J.L. Pawsey who headed
the Radiophysics radio astronomy group and who was responsible for my
joining the group in 1948.

The principal aim of the new interferometer was to provide
accurate positions, using the 60 m spacing to identify the correct central
lobe of the 270 m interferometer (lobe ambiguities were known to be a
problem even then). This was the first time I had used an interferometer
with such a small spacing as 60 m (20 λ); the solar interferometer had
resolved ambiguities using two slightly different larger spacings (large
enough to resolve out the quiet Sun). The problem with small spacings for
radio source work arose from the correspondingly long periods of the interference patterns; receiver instabilities were likely to limit the sensitivity achieved. Previously, a high-pass filter technique had been used to
reduce the effect of gain fluctuations, but this was not altogether
satisfactory. I had begun to look for a better alternative when I received
some unintended assistance from Cambridge. News came through the grapevine
that a revolutionary system had been introduced there but it was all very
hush-hush and no details were known; it was believed that it involved a
modulation of the interference pattern. This seemed to be just what was
needed. A little thought suggested modulation by interchanging maxima and
minima on the interference pattern by switching phase and using a synchronous detector, as in the Dicke system. The necessary equipment was built up
and it worked very well. Later I found that this was precisely the system
used at Cambridge, the only difference being their use of a hardware switch
in the antenna feed lines whereas I had used an electronic switch following
the preamplifier.

From almost the first observations with the interferometer the
results were surprising. "Point sources" detected by the close pair of
antennas often seemed to bear no relation to those detected with the wider
spacing, and when this occurred, the apparent flux densities on the close
spacing were usually larger. Obviously sources of large angular size were
involved, sources which were being resolved at a spacing of 90 λ. Further
indications that the "radio star" hypothesis might be incorrect came from
identifications which Bolton had suggested with three nebulae, at least

one of which (M 87), was known to be extragalactic (Bolton, Stanley & Slee
1949). The strong source Cygnus A had also been shown to have a possible
identification with a faint distant galaxy, although the peculiar nature
of this galaxy was not then known and Minkowski rejected the identification
(Mills & Thomas 1951). With more accurate positions from the three antenna
interferometer I confirmed Bolton's identifications and showed that the
possibility of the Cygnus identification was in fact increased (Mills
1952a). I also showed from an analysis of the 76 sources catalogued that
there were two major classes of radio source, the stronger sources being
closely confined to the Galactic plane (Class I) and the weaker sources
apparently randomly distributed over the sky (Class II) (Mills 1952b). The
latter class included known extragalactic members, some of which were of
appreciable angular size (as also were some of the galactic sources). It
could not be determined, however, whether this class was composed entirely
of galaxies like those identified, whether it included other classes of
extragalactic sources, or, indeed, whether it included a proportion of the
conventional "radio stars".

        At about this time I received my first intimations that the
Cambridge group did not necessarily accept Sydney results. Although I was
not involved, travelling members of the Sydney group like J.L. Pawsey and
J.G. Bolton reported that the "radio star" hypothesis was still alive and
well at Cambridge. But at some point the position must have radically
altered, possibly after Baade & Minkowski (1954) had positively identified
the Cygnus A galaxy (the identification previously rejected by Minkowski!),
the source Perseus A (with NGC 1275) and the galactic source Cassiopeia A,
all on the basis of Cambridge, rather than Sydney, position measurements.
Also, independent but almost simultaneous observations at Cambridge,
Jodrell Bank and Sydney had measured the angular sizes of some of the
hitherto unresolved northern sources and shown that they were certainly
not stellar (Hanbury Brown, Jennison & Das Gupta 1952; Mills 1952; Smith
1952).

        A photograph of the transportable antenna system used with a
radio link for the Sydney observations is reproduced in Figure 2. This
antenna system was moved between different locations in the vicinity of
Badgery's Creek to provide a variety of baselines with different spacings
and orientations. The responses with these baselines were used to construct two dimensional models of the radio sources (Mills 1953).

Figure 2.   The portable Yagi antennas used with the antenna of
Figure 1 to form a radio link interferometer.   The operating
frequency was 101 MHz.

THE 2C AND SYDNEY SURVEYS

For some time there was little contact between Sydney and
Cambridge, perhaps because both groups were busy with new instrumentation.
As a result of experience with the three antenna interferometer, my own
efforts were directed towards the development of a pencil beam system. By
then, I knew that collecting area was relatively unimportant, the important
thing was a large overall size to give high resolution. As a filled array
seemed to be wasteful, I first looked at various passive configurations
such as crosses and rings, but these all suffered from high sidelobe levels.
Suddenly it occurred to me that by combining the phase-switch, which I had
used on the interferometer, with a crossed array the sidelobe problem
would be substantially reduced. Thus was born the first "Cross" (Mills &
Little 1953). In the introduction to this paper we wrote, "It therefore
seems desirable to rely mainly on the use of pencil beam aerials of high
resolving power for future work and to reserve the use of interferometric

methods for special applications". By contrast, the Cambridge group were
constructing the large four antenna interferometer subsequently to be used
for the 2C and 3C surveys.

Our work was delayed somewhat by the scepticism of the powers-
that-be and I remember being informed with great authority that my proposed
"Cross" could not possibly work. It was first necessary to build a pilot
model to demonstrate that the principle of operation was sound and that
funds should be provided for an instrument of adequate size. This did
have some advantages, however, in enabling us to experiment with instrumental techniques and eventually produce a very reliable system. Also we
obtained one scientific result, the first detection of radio continuum
radiation from the Large Magellanic Cloud. A photograph of the pilot
model is shown in Figure 3; the dimensions were a mere 36 m.

Figure 3. The first "Cross" constructed as a pilot model for
experimental investigation of the technique at the Potts Hill
field station of the Radiophysics laboratory (Mills & Little
1953). The frequency was 97 MHz and the overall dimensions
of the arrays 36 m.

Following this successful demonstration, the construction of a
larger and more useful instrument was approved. However, there were some
complications. Firstly, another field station had to be established

because Badgery's Creek did not possess an adequate area of flat ground. Fortunately we knew of a disused air-strip only a few kilometres from Badgery's Creek and, after negotiations, we were able to lease a suitable area which formed the nucleus of the Fleurs Radio Observatory, used for many instruments during the next thirty years. Secondly, by the time a suitable design had been thrashed out I had the opportunity to spend six months in the U.S.A. at the California Institute of Technology and the Department of Terrestrial Magnetism. This was too good an opportunity to miss so I left the whole operation in the capable hands of Alec Little. I need have had no worries. On my return, in early 1954, construction of the antenna was nearly complete and shortly afterwards we were able to begin the process of making it work. The first good records were coming through before the end of 1954.

By then, however, the Cambridge 2C survey of radio sources (Shakeshaft et al. 1955) was already completed and 1936 sources catalogued. The Cambridge interferometer comprised four antennas placed at the corners of a rectangle measuring 580 m x 51 m at a frequency of 81 MHz. Our second Cross operated at a frequency of 85.5 MHz, and had overall dimensions of 450 m and a beamwidth of 48 arcmin. (e.g., Mills et al. 1958). A photograph of this "Cross" is shown in Figure 4. In Figure 5 a later photograph shows a group of "Fleurs" workers.

My first realization that all was not well with the new Cambridge survey came with a letter from Fred Hoyle. He was worried about the results of counting sources catalogued in the 2C survey, as reported by Ryle. For a homogeneous distribution of sources, not too far away, the cumulative log N - log S count should have a slope of -1.5. Ryle reported a slope of -3.0 and used this result to argue that the sources were placed at a great cosmological distance in an evolving Universe (for further discussion, see McCrea's article elsewhere in this volume). The steady-state Universe was inadmissable and this, of course, is what worried Hoyle. Although we had just begun observations with the Cross and could not provide detailed results, it was nevertheless clear from a simple inspection of available records that the slope of the source counts could not possibly be as steep as this. Somewhere the Cambridge group must have gone wrong. I replied to Hoyle with this reassuring news.

A preliminary account of our actual statistics was given by Pawsey at the 1955 Jodrell Bank Symposium on radio astronomy (Pawsey 1957).

Figure 4. The second "Cross" nears completion in early 1954
at the newly established Fleurs field station. The frequency
was 85.5 MHz and the overall dimensions 450 m (Mills _et al_.
1958). For a full aerial view, see Fig. 7 of Bowen's article
in this volume.

Figure 5. A group associated with the Fleurs field station
(c. 1956). From the left they are: K. Hawkins, C.A. Shain,
O.B. Slee, B.Y. Mills, K.V. Sheridan, A.G. Little and
H. Rishbeth. C.A. Shain had by then established a "low-
frequency Cross" at the field station.

My first impressions had been confirmed, and counts of about 1000 radio sources of rather roughly known flux densities in some selected areas gave a slope slightly steeper than -1.5, but nowhere near as steep as the Cambridge counts. Interchange of correspondence and results began between Pawsey, me and Ryle in an attempt to resolve the discrepancies. In addition, Rudolph Minkowski visited Sydney early in 1956 and, after thorough study of our results and discussions on their significance, continued on to Cambridge, reporting back on his general conclusions. It was now obvious that there was little agreement between the sources catalogued at the two places and the discordances arose at a level where the 2C counts began to show a steepened slope. I had found that this was also the level where confusion effects in the 2C catalogue could be expected to dominate.

So far there had been a free interchange of opinions and observational results, but the battle lines were now drawn. Ryle (1956) published a popular account in *Scientific American* of the Cambridge observations and the sweeping conclusions that he drew from them. Since we now had good reason to believe that the 2C catalogue was seriously wrong and it appeared that Ryle had ignored our private criticism, it was generally felt in Sydney that the time had come to make a public statement. Accordingly, I sent a letter to the Editor of *Scientific American* briefly describing the discordances we found with the 2C survey and giving reasons for them; this letter was answered by Ryle. Both letters were then published in the December, 1956 issue, and this was the end of any further co-operation. Our aim of publicising the uncertainties in Ryle's cosmological model had been achieved, but, in retrospect, it may have led to an easier resolution of our differences if I had ignored the article.

A full account of our checking of the 2C survey in a sample area was published as soon as possible (Mills & Slee 1957). Our conclusions were summarised as follows: "We have shown that in the sample area which is included in the recent Cambridge catalogue of radio sources there is a striking disagreement between the two catalogues. Reasons are advanced for supposing that the Cambridge survey is very seriously affected by instrumental effects which have a trivial influence on the Sydney results. We therefore conclude that discrepancies, in the main, reflect errors in the Cambridge catalogue and, accordingly, deductions of cosmological interest derived from its analysis are without foundation". These results

Figure 6. A direct comparison of the Sydney and 2C catalogues
from Mills & Slee (1957). The primary beamwidths have been
indicated to show the effects of resolution on the surveys.

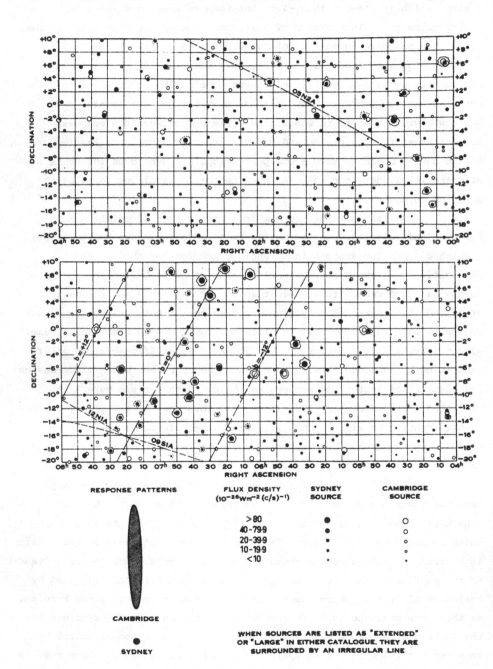

RESPONSE PATTERNS

CAMBRIDGE

SYDNEY

FLUX DENSITY (10⁻²⁶Wm⁻²(c/s)⁻¹)	SYDNEY SOURCE	CAMBRIDGE SOURCE

$(10^{-26} \text{Wm}^{-2} (c/s)^{-1})$

Flux Density	Sydney Source	Cambridge Source
>80	●	○
40-79.9	●	○
20-39.9	●	○
10-19.9	●	○
<10	·	·

WHEN SOURCES ARE LISTED AS "EXTENDED"
OR "LARGE" IN EITHER CATALOGUE, THEY ARE
SURROUNDED BY AN IRREGULAR LINE

were demonstrated in the diagram reproduced in Figure 6, which shows a
direct comparison of the sources in the 2C catalogue and those found at
Sydney. The reception areas of the primary beams of the two instruments
were shown in order to illustrate graphically the difference in the
confusion associated with the surveys.

The source counts obtained using this catalogue were in agreement with our preliminary results reported at the Jodrell Bank symposium.
A slope of -1.8 with a probable error established as 0.1 was obtained.
The result was marginally steeper than -1.5 but our analysis showed that
the instrumental effects of random noise and blending combined to increase
the apparent slope and reduce the statistical significance of the excess.
We commented as follows: "We find only a small excess of faint sources
which, when account is taken of instrumental effects, is found to be
insignificant as far as any cosmological evidence is concerned. A real
excess is possible but that suggested by Ryle and Scheuer is impossibly
large". Note that, although in relation to Ryle's model we referred to
an excess of *faint* sources, the statistical uncertainty actually arises
from the small numbers of *strong* sources. The high flux density end of
the source counts is the statistically uncertain part, as I emphasised
later.

THE P(D) ANALYSIS

If this had been all, the controversy would have had little to
fuel it. There was, however, another string to the Cambridge bow. The
Cambridge interferometer, with a large collecting area, had a high signal-
to-noise ratio and was completely confusion limited; an almost continuous
interference pattern was observed. Quite correctly, they concluded that
a statistical analysis of the envelope of the continuously beating interference pattern should give information about the distribution of radio
sources. The calculations were performed by Peter Scheuer on the assumption that the patterns were generated by a population of randomly
distributed point sources in a Euclidean Universe. This "P(D) analysis"
also pointed to a large excess of faint sources (Ryle & Scheuer 1955;
Scheuer 1957).

In Sydney we knew from the 1955 paper that this work had been
done, but had no details. We had independently carried out a similar,
but much simplified, analysis and applied it to our own results (Mills &

Slee 1957). We did not find evidence for an excess of faint sources, but rather for a cut-off in the population which we attributed to the effects of redshift below the limits of our survey.

As the 2C survey was progressively undermined, the Cambridge group came to rely increasingly on the P(D) analysis and appeared to believe that we did not understand its significance. Indeed we did, but we had a different concept of what was in the sky and believed that there could be a different explanation for their results. From early in 1950 we had known that not all radio sources of high galactic latitude were of negligible angular size: Centaurus A was several degrees across, Fornax A about a degree, and there were many other resolved sources. There were no such *very* large sources in the northern sky to impress themselves similarly on the Cambridge group, and their efforts to discount the importance of extended sources using the 2C antennas did not convince us. There was also practically no direct information on the general luminosity function of radio sources, and it seemed quite possible that these might be relatively common extragalactic objects which partake in the general clustering of external galaxies.

Both angular size and physical clustering would have effects on the Cambridge P(D) curve like those actually observed. In addition, there was the question of "background fluctuations", either originating in the galactic halo or possibly from the remains of old evolved extra-galactic sources: a low-level complex background would give a continuous interference pattern representing the Fourier component of the distribution selected by the interferometer. We had no information about the flux density scale of the P(D) curve and could not quantitatively explore these possibilities. There was no reason to doubt the validity of the Cambridge measurements, but we did query the validity of their simplified model. If the 2C survey had been correct, the only possible interpretation would have been that given by Ryle, but the P(D) analysis seemed open to several interpretations. To use Minkowski's apt description, it was "low grade information".

THE "EXTENDED SOURCES"

An important feature of the Sydney catalogues was the inclusion of "extended sources" at high galactic latitude which were not associated with nearby galaxies like Centaurus A or Fornax A. These were described by Mills & Slee (1957) as follows: "Although the majority of sources have

angular sizes much less than the beamwidth of the aerial and thus cause no
widening of the response pattern, there are some for which the response is
appreciably widened. These may be either sources of large angular size or
blends of two or more small sources; in general it is not possible to
distinguish between these possibilities."

It was believed that, even though their nature was unclear,
they should be included in the counts with their integrated flux density.
In this way "similarity" would be preserved, that is, close and large
extended sources would be counted the same way as similar unresolved,
more distant sources, and the counts would not be seriously affected by
the resolution of the instrument. This would be true whether these sources
were physical clusters of small sources or extended regions of low brightness. This concept was often misunderstood. For example, a typical
statement was: "If non-statistical associations of sources are counted as
single sources, it is evident that weak sources are removed and intense
sources added, and the apparent slope of the log N - log S curve is therefore decreased" (Ryle 1962). Clearly no thought had been given to similar
more numerous weak sources below the survey limit which, when clustered
together and observed with an instrument of finite resolution, would
produce significant numbers of apparently single sources above the limit
to replace those moved to higher intensities. In fact the inclusion or
omission of the extended sources had no really significant effects on our
source counts, but it was noted that they were more numerous than expected
on the basis of our analysis of chance blending.

It was evident, however, that if physical clustering effects or
large sources were common, they could have a gross effect on the Cambridge
interferometer statistics and would probably vitiate their interpretation.
Further controversy therefore arose around the reality of these sources.
Could they be predominantly side-lobes? Our position was clear: "It
therefore seems safe to assume that practically all our sources represent
real concentrations of radio emission, the majority probably being
physically discrete. An analysis of the catalogue should therefore give
meaningful results" (Mills & Slee 1957).

With the publication of our first full-scale catalogue, between
declinations of $+10^\circ$ and -20° (Mills, Slee & Hill 1958), these sources
assumed greater importance. In this area there were what later proved to
be exceptionally large numbers of quite strong extended sources. In a

high latitude area we chose to analyse, there were 20 extended sources
with integrated flux densities above 40 Jy. Our analysis of chance blends
had indicated only two such sources. This impressed us with the possibility that it was more than a chance phenomenon and we wrote: "Thus on
present evidence it would seem quite possible that individual radio
galaxies should appear relatively frequently in physically related groups
of two or more, and that close clustering of galaxies can create vast
radio sources of large angular size..."

      An opportunity to check the nature of the extended sources
arose shortly after the commissioning of the Parkes 64 m reflector in 1961.
By then, however, I had left the Division of Radiophysics to set up a radio
astronomy group in the University of Sydney. My application was rejected
and it was made clear that the new instrument was for the use of Radiophysics staff and visitors. One such visitor was Peter Scheuer who, in
fact, made such a check in 1963. The results were inconclusive (Scheuer
1965).

      Over the years, some of these sources have also been examined
at the Molonglo Radio Observatory, but no systematic work has been done.
It now seems clear, however, that chance blending effects were more than
we had originally thought or, put another way, it was not as easy to
separate close sources as we had assumed in our analysis. Also, the
effects of numerous weak sources blending with the stronger sources had not
been included. It now seems that most of the high latitude extended
sources are likely to be chance blends, many are known to be double sources
or associated with clusters, some are very large radio galaxies, and some
are probably associated with background irregularities;· only a small
proportion are likely to be sidelobes or real sources affected by sidelobes.
The logic of including them in our source counts to reduce instrumental
effects can hardly be questioned, but their effect on the Cambridge P(D)
analysis of the time seems to be an open question. Clearly those extended
sources which resulted from chance blending of small diameter sources
would have had no effect.

        THE LUMINOSITY FUNCTION

      The main point of the initial Mills & Slee paper was that,
while we found a small apparent excess of faint sources, we did not regard
it as significant. The full catalogue of the sky between declinations of
$+10°$ and $-80°$ was later presented in three papers (Mills, Slee and Hill

1958, 1960, 1961) in which some 2200 radio sources were catalogued; the
results were similar. A further instrumental effect, a systematic over-
estimation of flux densities near the survey limit, was also uncovered
and this caused us to continue to doubt the physical significance of the
"excess", which by now had become quite significant in a statistical sense.
Continued work at Cambridge on the 3C survey, at double the 2C frequency
(Edge et al. 1959), and eventually on the 4C survey with a much larger
interferometer (Scott & Ryle 1961) meanwhile had progressively reduced
the slope of the source counts obtained with their interferometer systems
to the same as our uncorrected value, -1.8. A good agreement between
individual sources was also usually obtained. The disagreement over the
form of the source counts had largely vanished, but a basic difference
remained over their interpretation. It is significant that the 3C survey,
as revised by Bennett (1962) with four times the resolution and one
quarter the number of sources catalogued in the 2C survey, eventually
became the vade mecum of northern astronomers. Cambridge had learnt the
lesson!

There was a prevalent view among cosmologists that even a
slope of -1.5 carried very significant cosmological information as the
effect of redshift was to reduce the steepness for faint sources in any
cosmology involving normal geometries and no evolution (see the article
by McCrea elsewhere in this volume). My view was that, although this was
correct, the level at which redshift effects became important and reduced
the steepness could not be predicted until one knew the radio luminosity
function. It was clear that extragalactic radio sources could not be as
common as ordinary galaxies or their integrated emission would exceed the
observed background brightness, but there was a large possible range which
had to be tied down before one could say anything significant about
cosmology. This was particularly important because of the likelihood of
physical clustering if the sources were relatively common, and the strong
instrumental artefacts which this would induce.

The only prospect for progress seemed to lie in an unambiguous
determination of the luminosity function by direct identification with
galaxies whose redshift could be measured. In the late 1950s I pursued
this idea, probably as far as possible at the time. By 1957 we had
installed a radio link interferometer with spacing of 10 km to operate
in conjunction with one arm of the Cross (Goddard, Watkinson & Mills 1960).
This was one of the "special applications" for interferometers envisaged

by Mills & Little (1953) and was based on my original portable radio link
system (Mills 1953). With a lobe spacing of just over one arc minute, it
could be used to separate the stronger sources into the categories "small"
and "large" and obtain some indications of actual sizes for sources in the
range 10 to 60 arc seconds. Armed with preliminary data from this interferometer and the new Cambridge 3C survey, which helped to improve the
right ascension for many of our sources, I spent a few months in late 1958
examining original Palomar Sky Survey plates at the California Institute
of Technology. Under the tutelage of Minkowski, I was introduced to the
arcane mysteries of the Hubble classification scheme and the subtle
differences between "normal" and "peculiar" galaxies.

It was soon apparent that, within the area of positional
uncertainty of any source, there were usually numbers of possible galaxy
identifications and it was impossible to choose between them, even with
the additional clues provided by the angular size data. By limiting the
search to (i) bright galaxies without obvious peculiarities, (ii) fainter
"peculiar" galaxies, and (iii) even fainter "double" galaxies, some degree
of statistical validity could be obtained for the identifications. Out of
400 radio sources in a defined area I was able to suggest 46 possible
identifications, about half of which were described as being likely, and
some of which had already been suggested by earlier workers (Mills 1960).
In a "... first step towards establishing the radio luminosity function..."
I found that a subsample of 35 of these identifications gave a luminosity
function which could reasonably account for about half the observed high
latitude radio sources. I was concerned about the possibility that substantial numbers of quite normal looking galaxies could be radio sources,
but there was then no way of finding out. It eventually required much
more accurate positions obtained with the new radio telescopes of the
1960s and 1970s.

Thus my own work in cosmology ended in this frustration. For
most of the next decade I was involved with building a new instrument and
forming a new radio astronomy group. By the time our results were coming
in, most of these problems had been resolved, at least qualitatively, by
the recognition of large numbers of identifications with ordinary looking
galaxies and an unsuspected new extragalactic class, the quasars.

EPILOGUE

What would be the reaction of astronomers of the 1950s if

faced by the source counts as we know them today, but with a complete lack of all other modern knowledge? I have little doubt that, given this fanciful situation, the really significant feature would be seen as the cut-off at low flux densities, evidence for redshift at work. The comparative shortage of strong sources could hardly be seen as significant in itself, two or three standard deviations less than the numbers expected with a slope of -1.5, but not enough to launch a new cosmology. Clustering models could further reduce the statistical significance to a negligible amount. By contrast, the relative sharpness of the redshift cut-off would be seen as very significant because no simple cosmological models predict it. Evolution of distant sources would certainly be one explanation, but so would be the instrumental effects of clustering. One might even be able to invent a geometrical explanation. It is hard to imagine a violent controversy arising over the respective merits of the explanations. The actual controversy in the 1950s was triggered by the 1600 or more fictitious sources firmly listed in the 2C catalogue.

It now seems clear that the source counts by themselves present little cosmological information. The crucial data needed to interpret the source counts are the identifications and their redshifts which lead to the luminosity function. For instance, we know today that the flat-spectrum sources have counts with an initial slope hardly distinguishable from that expected in a static Euclidean Universe. Nevertheless, it is believed that these are predominantly distant quasars with strong evolution, as shown by application of the V/V_m test to identified members. It is implied that the evolutionary process almost exactly counterbalances the effects of redshift on the counts of the closer sources, a curious situation to ponder.

The 1950s were stimulating and exciting for both radio and optical astronomers. For the first time it became clear to all of us that radio astronomy promised a great extension to the powers of optical astronomy for exploring the distant Universe. The Sydney-Cambridge controversy eventually concerned only the degree to which the radio data stood alone and could be interpreted by simple models.

REFERENCES

Baade, W.A. & Minkowski, R.L. (1954). Identifications of the radio sources in Cassiopeia, Cygnus A and Puppis A. Astrophys. J., 119, 206-214.

Bennet, A.S. (1962). The revised 3C catalogue of radio sources. Mem. R. Ast. Soc., 58, 163-172.

Bolton, J.G., Stanley, G.J. & Slee, O.B. (1949). Positions of three discrete sources of galactic radio frequency radiation. Nature, 164, 101-102.

Edge, D.O. & Mulkay, M.J. (1976). Astronomy Transformed. Wiley.

Edge, D.O., Shakeshaft, J.R., McAdam, W.B., Baldwin, J.E. & Archer, S. (1959). A survey of radio sources at a frequency of 159 Mc/s. Mem. R. Ast. Soc., 68, 37-60.

Goddard, B.R., Watkinson, A. & Mills, B.Y. (1960). "An interferometer for the measurement of radio source sizes. Aust. J. Phys., 13, 665-675.

Hanbury Brown, R., Jennison, R.C. & Das Gupta, M.K. (1952). Apparent angular sizes of discrete radio sources. Nature, 170, 1061-1063.

Little, A.G. & Payne-Scott, R. (1951). The position and movement on the solar disc of sources of radiation at a frequency of 97 MHz, Pt. I - Equipment. Aust. J. Sci. Res. A., 4, 489.

Mills, B.Y. (1952a). The positions of six discrete sources of cosmic radio radiation. Aust. J. Sci. Res. A., 5, 456-463.

—   (1952b). The distribution of the discrete sources of cosmic radio radiation. Aust. J. Sci. Res. A., 5, 266-287.

—   (1952c). Apparent angular sizes of discrete radio sources. Nature, 170, 1063-1064.

—   (1953). The radio brightness distributions over four discrete sources of cosmic noise. Aust. J. Phys., 6, 452-470.

—   (1960). On the identifications of extragalactic radio sources. Aust. J. Phys., 13, 550-577.

Mills, B.Y. & Little, A.G. (1953). A high-resolution aerial system of a new type. Aust. J. Phys., 6, 272-278.

Mills, B.Y., Little, A.G., Sheridan, K.V. & Slee, O.B. (1958). A high resolution radio telescope for use at 3.5 m. Proc. I.R.E., 46, 67-84.

Mills, B.Y. & Slee, O.B. (1957). A preliminary survey of radio sources in a limited region of the sky at a wavelength of 3.5 m. Aust. J. Phys., 10, 162-194.

Mills, B.Y., Slee, O.B. & Hill, E.R. (1958). A catalogue of radio sources between declinations of +10° and -20°. Aust. J. Phys., 11, 360-387.

Mills, B.Y., Slee, O.B. & Hill, E.R. (1960). A catalogue of radio sources between declinations of -20° and -50°. Aust. J. Phys., 13, 676-699.

— (1961). A catalogue of radio sources between declinations of -50° and -80°. Aust. J. Phys., 14, 497-507.

Mills, B.Y. & Thomas, A.B. (1951). Observations of the source of radio-frequency radiation in the constellation of Cygnus. Aust. J. Sci. Res., A4, 158-171.

Pawsey, J.L. (1957). Preliminary statistics of discrete sources obtained with the Mills Cross. I.A.U. Symposium No. 4: Radio Astronomy, ed. H.C. van de Hulst. Cambridge University Press, pp.228-232.

Ryle, M. (1956). Radio galaxies. Scien. Amer., 195, No. 3, 205.

— (1962). In discussion. I.A.U. Symposium No. 15: Problems of Extragalactic Research, ed. G.C. McVittie. McMillan, N.Y., p.344.

Ryle, M. & Scheuer, P.A.G. (1955). The spatial distribution and nature of radio stars. Proc. Roy. Soc., A., 230, 448-462.

Scheuer, P.A.G. (1957). A statistical method for analysing observations of faint radio stars. Proc. Camb. Phil. Soc., 53, 764-773.

— (1965). Extended radio sources II. Aust. J. Phys., 18, 77-84.

Scott, P.F. & Ryle, M. (1961). The number flux density relation for radio sources away from the galactic plane. Mon. Not. R. Ast. Soc., 122, 389-397.

Shakeshaft, J.R., Ryle, M., Baldwin, J.E., Elsmore, B. & Thomson, J.H. (1955). A survey of radio sources between declinations -38° and +83°. Mem. R. Ast. Soc., 68, 97-154.

Smith, F.G. (1952). Apparent angular sizes of discrete radio sources. Nature, 170, 1065.

Star trails above the 17 ft dish at Dover Heights in Sydney in 1952 (courtesy CSIRO Radiophysics Division)

EARLY WORK ON IMAGING THEORY IN RADIO ASTRONOMY

R. N. Bracewell
Radio Astronomy Institute
Stanford University
Stanford, California 94305 U.S.A.

By the end of the 1950s, radio astronomy was generating a
recognizable sub-discipline which was to prove influential in the later
development of antennas and radio interferometry and also in information
theory as applied to optics. Radar and radar astronomy, optics, electron
microscopy, and, unexpectedly, image reconstruction in X-ray tomography
were all to be influenced; in fact, a whole range of mathematical ideas
connected with imaging in radio astronomy began in the 1950s to be pulled
together and to crystallise as a whole.

This chapter will explain how these ideas developed as I saw
them over years spent in Sydney, Cambridge and California and on visits
to France. In addition, notes on individuals encountered along the way
round out the human aspects that tend not to be chronicled in the terse
journals.

Fortunately I had a good grounding in mathematics at Sydney
University. T. G. Room had just taken over from H.S. Carslaw, whose
book <u>Fourier Series and Integrals</u> was studied like the Bible. My other
mathematics teachers were R.J. Lyons, W.B. Smyth-White, H.H. Thorne and
E.M. Wellish. In physics I was much influenced by V.A. Bailey, a mathe-
matically powerful Oxford experimentalist.

I first encountered Joseph L. Pawsey (Lovell 1964) in a course
of lectures on transmission lines and aerials that he gave at the Radiophysics Laboratory, Sydney, in 1942. The contrast with the orthodox text
of Bewley (1933) offered food for serious thought and the lectures introduced me to the marvelous duality of physical vis-à-vis mathematical
thinking. I have kept my lecture notes. Later when I followed Pawsey's
footsteps to the Cavendish Laboratory I found the same penetration to the
physics of a matter exemplified by J.A. Ratcliffe, who inspired me and
other disciples to problem-solving employment of the Fourier transform.

I first heard of Ratcliffe in 1940 when his Methuen Monograph <u>The</u>
<u>Physical Principles of Wireless</u> (Ratcliffe 1934), a model of clarity, was
assigned by Bailey.

Fig. 1.   Joseph Lade Pawsey (1908 - 1962).

        At Radiophysics Laboratory from 1942 to 1946 I worked on
speech modulation of magnetrons, developed hardware prototypes of various
components for L-band and S-band radars, and solved electromagnetic
boundary value problems by numerical methods. F.W.G. White, Chief of the
Division at the time of my arrival, had written a Methuen Monograph
<u>Electromagnetic Waves</u> (White 1942) which commanded my attention because
the subject was one I wished to master. I did not think the Chief
noticed my existence but years later he remembered that I had conjured
500 milliwatts of L-band power out of the new lighthouse triode by
embedding it in a pair of TE010 cavities. Not long after, by which time
E.G. Bowen was chief, I contributed the theoretical chapter on waveguides
and resonators to the <u>Textbook of Radar</u> (Bowen 1947). My first published
paper (Bracewell 1947) dealt with cavity resonators and I was gratified
to receive my first request for a reprint from the Microwave Laboratory,
Stanford, where W.W. Hansen had originated the application of resonators
to the klystron oscillator.

        Professor Balthasar van der Pol influenced the birth of
imaging theory. He had a great interest in applied mathematics and had
created Commission VI of URSI around 1950. By the time of the 13th
General Assembly in London in 1960, Commission VI was able to conduct a
series of sessions on "aerials and data processing" which were written up

and published under the editorship of Professor Samuel Silver (1963). I
first met van der Pol and Silver in 1952 at the 10th General Assembly of
URSI, of which I was Organizing Secretary, by which time I had already
been guided by their writings (van der Pol & Bremmer 1955; Silver 1949).
I had had the occasion to edit the manuscript of van der Pol and Bremmer
for Cambridge University Press around 1948 and was struck by the attention given to questions of notation. J. C. Jaeger (Patterson 1981), in
his 1944 lectures on the Laplace transform at the National Standards
Laboratory, Sydney, also discussed minutiae of notation. Since then I
have always regarded symbology, terminology and graphical presentation as
worthy of close attention.

Jaeger inhabited a small office upstairs from me and when I
would go to consult him on mathematical questions he would always be
turning the crank of a calculating machine. Later he had an electrical
model. Intrigued by this I undertook an ambitious computation of discon-
tinuity capacitances in radial transmission lines, a problem generated by
the lighthouse triode oscillator (Bracewell 1954).

In Cambridge from 1946 to 1949 I wrote many papers on very
long wave probing of the ionosphere, proved that there are two D regions
(Bracewell and Bain 1952), and found that the wave admittance Y(h) looking
upward through the ionosphere at height h obeyed the same Riccati equation $dY/dh = i\omega\mu_0 Y^2 - y(h)$, where y(h) is the shunt admittance of the
medium per unit distance, that governed the nonuniform radial transmission line. I became interested in solar physics (my first publication
from Cambridge went to Monthly Notices), and began to haunt solar observatories. Pawsey commissioned me to visit Paris, where I had already
become acquainted with M. and Mme d'Azambuja, to look into the acquisi-
tion of a Lyot filter for Radiophysics (Fig. 2). On returning to Sydney
I began further investigations into D-region, which were interrupted by
duties as Organizing Secretary of the forthcoming 10th General Assembly
of URSI in Sydney and Canberra. My activities included soliciting funds
from company managers, creating a complex staff, and winetasting in
preparation for social events.

What an experience the Sydney URSI meeting was. It opened in
the Great Hall of Sydney University with the "Ride of the Valkyries,"
with Eugene Goossens conducting the Sydney Symphony Orchestra. There had
been differences of opinion about the appropriateness of Wagner, but the

bloodcurdling reverberations in that Gothic stone cavern, first thing in
the morning, proved to be unforgettable. After some more conventional
items we listened to John Antill's "Corroboree," a marvelous ballet suite
with bullroarers in the background that solved the problem of how to
introduce something Australian without inviting comparison with
Beethoven. These and many other programme details were meticulously
planned by D.F. Martyn (Massey 1971) who had initiated the invitation to
Australia and then masterminded the organization and funding. The
organizing committee was chaired by my old electrical engineering
professor J.P.V. Madsen and afforded educational insights into commit-
teemanship which stood me in good stead when I duly rose to be a chairman
myself, drawing complex assignments such as the successive Arecibo
Committees (1962-1976) and other hot potatoes. A great group of radio-
scientists attended, some spending a month at sea crossing the Indian
Ocean to reach Australia. I designed a badge for the participants and an
URSI flag, which Frank Kerr had made. Appleton (Ratcliffe 1966) raised
the flag for the duration of the meeting and it has continued to grace

Fig. 2. R.N. Bracewell, R.G. Giovanelli, Bernard Lyot and
L. d'Azambuja at Meudon Observatory, 1949.

subsequent General Assemblies. I astonished Sydney Chapman (Cowling 1971) by handing him the manuscript of a popular talk on geomagnetism for him to read on an ABC radio program. After recognizing that the piece was condensed from his own work he graciously broadcast it (Chapman 1952).

The immediate inspiration to write my book on the Fourier transform came from Pawsey, who wanted to see a Pictorial Dictionary of transforms. Pawsey did not have much of a mathematical background – he once asked me what variance meant – but thought in physical terms. He was concerned about the reliability of brightness distributions obtained by Fourier transformation of measured visibilities and proposed what he called the Sausage Theorem: "If the error bars on the visibility measurements fit inside a certain sausage then the calculated source distribution runs down the middle of another sausage." Pawsey very reasonably wanted to know how fat this other sausage was and my job was to find out. It is a very good question.

My paper at the 1960 General Assembly of URSI (Silver 1963) is a synopsis of the topics whose origins are discussed below.

First there is restoration, which I first heard about from Pawsey at Radiophysics. In 1945 and 1946 Pawsey, together with Ruby Payne-Scott and Lindsay McCready, was observing solar noise as an exercise connected with antennas intended for radar use.(See the contribution by Christiansen elsewhere in this volume.) The term radio astronomy did not then exist. (I first heard Pawsey use it in 1948 in a talk he gave at the Cavendish Laboratory while on his way to the Stockholm URSI Assembly.) Pawsey was interested in deducing the diameter of the sources of radio emission on the sun, at first with a view to deciding whether the whole sun was active or whether the sources were small. Observation settled the association with sunspots immediately, a remarkable achievement considering that the antennas had beamwidths around 20 degrees. Then, however, there were the more refined questions of the size and shape of the intensity distribution of the solar sources. For the diameter, Pawsey had thought in terms of an equivalent rectangular source, an idea borrowed from emission-line spectroscopy. The question of the actual shape of the source distribution, however, was first faced in connection with cosmic noise (from the Galaxy) where it was apparent that the observed distribution could not be the "true" distribution. The notion that the observed distribution bore the same

relationship to the "true" distribution that the output of a low-pass
filter bears to the input waveform caused a lot of excitement. Of
course, one only had to apply the convolution theorem to the known
convolution integral relating observed distribution to "true"
distribution to discover this result. But in those days convolution was
not a household word at the Radiophysics Laboratory, nor was it anywhere
else. Phrases peculiar to each field abounded, such as Faltung,
composition product, superposition integral, Carson's integral. A glance
at standard textbooks of the time, for example Bush, Jeffreys and
Jeffreys, Stratton, Titchmarsh, and Watson, reveals how much has been
gained as convolution has come to be appreciated as a pervasive concept
for which we have now one name that is universally understood. Even less
of a houshold notion in electrotechnical circles was the convolution
theorem. The idea was certainly in existence, but again as a special
concept in each field of application. For example, the array-of-arrays
rule for antennas is an embodiment of the convolution theorem (see T.R.
Kaiser's chapter on aerials (Bowen 1947) for the general level of
sophistication). In statistics, characteristic functions also have a
property which exemplifies the convolution theorem.

        Thus the reduction of the antenna mapping problem to a
familiar low-pass filtering action on waveforms was surprising and
pleasing. I had learnt about the convolution theorem from Doetsch's
Theorie und Anwendung der Laplace-Transformation and later from van der
Pol and Bremmer, but the concept lay dormant in my case until the true
connection between the Laplace and Fourier transforms dawned on me.

        Meanwhile, J.S. Hey et al. (1948) hit upon a reassuring method
of restoration (independently known in the study of integral equations as
the method of successive substitutions) that proved to be perfectly
practical for numerical work in two dimensions, even on a spherical
surface, and was thought for a time to be the answer to the problem of
finding the "true" distribution (Bolton & Westfold 1950). I had a
lurking suspicion, however, that the sequence of successive approximations might not converge or, if it did, that it might not converge to the
true distribution. Hand computation showed divergence apparently associated with noise in the data or with round-off error; also there was a
philosophic doubt whether observations of finite resolving power could
get at the full truth.

After the URSI meeting Pawsey decided that the time was ripe for a textbook on radio astronomy (Pawsey & Bracewell 1955) and that he should strike while the iron was hot. He invited me to join him on the grounds that my work had been disrupted by preparations for URSI, and he passed over other conceivable coauthors who were fully engaged in researches. A hidden agendum, I now suspect, was to wean me from the ionosphere and indeed my assignment to URSI may have been the first step. Later that year someone threw away several notebooks containing D-region data I had accumulated all over New South Wales, an event that put a dent in my list of publications and kept me in radio astronomy. After laying out the sequence of chapters for the book we wrote first drafts which we repeatedly exchanged for heavy revision, meeting almost daily. You can no longer tell, in most cases, who wrote the first draft of each chapter. Pawsey liked to write furiously with copious blanks and call for frequent and innumerable retypings while I liked to write slowly in quasi-final form; so I learnt a lot about technique. Some later authors flattered us by copying the illustrations. In our Fig. 1, which exhibits the concept of optical and radio "windows," there is a procession of clouds one of which, somewhat whimsically, is shown as opaque at one end and transparent at the other. This peculiar cloud has developed a self-replicating life of its own.

In 1961 we began rewriting for a second edition and in May 1962 Paul Wild signed up as a third author. From Massachusetts General Hospital Pawsey wrote to me in Sydney saying, "I congratulate you on talking Paul into taking this on. He will be a tower of strength." Upon Pawsey's sad demise, not long after, Wild withdrew. I could not bring myself to abandon the project--it just faded away.

In writing up the low-pass filter analogy for Chapter 2 of Radio Astronomy, I showed the triangular "filter characteristic" for a uniformly illuminated antenna. The concept of spatial frequency has to be understood to appreciate this triangular spectral sensitivity function and one has to accept that a brightness distribution on a spherical sky can possess sinusoidal Fourier components. Not everyone found these ideas obvious. In particular, the existence of a cut-off spatial frequency presented difficulties and was contested in the literature. Two short papers (Bracewell 1962a,b) some years later were elicited by continuing doubts.

To conserve the time I was devoting to our textbook, Pawsey assigned Jim Roberts to assist with a separate paper on aerial smoothing (Bracewell & Roberts 1954). The many long discussions Roberts and I had over tricky points, especially sinusoidal components on the celestial sphere, probably did not result in any saving of time but did lead to a very good paper which became required reading for beginners. The paper contained the first use of the famous shah function III(x), which has since caught on widely, possibly because it can be typed, and therefore printed, without special instructions. The shah function, an infinite string of unit delta functions at unit spacing, is not a function at all in the realm of mathematical analysis but it can be handled as a generalized function in the sense of Temple. The symbol is named after the Russian letter shah. St. Cyril modelled it after the Hebrew letter shin, which in turn traces back to a charming Egyptian hieroglyph depicting a row of papyrus buds and flowers. The draft of our paper contained the proof that the shah function is its own Fourier transform, III(s) = F[III(x)], a sophisticated business because the shah "function" violates the two main conditions for even possessing a Fourier transform! The approach via sequences of transformable functions, as expounded by M.J. Lighthill (1959), is the way to think about the shah function. To my astonishment, this beautiful proof was deleted, in my absence, to reduce the length of the paper; so I added it to the two-dimensional sequel (Bracewell 1956b). But in two dimensions the proof looks rather formidable, and perhaps for this reason it is never cited.

The (u,v)-plane, that indispensable tool for thinking about radio interferometers, also made its debut in this 1956 paper. Authors have since experimented with other symbols but the (u,v)-plane is now firmly entrenched in radio astronomical jargon.

In the 1954 paper one encounters for the first time invisible distributions, the principal solution, and the answer to the question whether restoration by successive substitutions converges to the "true" distribution. The answer, which is rigorous and transparent, shows that convergence does occur, in the absence of errors, to the principal solution, a novel concept and one that was harmonious with qualitative notions of finite resolving power. The principal solution is one of an infinity of distributions which, if scanned by the antenna, would yield the observed distribution. It is unique in that its spatial frequency

spectrum agrees with the spectrum of the true distribution in the range
of spatial frequencies where the antenna is responsive, and is zero elsewhere.

Once the convolution relation is accepted, restoration is seen
to be analogous to equalization in telephony. In telephone systems
equalization of spectral components affected by frequency-dependent
attenuation is practised with confidence but practice in radio astronomy
has wavered. Authors hesitated to publish principal solutions that
exhibited even slightly negative values and tended towards abandoning
restoration even though the errors in the raw antenna output might exceed
the errors of the principal solution. Furthermore, extrapolation of a
Fourier transform beyond the cutoff due to the antennas was repugnant to
many authors then, but de facto extrapolation by maximum entropy and other
techniques became common practice in publications twenty years later.

With the understanding of integral equations which I gained
from thinking about restoration in radio astronomy, I was able to deal
with other problems such as the inversion of running means (Bracewell
1955a) and correction for grating response (Bracewell 1963a) where the
existence of divergent but nevertheless useful series of approximations
is demonstrated. Other papers on X-ray diffraction and medical X-ray
techniques which then resulted emphasize the unpredictability of basic
science (see below).

Two discrete-interval theorems related to Shannon's sampling
theorem were established (Bracewell 1958a). One, relating to pencil-beam
mapping, specified the permissible coarseness of antenna beam spacing in
terms of the spatial cut-off frequency of the spectral sensitivity
function and was important for pencil beam mapping in raster mode. The
other, relating to Fourier synthesis, specified the permissible
coarseness of spacing of fixed or transportable interferometer elements
in terms of source diameter.

At the Paris Symposium on Radio Astronomy in 1958 a great
conclave of radio astronomers came together for an exciting interchange
covering every field from the moon to the cosmos. It was recognized that
this might never be possible again, although in fact the dramatis personae
strove to reconvene for years after. At that time antennas on the one
hand and receivers on the other were still perceived by many as distinct
from each other and as mere adjuncts to the astronomy, and no session was

planned for these earthy technicalities. Time after time, however, the
discussion revolved around mapping technique and the perils of interferometry and it became apparent that the explosive progress in image-
forming systems formed a common bond between radio astronomers of otherwise diverse interests. Subsequent meetings, up to the present day, have
often recognized these instrumental concerns. As Chairman of the Organizing Committee, Pawsey made me editor of the proceedings, which took
eight months out of my life. To ensure uniformity with earlier IAU
symposium volumes, D.H. Sadler gave me the strictest instructions,
including a requirement that the spine must read from bottom to top.
Stanford University Press Director Leon Seltzer, a most reasonable man,
told me however that no book published in America could conceivably be
titled upside down and that it was no business of his if England did not
observe a standard. This serious contretemps was bypassed when the
612-page volume proved to be so thick that the spine could be lettered
horizontally.

After Paris we all went to Moscow for the 10th General
Assembly of the International Astronomical Union, visiting Pulkovo on the
way to see the distinctive multiple-mirror radio telescope. Meeting the
Soviet radio astronomers for the first time was interesting (Fig. 3).

Fig. 3. Self portrait of I.S. Shklovsky with pig, executed in
the Georgievsky room of the Great Kremlin Palace on August 16,
1958.

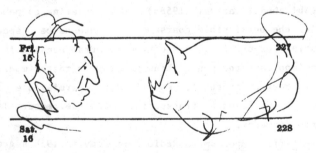

In January 1961 Swenson and I returned to the Soviet Union
under a new agreement to foster exchange of scientists between the U.S.
and U.S.S.R. and made a more extensive visit to radio astronomy sites
(Swenson & Bracewell 1961). When the National Academy of Sciences
undertook to have the Soviet visas entered in our passports the Academy

was bemused to find that of their first two exchange visitors one was a
British subject. I think the Russians were baffled too.

Meanwhile, the literature of what would in time be seen as
contributory steps toward image processing was accumulating. A paper
"Restoration in the presence of errors"(Bracewell 1958b) led to results
resembling Wiener's theory of linear prediction. A bothersome problem
that still worries many people who work with observational data is that
their signals are not stationary processes; and the statistics of these
postulated processes are not known to them. Wiener asks them to apply a
filter that they do not know to an entity which is not the one they have
observed. The classical distinction between experimental science and
observational science helps us to understand this. Practitioners of
Wiener filter theory must be experimentalists. Observers, on the other
hand, write papers based on single-shot observations and cannot afford to
obtain ensembles. They could, however, perhaps do a little better by
attending to the statistics of their errors.

How these imaging ideas came to be disseminated beyond the
radio-astronomical fraternity is interesting. Cambridge, Manchester,
M.I.T., Paris and Sydney were the important centres of diffusion. By the
late 1950s phase-switched interferometers had entered conventional radio
engineering for precision location of earth satellites, and many
classified applications must have been developing. Indicative of the
interest in the instrumental aspects of the explosion in radio astronomy
is the fact that of all my papers, "Radio Astronomy Techniques," a
lengthy chapter in vol. 54 of Handbuch der Physik, was more widely
requested, by more than an order of magnitude, than any other.
Commissioned in 1956, the final manuscript was delivered to meet an April
1957 deadline. Starting in 1957, I gave away several hundred preprints
and Springer Verlag later supplied 500 reprints (partly as reparations
for a five-year delay in publication) all of which were distributed to
correspondents, only some of whom could have been radio astronomers. The
success of Christiansen and Högbom's (1969) monograph Radiotelescopes
confirms the technological interest in interferometry techniques as
developed for exploring the radio sky.

Another key expository paper of the time, delivered in 1959 in
the U.S.A. to URSI Commission V on Radio Astronomy, laid emphasis on two
dimensions (Bracewell 1961a). In two dimensions the spectral sensitivity
function, the spatial analogue of the filter characteristic of a linear

time-invariant electrical transducer, is particularly helpful in understanding radio interferometry; the result that a <u>tee</u> <u>has</u> <u>the</u> <u>same</u>
<u>resolution</u> <u>as</u> <u>the</u> <u>cross</u> containing it, while having only three-quarters as
many elements, was discovered by thinking in the (u,v)-plane, an approach
that is now a standard thinking tool.

A type of <u>compound</u> <u>interferometer,</u> in which additional antennas
are deployed on the line of an existing east-west array and ingeniously
positioned to cancel odd-order grating lobes, led to the achievement of
antenna beamwidths narrower than one minute of arc by the end of the
decade (Bracewell 1961b). In the old view an interference null was
manoeuvred onto the grating lobe; now one could think alternatively and
with profit of minimizing ripple on the spectral sensitivity function
(Bracewell 1963b). Other configurations were explored.

The term <u>spectral</u> <u>sensitivity</u> <u>function</u> is still seen
occasionally but has been largely superseded by <u>transfer</u> <u>function</u>, the
usual term in electric circuit theory for the analogous one-dimensional
temporal quantity. It is routinely regarded as complex. In optics, the
terms <u>optical</u> <u>transfer</u> <u>function</u> and <u>modulation</u> <u>transfer</u> <u>function</u> also
exhibit a unification of terminology but the retention of two phrases,
one for the complex and one for the real quantity, reminds us that the
concept of spatial phase is still not quite commonplace.

The <u>complex</u> <u>visibility</u> V of interference fringes, which is now
so well understood that it is no longer necessary to mention the word
"complex," evolved from the concept of fringe visibility V in optics.

Fig. 4. Two-element interferograms (above) showing that the
envelope (broken curve), from which visibility magnitude V is
determined, is resistant to phase errors (below).

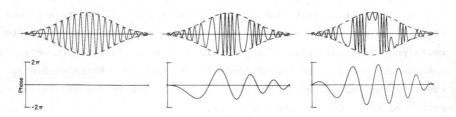

The quantity V, which is defined as a modulation coefficient in optics, is the absolute value of V. But the generalization from the real to the complex was not a natural step in optics, where "seeing" had to be accepted with resignation and the spatial phase of an interference pattern was not a physically measurable quantity. Thus when the Hanbury Brown and Twiss intensity interferometer startled the optical world, it was possible to claim as an advantage that stellar diameters were measurable by an intensity interferometer in the presence of atmospheric phase fluctuations (because intensity ignores carrier phase) and thereby engender deep controversy in the world of physics (see Hanbury Brown's own contribution elsewhere in this volume). And yet the familiar Michelson stellar interferometer also conserves diameter information in the presence of phase fluctuations (Fig. 4), as is clear to anyone understanding the Fourier transform relation between a source distribution and the observable complex visibility.

The formulation of the basic two-dimensional Fourier transform relationship of radio astronomy, connecting the celestial brightness distribution to the measurable quantity complex visibility (Bracewell 1958b) appeared in a special volume on radio astronomy edited for the Institute of Radio Engineers by Fred Haddock and now of historical value. The precise interpretation of u (and v) as a spatial frequency measured in cycles per direction-cosine unit was not always as obvious as it is today, and the alternative interpretation as a dimensionless baseline length measured in wavelengths still puzzles many.

McCready et al. (1947) were the first to describe the possibility of Fourier synthesis of an arbitrary one-dimensional source distribution. They utilized an interferometer which had been developed for finding the height of approaching aircraft, a vital parameter but one not readily yielded by an ordinary radar, which is best suited to rangefinding. By siting a single antenna on a cliff overlooking the ocean one obtained interference between a direct ray and a ray specularly reflected at almost grazing incidence from the ocean. The result was to dissect the antenna beam into fine lobes stacked regularly in angle of elevation. Aircraft height was deducible from the ratio of echo amplitudes received on the primary antenna and on a secondary antenna mounted above the first. McCready, Pawsey and Payne-Scott presented an equation for the power received, as the source rose above the horizon and swept through the interference fringes, in terms of a Fourier cosine

transform, which suggests lack of generality. The basic relationship
between complex visibility V and brightness, which requires the
exponential transform (and a redefinition of brightness), does not appear
in this paper. Even in its original (and current) meaning in optics as
"depth of modulation of an interference pattern," visibility V was not
mentioned. Instead, a related contrast ratio R = (1 − V)/(1 + V) of
minimum to maximum power was introduced and was still current a decade
later. The Fourier exponential transform relationship between brightness
and complex visibility V, as a function of antenna spacing, is quite
distinct from the (perfectly correct) Fourier cosine transform
relationship between brightness and the received power as a function of
source position while the antenna spacing remains fixed. This basic
paper appeared as an internal report of the CSIRO Radiophysics Division
on 16 June 1946, was received in London at the Royal Society on 22 July
1946, but was not published until August 1947. Naturally, the
unbelievable delay gave rise to apprehension.

  Although Fourier synthesis was clearly described as early as
1946 by McCready et al. and was impressively demonstrated by Ryle's group
during the 1950s (Ryle & Hewish 1960), to many people the concept
remained arcane. Up until the Pierce report on radiotelescopes (Keller
1961) Fourier synthesis was widely regarded with dubiety, especially at
the National Radio Astronomy Observatory, Green Bank. To allay concerns
a footnote to the Pierce report explained that "Aperture synthesis means
combining observations taken at different times by interferometry in such
a way as to simulate the observation that could be made at one time with
a large-aperture configuration." As my colleague A.R. Thompson drily
comments, "Fourier synthesis leads to transfer functions that can never
be simulated by single large antennas." Such synthesis invariably leads
to physically unrealizable point response functions that go negative. In
fact, Fourier synthesis yields (Bracewell 1962c, p. 123) the dreaded
principal solution.

  The Fourier cosine factor $\cos\Delta$ or $\cos(4\pi h\theta/\lambda)$ of McCready et
al. for small angles of elevation θ would become $\cos[4\pi(h/\lambda)\sin\theta]$ for
greater angles, as the authors clearly understood. But the recognition
of $\sin\theta$ as a direction cosine had to await explicit treatment of
two-dimensional source distributions. Brightness $b(\theta,\phi)$ as a function of
two directional coordinates was usually written $b(x,y)$ as though use of
Cartesian coordinates on the celestial sphere needed no comment. The

modified quantity measured in W m^{-2} Hz^{-1} (direction cosine unit)$^{-1}$, let
us call it b'(1,m), where (1,m,n) are direction cosines, has never
received a name or distinctive symbol. It may be defined by reference to
brightness b(1,m), measured in W m^{-2} Hz^{-1} sterad^{-1}, by b'(1,m)dldm =
b(θ,ϕ)dΩ. Only much later did the practical distinction expressed by
b(θ,ϕ) = nb'(1,m) become significant as long-baseline arrays developed.

Derivation of the two-dimensional Fourier transform
relationship

$$V(u,v) \propto F[b'(1,m)]$$

turns out to be rather sophisticated. Emil Wolf told me that he did not
think my derivation in Haddock's volume could be rigorous because the
corresponding result in optics involved an approximate argument from
Huygens' Principle. My derivation does make an assumption of ergodicity,
but this is surely acceptable for astronomical sources exhibiting no
parallax. I think the source of Wolf's objection lies in the fact that
in optics one thinks of diffraction in terms of waves incident on
apertures, whereas in radiofrequency electromagnetism one starts from
aperture distributions that are imposed. Certainly it is absolutely
rigorous to say that a cosinusoidal field distribution E$_o$ cos 2πlu·
cos 2πmv will launch plane waves in the directions ($\pm$l, $\pm$m, n). But it
was this fundamental Fourier transform property of electromagnetism
which, in my derivation of the V - b' relationship, gave rise to the
result that V $\propto$ F[b'].

The approach to a full theory of the curved sky contained an
intermediate cylindrical stage propounded by J. Arsac (1957). Arsac sent
his manuscript to Pawsey asking whether it would be suitable for the
Australian Journal of Physics and I got the job of returning detailed
comments. I had written an internal memorandum in 1953 on the   " $\overline{ZK}$
Array". In Morse Code a digraph with vinculum stood for a special long
character, in this case ▬ ▬ •• ▬ • ▬ , a mnemonic for an
arrangement of four antennas (dashes) that could be formed from a
seven-element equispaced array by removing the third, fourth and sixth
(dots). Another representation would be the binary number 1100101. The
interesting thing about this arrangement, or "unfilled" array, was that
the resolution should be as good as that of the seven-element filled

array.   This deduction followed from the respective spectral sensitivity
functions which, I found, could be calculated instantly by base-10
multiplication of the binary representation by its reverse. Thus 1100101
× 1010011 = 1111114111111. This signifies a sensitivity of 4 to spatial
d.c. (uniform background), and unit sensitivity to all spatial
frequencies, positive and negative, out to a maximum spatial frequency of
six units. For comparison, the filled seven element array 1111111 has a
spectral sensitivity function 1111111 × 1111111 = 1234567654321, which is
graded differently, due to redundancy, but covers exactly the same set of
spatial frequencies, no more, no less. We now learnt that Arsac (1955)
had actually built a "$\overline{ZK}$ array". I believe this idea was also known to
N.F. Barber in connection with underwater acoustic arrays. The whole
subject of minimum redundancy arrays started from the Arsac array.

        Today the Fourier analysis of brightness distributions
continues to wrinkle brows. It is probably fair to say that, when the
total set of baselines is <u>noncoplanar</u> and a third coordinate w is needed
to represent those baselines that lie out of the (u,v)-plane, many radio
astronomers accept the relation

$$V(u,v,w) \propto \iint_{\text{source}} b'(l,m) \, \exp[-i2\pi(lu + mv + nw)] \, dldm$$

for a compact source as being exact; but various inversion formulas have
been published, and a generally accepted standard has not appeared.

Fig. 5. H.C. Minnett, W.N. Christiansen and R.N. Bracewell
(left to right) at Radiophysics Laboratory around 1953.

Fan beam scans, as perfected by Christiansen by 1952, led to interesting developments. Christiansen maintains that reconstruction from fan beam scans in different orientations was an early example of earth rotation synthesis, or supersynthesis, to use Ryle's term (see Christiansen's and Scheuer's contributions elsewhere). It is true that the scans were first symmetrized by subtraction of active regions so that cosine transforms would suffice, but it is unreasonable to deny the claim on the grounds that the antennas were not transportable. The actual reconstruction was at first carried out by a laborious numerical-graphical procedure and in 1954, at a meeting of the radio astronomers at the Radiophyics Laboratory, Sydney, fan beam reconstruction was an item listed in the minutes as requiring attention. Pawsey assigned this project to me.

Christiansen shared a small office with Harry Minnett, later Chief of the Division, and myself (Fig. 5). The day he moved in, he inspected the scene briefly and left, but soon returned with a lightmeter and proceeded to measure light levels at desk height all over the room while Minnett and I busied ourselves with our work. After this unnerving experience Minnett and I moved over a little. A succession of Christiansen's collaborators worked in an adjacent office on the graphical computations and I used to consult with them. A huge sheet of graph paper was spread on a sloping drawing table and myriads of pencilled numbers ran across the paper at assorted angles. My method of chord construction (Bracewell 1955b), which was adopted by Swarup and Parthasarathy for restoration of the scans, was in keeping with the graphical spirit of the approach. The distinction between restoration and reconstruction dates from those times and has spread to other fields such as radiology, where the term reconstruction has been taken into computerized tomography.

The outcome of my assignment at the 1954 planning meeting was a readable paper on strip integration (Bracewell 1956a) expounding among other things the projection-slice theorem, a theorem that has since become famous. At the time the relation seemed so obvious that no one thought of giving it a name; it was well-known to a handful of radio astronomers in Sydney and Cambridge – to X-ray crystallographers it would have been a trifle. Nevertheless, one key illustration was effective in the spread of the theorem; the distinctive spatial organization of the diagram (Fig. 6) enables its descendants in the modern literature of

Fig. 6. An early illustration (1956) of the famous
"projection-slice" theorem in which Fourier transformation is
organized horizontally and line integration vertically.

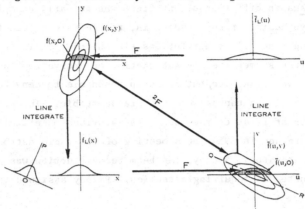

optics and image formation to be recognized at a glance. Just as
intriguing to me was the cyclic relation between the Fourier, Abel and
Hankel transforms, a tool that I have since frequently used, for example
in discovering that the cross-section of the grating ringlobe generated
in rotation synthesis is the half-order derivative of a sinc function!
The Fourier-Abel-Hankel cycle of transforms is only now beginning to
appear in textbooks of image processing. Practical methods of solving
the reconstruction problem were described in the strip integration paper,
including an iterative method. In a later paper (Bracewell & Riddle
1967) I gave the space-domain inversion procedure which came to be
universally used in computerized tomography for inverting line integrals
of X-ray absorption. As a result of medical interest these items became
the most extensively mentioned of my astronomical papers in the Science
Citation Index, another indication of the significance of radio
astronomical spinoff in the everyday world. Alan Cormack had also solved
the reconstruction problem in 1963 (as reported in Physics Today, 32, 19,
1979) and received half the Nobel Prize for Physiology and Medicine in
1979 for his contribution to the mathematical side of X-ray tomography.

The history of this unexpected contribution of an astronomical
algorithm to medical computing is touched on by Brooks and diChiro
(1976). To my surprise I received a visitor from the General Accounting
Office in Washington around 1976 asking how my medically-related

astronomy research could have been fostered by the National Institutes of Health and what they should do in future!

Fan beam reconstruction disappeared from solar radio astronomy as a result of the invention of the <u>crossed-grating</u> <u>interferometer</u>, often known as the Chris cross. The concept of this instrument occurred to me in 1954 and I mentioned it to Otto Struve (Cowling 1964), who had invited me to spend 1954-1955 lecturing on radio astronomy in the Astronomy Department at Berkeley, when he asked for a suggestion as to how Berkeley might enter radio astronomy. I presented Struve with a design for a 10 cm wavelength cross and also wrote to Pawsey describing the concept. This design, which is recorded in a report entitled "A Proposal for a Microwave Spectroheliograph", dated September 15, 1955, was ultimately implemented at Stanford and led to monthly publication of eleven years (1962-1973) of continuous daily microwave spectroheliograms (Graf & Bracewell 1975). The digital spectroheliograms came to be available worldwide by teleprinter the same day and received a commendation from NASA for contributing to the manned landing on the moon. Meanwhile the

Fig. 7. Signatures of famous radio astronomers carved with hammer and chisel on pier East-1 at Stanford.

Radiophysics Laboratory built an instrument suited to 21 cm wavelength under the direction of Christiansen, who has described its origins

elsewhere in this volume. The first published description of the
Chris-cross concept appeared in Bracewell (1957).

The Stanford instrument was the first radiotelescope to
produce its output digitally in <u>camera-ready form</u>. It attracted many
famous visitors, numbers of whom left their names carved with hammer and
chisel in the concrete piers (Fig. 7). It also incorporated the first
system of <u>phase-and-amplitude</u> <u>calibration</u> of transmission lines; the
installation used modulated gas-discharge tubes at each of the 32
antennas. Executing a regular raster scan on the moving target presented
by the sun required precision phasing of 16 antennas independently so
that a beam would be correctly formed and also lie at the desired zenith
distance (within 0.1 minutes of arc) for each scan line. Such absolute
pointing accuracy was far beyond what was required of the great optical
telescopes. Scanning was automatic and computer-controlled, at first
with a real time analogue device and later with a card reader.
Subsequent innovation in <u>Fourier synthesis</u> instrumentation at Stanford is
documented in a cumulative report (Bracewell 1980).

Designing waveguide components for the cross was a pleasant
reversion to old skills. Microwaves is still one of the subjects I try
to keep up with. This familiarity later helped me crack the puzzle of
the Arecibo line feed which, in its original square cross-section form,
was mysteriously inefficient. For years I travelled regularly to
Arecibo, Puerto Rico. One tropical evening in San Juan, as we leaned
over a railing and listened to the coquís chirping, Hanbury Brown
generously offered me his job! Not wishing to test his authority to
deliver, I cautiously said nothing.

Throughout the developments in data processing of the fifties,
the unifying theme has been the application of Fourier analysis. Even in
such an apparently unrelated subject as the surface tolerance of the
paraboloids used for radio astronomy it was shown that the degradation of
gain is characterised by the root-mean-square departure of the paraboloid
of best fit, not from the reflector surface itself, but from the surface
from which one has first removed the Fourier components of period less
than a wavelength (Bracewell 1960). Not only does the Fourier transform
control antenna design and the relation between source brightness
distributions and observable visibility but also, for different reasons,
relates the waveforms and spectra of our electronic circuits, and to cap
everything, has revolutionized the computing on which our image formation

now depends.   Lord Kelvin had an inkling of this when he wrote,
"Fourier's theorem is not only one of the most beautiful results of
modern analysis, but it may be said to furnish an indispensable
instrument in the treatment of nearly every recondite question in modern
physics".

Fig. 8.   Brush drawings of E.G. Bowen and R.N. Bracewell
(right) by cartoonist Emile Mercier.

The splendid atmosphere at the Radiophysics Laboratory under
the direction of E.G. Bowen and J.L. Pawsey and the stimulation of such
colleagues as W.N. Christiansen, B.Y. Mills and J.P. Wild kept the mind
occupied with good problems.   If I seem to have quoted lots of my own
papers, the fact is that while all the radio astronomers vigorously
discussed their data processing problems, they wrote mostly about
astrophysics and confined instrumental matters somewhat apologetically to
introductions and appendices.   A memorable exception would be J.P. Wild's
papers of the following decade discussing the principle of the circular
array at Culgoora.   Looking back now we see that no apology was required
for the instrumental and image formation aspects of radio astronomy in
the fifties.   Starting from modest beginnings and by small steps radio
astronomers took the separate fields of antennas, receivers and
information theory and welded them into image forming systems that have

improved by seven orders of magnitude in resolution, surpassing the optical telescope, and inspired other developments in fields as diverse as optics, acoustics, seismic probing and X-ray tomography.

REFERENCES

Arsac, J. (1955). Nouveau réseau pour l'observation radioastronomique de
la brillance sur le soleil à 9350 Mc/s. Comptes Rendus,
240, 942.
Arsac, J. (1957). Application of mathematical theories of approximations
to aerial smoothing in radio astronomy. Aust. J. Phys., 10,
16-28.
Bewley, L.V. (1933). Traveling Waves on Transmission Systems. New York:
Wiley.
Bolton, J.G. & Westfold, K.C. (1950). Galactic radiation at radio
frequencies. I. 100 Mc/s survey. Aust. J. Sci. Res., 3,
19-33.
Bowen, E.G., ed. (1947). A Textbook of Radar. Sydney: Angus and
Robertson; 2nd ed., 1954, Cambridge: Cambridge University
Press.
Bracewell, R.N. (1947). Charts for resonant frequencies of cavities.
Proc. I.R.E., 35, 830-841.
Bracewell, R.N. (1954). Step discontinuities in disc transmission
lines. Proc. I.R.E., 42, 1543-1547.
Bracewell, R.N. (1955a). Correction for running means by successive
substitutions. Aust. J. Phys., 8, 329-334.
Bracewell, R.N. (1955b). A simple graphical method of correcting for
instrumental broadening. J. Opt. Soc. Amer., 45, 873-876.
Bracewell, R.N. (1956a). Strip integration in radio astronomy. Aust.
J. Phys., 9, 198-217.
Bracewell, R.N. (1956b). Two-dimensional aerial smoothing in radio
astronomy. Aust. J. Phys., 9, 297-314.
Bracewell, R.N. (1957). Antenna problems in radio astronomy. I.R.E.
National Convention Record, 5, 68-71.
Bracewell, R.N. (1958a). Radio interferometry of discrete sources.
Proc. I.R.E., 46, 97-105.
Bracewell, R.N. (1958b). Restoration in the presence of errors. Proc.
I.R.E., 46, 106-111.
Bracewell, R.N. (1960). Antenna tolerance theory. In Statistical
Methods of Radio Wave Propagation, ed. W.C. Hoffman, pp.
179-183. New York: Pergamon.
Bracewell, R.N. (1961a). Interferometry and the spectral sensitivity
island diagram. IRE Trans. on Antennas and Propagation, AP-9,
59-67.
Bracewell, R.N. (1961b). A radiotelescope with a minute-of-arc beam.
Astronom. J., 66, 279.
Bracewell, R.N. (1962a). A corrected formula for the spectral
sensitivity function in radio astronomy. Aust. J. Phys., 15,
445-446.
Bracewell, R.N. (1962b). Electrical scanning and low pass filtering in
radio astronomy. Aust. J. Phys., 15, 447-449.
Bracewell, R.N. (1962c). Radio astronomy techniques. Handbuch der
Physik, 54, 42-129.
Bracewell, R.N. (1963a). Correction for grating response in radio
astronomy. Astrophys. J., 137, 175-183.
Bracewell, R.N. (1963b). Proposal leading to future large radio
telescopes. Proc. Nat. Acad. Sci., 49, 766-777.
Bracewell, R.N. (1980). Stanford Radio Astronomy Institute. Bull.
Amer. Astronom. Soc., 12, 369-372.

Bracewell, R.N. & Bain, W.C. (1952). An explanation of radio propagation
    at 16 kc/sec in terms of two layers below E layer. J. Atmos.
    Terr. Phys., 2, 216-225.
Bracewell, R.N. & Riddle, A.C. (1967). Inversion of fan beam scans in
    radio astronomy. Astrophys. J., 150, 427-434.
Bracewell, R.N. & Roberts, J.A. (1954). Aerial smoothing in radio
    astronomy. Aust. J. Phys., 7, 615-640.
Brooks, R.A. & diChiro, G. (1976). Principles of computer assisted
    tomography (CAT) in radiographic and radioisotopic imaging.
    Phys. Med. Biol., 21, 684-732.
Chapman, S. (1952). The mystery of the magnetic compass. The ABC
    Weekly, September 13, pp. 8-9.
Christiansen, W.N. & Högbom, J.A. (1969). Radiotelescopes. Cambridge:
    Cambridge University Press.
Cowling, T.G. (1964). Otto Struve 1897-1963. Biog. Mem. of Fellows of
    the Roy. Soc., 10, 283-304.
Cowling, T.G. (1971). Sydney Chapman 1888-1970. Biog. Mem. of Fellows
    of the Roy. Soc., 17, 53-89.
Graf, W. & Bracewell, R.N. (1975). Synoptic maps of solar 9.1 cm
    microwave emission from June 1962 to August 1973. Boulder:
    World Data Center A.
Hey, J.S., Parsons, S.J., & Phillips, J.W. (1948). An investigation of
    galactic radiation in the radio spectrum. Proc. Roy. Soc.,
    A192, 425-445.
Keller, G. (1961). Report of the advisory panel on radio telescopes.
    Astrophys. J., 134, 927-939.
Lighthill, M.J. (1959). An Introduction to Fourier Analysis and
    Generalized Functions. Cambridge: Cambridge University Press.
Lovell, A.C.B. (1964). Joseph Lade Pawsey 1908-1962. Biog. Mem. of
    Fellows of the Roy. Soc., 10, 229-243.
McCready, L.L., Pawsey, J.L. & Payne-Scott, R. (1947). Solar radiation
    at radio frequencies and its relation to sunspots. Proc.
    Roy. Soc. A, 190, 357-374.
Massey, H. (1971). David Forbes Martyn 1906-1970. Biog. Mem. of Fellows
    of the Roy. Soc., 17, 497-510.
Patterson, M.S. (1981). John Conrad Jaeger 1907-1979. Biog. Mem. of
    Fellows of the Roy. Soc., 28, 163-203.
Pawsey, J.L. & Bracewell, R.N. (1955). Radio Astronomy. Oxford:
    Clarendon Press.
Ratcliffe, J.A. (1934). The Physical Principles of Wireless, 3rd ed.
    London: Methuen.
Ratcliffe, J.A. (1966). Edward Victor Appleton 1892-1965. Biog. Mem. of
    Fellows of the Roy. Soc., 12, 1-21.
Ryle, M. & Hewish, A. (1960). The synthesis of large radio telescopes.
    Mon. Not. Roy. Astron. Soc., 120, 220-230.
Silver, S., ed. (1949). Microwave Antenna Theory and Design. New York:
    McGraw-Hill.
Silver, S., ed. (1963). Monograph on Radio Waves and Circuits.
    Amsterdam: Elsevier.
Swenson, G., Jr. & Bracewell, R.N. (1961). Some Russian radio
    telescopes. Sky and Telescope, 22, 77-80.
van der Pol, B. & Bremmer, H. (1955). Operational Calculus Based on the
    Two-Sided Laplace Integral. Cambridge: Cambridge University
    Press.
White, F.W.G. (1942). Electromagnetic Waves. London: Methuen.

SECTION THREE: *England*

The development of radar in England was perhaps the greatest technological contribution to the survival of that nation during the darkest days of World War II. This was achieved by concentrating the very best of British physicists in a few key laboratories (in particular the Telecommunications Research Establishment) where for six years they worked under tremendous pressure, often involving combat testing of the latest radar gear. Bernard Lovell and Martin Ryle emerged from this crucible not only as experts in radio techniques, but as men who knew how to get a job done. They founded radio groups at Manchester and Cambridge Universities amidst auspicious conditions for scientific and technical success: skilled and enthusiastic teams largely composed of wartime colleagues, "connections" to the military invaluable for obtaining surplus equipment, strong support from mentors (P.M.S. Blackett and J.A. Ratcliffe), and a subject ripe for exploitation.

The articles in this section recall those times. Jodrell Bank specialized at first in the radar study of meteors and the ionosphere, but shifted in the 1950s to pioneering work on galaxies and discrete sources. Large paraboloids and long baseline interferometry became specialties. At Cambridge, the Cavendish Laboratory group initially concentrated on solar observations, but then broadened to include the new and mysterious "radio stars." Their relentless pursuit of the techniques of interferometry and the cataloguing, mapping, and interpretation of radio sources led to a variety of fundamental contributions.

One of the portable antennas used in the first intensity inter-ferometer at Jodrell Bank in 1952 (courtesy Roger Jennison)

THE ORIGINS AND EARLY HISTORY OF JODRELL BANK

Sir Bernard Lovell
University of Manchester
Nuffield Radio Astronomy Laboratories
Jodrell Bank

WORLD WAR II INFLUENCES

The observatory at Jodrell Bank emerged in a most unusual manner from my research work as a young man on cosmic rays. In the years immediately before the outbreak of World War II, I was on the staff of the Physics Department of the University of Manchester as an Assistant Lecturer. Originally I continued my researches on the electrical conductivity of thin films, which I had worked on as a postgraduate student in Bristol, but when P.M.S. Blackett succeeded W.L. Bragg as the Professor of Physics in 1937 I eagerly seized the opportunity of joining him in his cosmic ray researches. I took over from a young Chinese, Hu Chien Shan (who had returned to China to help in the war against Japan), and modernised an automatic counter-controlled cloud chamber. With this device I first studied the cosmic ray showers produced by the newly discovered particles, the 'heavy' electrons, then called mesotrons, later mesons and now recognised as the mu-meson or muon. With L. Jánossy (Jánossy & Lovell 1938) and J.G. Wilson (Lovell & Wilson 1939) I was using this device to investigate the extensive cosmic ray air showers in 1939. In the later stages of this work we were using two cloud chambers separated by 20 metres to photograph showers which we estimated to have energies approaching 10^{16} eV.

At that time the energy spectrum had been established only to about 10^{10} eV, and there was little information about the spectrum at higher energies or about the upper limit of the energy range. Since these parameters were important matters concerning the nature and origin of the primary cosmic ray particles and since our two chamber equipment recorded these energetic events only rarely, we were naturally seeking means of improving the statistics. In pursuit of this aim I had equipped a lorry with suitable apparatus which I planned to use at the high altitude Pic du Midi observatory, but during the preparations Blackett advised me not to start on the journey because of the critical situation in Europe. In fact, almost immediately, in

August of 1939, groups of University staff were sent to operational Air
Ministry stations to be introduced to the secret RDF ("Radio Direction
Finding") equipment, subsequently to be known as Radar.

I was sent with Wilson to an operational Air Ministry station at
Staxton Wold in Yorkshire, a few miles inland from Scarborough on the north-
east coast of England, the site of one of the chain of powerful long wave
(20 to 55 MHz) coastal defence radars known under the code CH (Chain Home).
This chain of radars was in continuous day and night operation plotting
every aircraft that approached Great Britain and reporting the information
to the headquarters of RAF Fighter Command near London. Of course, there
were no computers in those days and the operators had before them a large
cathode ray tube (c.r.t.) display with a simple range-amplitude time base,
from which they measured the range of the aircraft echo. These radar echoes
from the aircraft, although fluctuating in amplitude, persisted for long
periods as the aircraft moved through the rather wide beam of the radar. I
was in the control room of the Staxton Wold radar on the Sunday morning,
3rd September 1939, when Prime Minister Chamberlain broadcast to the nation
that we were at war with Germany. I expected to see immediately large numbers
of echoes on the c.r.t. signalling the advance of the German bombers. There
were, indeed, a plentiful supply of echoes, very often short lived, but there
were no German bombers and the operators explained to me that these echoes
were "ionosphere".

Although I did not realise it then, this was the moment when
Jodrell Bank was to have its origin. We could understand why the ionosphere
would give a fairly steady echo on the radar, but we were mystified by the
many echoes which were transient - lasting only for a second or so. Wilson
and I turned to one another and simultaneously remarked "echoes from cosmic
ray ionization". We might, it occurred to us, have the powerful new technique
for studying the extensive cosmic ray showers which we were seeking. But the
war had begun and I was despatched to help the group developing airborne
radar, first at Scone airport near Perth in Scotland and then to St. Athan
in South Wales. Although my main task was to lie in the development of the
centimetric airborne radars, for the first months of the war I was concerned
with the 1½ metre Airborne Interception (AI) radar then being fitted in
Blenheim aircraft. Again I occasionally saw these transient echoes when
flying at 10,000 ft or more. When Blackett visited the aerodrome in South
Wales as a member of the Tizard Committee, I described the phenomenon to him

and suggested that the echoes might be radar reflections from the ionization
caused by extremely energetic cosmic ray showers.

At that time I was on an isolated aerodrome working under severe
wintry conditions in an aircraft hangar. Since there was no library and poor
communications with the outside world, I was unaware that Blackett had
delivered the Guthrie Lecture of the Physical Society in London on 26 February
1940. In this lecture he had considered the implication of the energy spectrum
of cosmic rays - a simple inverse exponent law applicable over an energy
range of at least 10^{18} to 1. Blackett concluded that such a simple form of
spectral distribution over this enormous intensity range must have some deep
cosmological significance in that it might be related to the structure of the
Universe (in the same class as the cosmological expansion). His recent
thoughts on this problem no doubt excited his interest in my suggestion that
we might, through radar, have a new means of investigating the high energy
region of the cosmic ray spectrum. In any event he demanded that I should
do the calculations properly and send him the draft of a paper. Under those
conditions that was not an easy matter, but when we moved to the south coast
of England in the spring of 1940 to rejoin the main radar establishment
(later to be known as the Telecommunications Research Establishment (TRE)
and now the Royal Signals and Radar Establishment (RSRE)) there was a modicum
of a library from which, in what time I could spare, I studied the ionospheric
literature and wrote the draft of a paper in which I set down the elementary
calculations to show that the ionization produced in the atmosphere by large
cosmic ray showers should be detectable by the modern high power pulse transmitters. Blackett, when he received the draft of my paper in London, re-
wrote it whilst sheltering from the night bombing of London, and communicated
it to the Royal Society in October 1940. This paper, "Radio echoes and cosmic
ray showers" (Blackett & Lovell 1941) was eventually published in 1941 but by
that time neither Blackett nor I had any time or opportunity for further
thoughts on this problem except to agree that if ever peace came and we
survived the war, then we would look into this possibility again.

THE ORIGIN OF JODRELL BANK

As we emerged from the conflict five years later Blackett returned
to his post in Manchester and I followed him as a lecturer on the staff of
the Physics Department of the University. Clearly Blackett expected me to
start work on this problem of the transient echoes. I wanted a long wave,
high-powered radar and not the airborne centimetre equipment I had been
concerned with in recent years. With the assistance of J.S. Hey, who was

then with the Army Operational Research Group, I borrowed a 4.2 m wavelength, mobile radar with Yagi aerials which had been used to direct the anti-aircraft guns. There were three trailers containing the transmitter, receiver and a diesel generator respectively, and the army delivered them to me in the yard outside the Physics Department of the University. Soon I had this operating, but to my dismay found that the c.r.t. display was almost entirely obliterated by interference. The origin of this was quickly identified as the electric trams which regularly ran along the adjacent street and worked off the city's direct current supply. The only hope of using this equipment seemed to be to find a location outside of the city. After a few enquiries and visits I was directed to the Botanical Grounds of the University about 25 miles south of the city. This was a small area known as Jodrell Bank; about 11 acres had been purchased before the war as an extension to the University's main botanical grounds near the University. I was given permission to move my trailers there for a few weeks and in mid-December 1945 I was deposited there with snow on the ground and with the nearest public electricity supply many miles distant. The staff of these botanical grounds consisted of two gardeners, who helped to start the diesel. By an extraordinary coincidence, I first operated the radar in mid-December when the Geminid meteor shower was at maximum (see Figures 1 and 2). There were

Figure 1 : The first day at Jodrell Bank - mid-December 1945. The receiver cabin of an Army 4.2 metre radar. The Yagi aerials are on the roof. The wooden huts contained the fertilisers and tools of the two gardeners working in the botanical grounds.

Figure 2 : The first day at Jodrell Bank - mid-December 1945. The diesel generator (stuck in the mud) and in the background the transmitter cabin of the Army 4.2 metre radar.

many of the transient echoes of the type we had noticed on the coastal radars six years earlier. This was immensely encouraging, but at that time we still did not know that they were associated with the meteor ionization.

THE EARLY POST-WAR YEARS AT JODRELL BANK 1946-1950

The study of meteors

In the early months of 1946 when I was alone at Jodrell Bank with this ex-Army radar equipment, a number of diverse influences led to our ultimate destiny. Hey gave me a report which he had written as a secret memorandum about his investigation of the attempts to detect the German V2 rockets with this type of army radar when the V2 bombardment of London commenced in September 1944. The operators of these radars had seen frequent echoes and therefore gave warnings of the approach of rockets which did not arrive. Indeed, intelligence correlation showed that at the times reported no rockets had been launched by the Germans. As a member of the Army Operational Research Group, Hey had been given the task of investigating this peculiarity and had concluded that the transient echoes probably had an ionospheric origin and that some of them could be associated with the radar reflections from the ionized trails of meteors. I knew nothing about meteors at that time, but soon discovered that the pre-war literature was

plentifully scattered with proposals that the trail of electrons left by a
meteor burning up 100 km high in the atmosphere scattered radio waves back
to earth to give a detectable short-lived echo (for a summary see Lovell
(1948)). I could find nothing very helpful in the textbooks about meteors
and neither did the professional astronomers seem interested. By good
fortune Blackett had at this time in his laboratory a young Norwegian,
Nicolai Herlofson (later to be Professor at the Royal Institute of Technology
in Stockholm). Herlofson had been a meteorologist during the war and he knew
that the authorities in the U.K. on meteors were the amateurs. The director
of the meteor section of the British Amateur Astronomical Association was
J.P.M. Prentice, a solicitor. [Prentice discovered Nova Herculis in 1934 .
He died in 1981 and for an account of his life and work see Lovell (1982).]
Liaison between us was soon established and Prentice offered to come to
Jodrell during the Perseid meteor shower in August 1946 in order to make
visual observations for correlation with the radar echoes. Our paper on
that work, published a year later (Prentice et al. 1947), showed that we
had good correlation with radar echoes lasting more than 0.5 seconds, but
that the correlation with echoes of short duration was very poor. Prentice
then informed us that in October the earth would cross the orbit of the
Giacobini-Zinner Comet near the head of the comet, and that he expected a
major, although short lived, meteor shower. J.A. Clegg had joined me in the
summer and had much wartime experience on Yagi aerials. We borrowed an Army
searchlight and on this Clegg built a beautiful Yagi array which we could
direct easily to any part of the sky (Figure 3). With this and the 4.2 m
radar we observed an astonishing meteor shower during the night of 9-10
October 1946. At the peak of the shower we were observing 168 echoes per
minute, but when the aerial beam was directed into the radiant point of the
meteors, the rate fell almost to zero (Lovell et al. 1947). This decisive
proof that the transient echoes were from meteors and that the trails were
specularly reflecting was an important event in the history of Jodrell Bank.

        Space does not allow me to pursue in detail the subsequent
history of our work on meteors at Jodrell Bank. I will mention, however, that
most important in our growth was the arrival of two students - J.G. Davies
in 1947 and a visitor from New Zealand, C.D. Ellyett, in 1946. Important
landmarks were the discovery of the daytime meteor streams in 1947, the
development of the diffraction method for the accurate measurement of
meteor velocities by Davies and Ellyett, the application of this technique
to settle the argument about the origin of the sporadic meteors, and the 3-

Figure 3 : A view of Jodrell Bank in 1951 showing the Yagi aerial array
built in 1946 on an Army searchlight mount. In the background is another
array of Yagi aerials, used for the radio echo determination of meteor
radiants, a 30 ft paraboloid used for the observation of radio star
scintillations, and one of the prefabricated buildings which were replacing
the trailers at that time. The men in the foreground are (left to right)
C. Hazard, R. Hanbury Brown, and J.G. Davies.

station technique devised by Davies by which it became possible to determine
the orbits of individual meteors. One of the last results published (Davies
& Gill 1960) by Davies on his meteor work, before he became diverted to
other matters, showed that the orbits of short period sporadic meteors are
nearly circular and have inclinations to the ecliptic near 60 deg. This
presented a solar system problem which has not yet been solved. Much
research was also carried out on the physical processes in the meteor trails
and on the determination of upper atmospheric winds from the study of the
meteor trails. The same radar equipment was used also in some of the early
researches on the reflection of radio waves from the aurora borealis.
Further details regarding the meteor radar research at Jodrell Bank in the
1950s can be found in Lovell (1954, 1957) and in Greenhow and Lovell (1960).

The construction of large aerials

Although the work on meteors had helped to establish our post-
war scientific reputation, it was an accidental by-product of the reason
for my possession of the radar equipment and for my presence at Jodrell
Bank. My aim was to investigate extensive cosmic ray showers by the radar
technique and another curious sequence of events shaped our future.

During the early months of 1946, before the decisive correlation of the transient echoes with meteors had been established, I was still of the opinion that at least some of these could have their origin in the reflection from the ionization left by the large cosmic ray showers. I was beginning to look again into the war time calculations of the Blackett-Lovell paper when Blackett showed me a letter which he had received from T.L. Eckersley, an authority on ionospheric problems. Eckersley had remarked that we had given the cross section for scattering of the electrons as $(e^2/mc^2)^2$ omitting the factor $8\pi/3$. Although Blackett was annoyed at this carelessness on my part, the factor had little effect on the calculations. Eckersley also queried the possible effect of the damping factor, which we had not even mentioned in the war-time paper. When I looked into this I discovered to my alarm that this was, indeed, a very large factor which would certainly invalidate the calculation that we had enough sensitivity with the radar equipment to detect the cosmic ray ionization. Fortunately, by that time Blackett could see for himself the transient echoes on the c.r.t. and he became far more concerned with the attempts to discover their origin than with recrimination over past errors. Therefore, he allowed and encouraged me to carry on at Jodrell rather than return to the laboratory in Manchester.

The other fortunate circumstance was that during this critical period Clegg joined me. I have referred to the Yagi aerial system he built on the searchlight mount for the study of the Giacobinid meteor shower. Of greater ultimate significance were my discussions with him about the cosmic ray problem. The inclusion of the damping factor in the calculations implied that we must develop a far more sensitive equipment. We had no money to obtain a more powerful transmitter at that time and on those long wavelengths we could not improve the receiver, so we concluded that we must build a large aerial system. Our first idea was to erect a tower out of scaffolding tubes and construct on this a large horizontally directed array. We had no significant help and the concrete foundations and the erection of the scaffolding proved to be such an arduous task that by the time the tower was 20 ft or so above the ground we abandoned it. We decided it would be easier to build a large paraboloid, which would be more flexible in changing wavelengths and also easier to feed with a powerful transmitter.

That was the origin of the 218 ft aperture transit telescope. It was started by Clegg and myself, but by the time we finished it in the summer of 1947 we had acquired a few more young people to help us. The aperture was determined by the amount of space we could find in the field. We fixed 24 perimeter posts of scaffolding tube 24 ft high into concrete blocks. The main framework of the bowl was then formed by ⅜ inch steel wires radiating from the centre to the top of these posts and strained to the ground in more concrete outside the perimeter. These wires were held to the ground by more vertical posts along the radius. On this framework we wound many miles of thin 16 gauge wire to form the reflecting screen. The focal length was 126 ft and we had to devise some means of suspending the feed at this height at the centre of the reflector. Initially we borrowed one of the wooden towers used on the CH radar chain, but after beginning the erection we decided that we did not have the experience to erect this mast. But by that time our work on meteors was becoming known to the grant giving agencies, and I secured a grant of £1000 from the Department of Scientific and Industrial Research to engage a contractor to supply us with a 126 ft steel tube, pivoted at the base and held by guys (Hanbury Brown & Lovell 1955).

This transit telescope was a great success - but not in the research for which it was built. Early in the autumn of 1947 we connected the transmitter to it in a brief search for the echoes from cosmic ray showers. But out interests quickly shifted when we realised that on a wavelength of a few metres the telescope had a beamwidth at least six or seven times smaller, and a collecting area much larger, than any other instrument which had so far been used in the study of the radio emissions from space (for a photograph of this aerial, see the article of Hanbury Brown elsewhere in this volume).

The first researches on cosmic radio waves at Jodrell Bank

It will be seen that we had, in a circuitous manner and without any intention of doing so, produced a first class instrument for the study of the cosmic radio waves. Although the paraboloidal wire bowl was fixed to the ground, the beam could be moved several degrees either side of the zenith by the adjustment of the guys holding the 126 ft steel mast, and thus over a period of time a strip of sky several degrees wide passing through the zenith at Jodrell Bank could be investigated. Hanbury Brown had joined us in September 1949 as a Ph.D. candidate, and his deep experience of war-time electronics soon became a major influence on our development. In fact, it was Hanbury Brown with Cyril Hazard (who also came to us in October 1949 as a postgraduate student) who made the first significant observations with the transit telescope. Using this instrument in the early months of 1950 on a frequency of 158.5 MHz they were the first to provide unambiguous evidence that M31, the Andromeda Nebula, was a radio source (Hanbury Brown & Hazard 1950). It is very hard to realise now that in those days most people believed firmly that the radio emission all originated in the Milky Way. Indeed, in their paper Hanbury Brown and Hazard referred to the work of Bolton and Stanley in Australia and of Ryle and Smith in Cambridge who with their interferometers had demonstrated the existence of an isotropic distribution of sources less than 6 minutes of arc in diameter; Hanbury Brown and Hazard remarked, "Although the majority of these sources are unidentified with visual objects, their distribution indicates that they lie in our own galaxy. Attempts to detect radio emissions from specific extragalactic objects have hitherto been unsuccessful." For further details of this work, see the article by Hanbury Brown elsewhere in this volume.

This paper on M31 was published in November 1950. By that time 40 other papers had originated from the Jodrell Bank group (Figure 4), 38 of which concerned work on meteors or the aurora borealis, and only two concerned the reception of radio waves. The first of these described the solar outburst of 25 July 1946, which had forced itself on our attention when we were studying the transient echoes (Lovell and Banwell 1946). The second was a more deliberate investigation by Gordon Little and myself.

Figure 4 : The Jodrell Bank group in 1951 photographed in front of the "searchlight aerial". From left to right: M. Almond, C. Hazard, R.C. Jennison, (the late) J.S. Greenhow, G.S. Hawkins (partially hidden), J.G. Davies, A.C.B. Lovell, C.G. Little (behind Lovell), J.A. Clegg, A. Maxwell, I.B. Hazzaa (a visiting Egyptian), T.R. Kaiser, W.A.S. Murray, R. Closs and M.K. das Gupta. R. Hanbury Brown had joined the group in 1949, but was absent at the time of the photograph.

Little had joined me to study for his Ph.D. in 1948 and together we tackled the problem of the origin of the fluctuations in the intensity of the radiation from the source in Cygnus which Hey and his colleagues had discovered in 1946. Hey had assumed that the intensity variations must arise in the source and hence be evidence that the source was of small angular extent. In an attempt to settle this question unambiguously, we used spaced receivers on a wavelength of 3.7 m to observe the fluctuations simultaneously at separations of 100 m, 3.9 km, and, in conjunction with Ryle and Smith in Cambridge, 210 km. The complete time correlation of the fluctuations at 100 m, the partial correlation at 3.9 km and the complete lack of correlation at the 210 km spacing seemed clear proof to Little and myself that the disturbances had an ionospheric origin and that the radiation from the source was steady. However, Ryle, who had recently published a paper (Ryle 1949) in which he derived the maximum physical dimensions of the Cygnus source from the period of the fluctuations, would not agree, and so the work was published separately (Smith 1950, Little & Lovell 1950).

THE DECADE 1951-1960

The success of the transit telescope soon led to a marked change in our researches. Whereas to the end of 1950 the major research effort had been radar based, during the next five years our effort was almost evenly distributed between the radar studies and the subject, which by this time, became known as radio astronomy. In that five-year period 36 of our research publications were based on radar and 28 on radio astronomy. For the next five years to 1960 the bias shifted markedly in favour of radio astronomy, 49 to 29. In the decade following the end of World War II, foreign exchanges - especially with the Soviet Union - remained difficult. We were therefore fortunate that many distinguished astronomers and other scientists visited Jodrell Bank during those years, and in various respects encouraged and influenced the new developments. Amongst the more significant of these meetings may be mentioned the meetings in September 1948 on meteor astronomy (Anon 1948), the meeting of the Royal Astronomical Society in June-July 1949 (Anon 1949); the symposium on radio astronomy in July 1953 attended by many foreign astronomers (Figure 5) (Hanbury Brown 1953), the symposium on meteor physics in July 1954, attended by some of the first Soviet scientists to visit the West after the war (Kaiser 1954), and the IAU symposium on radio astronomy 25-27 August 1955 (van de Hulst 1957) immediately preceding the ninth General Assembly of the IAU in Dublin.

Figure 5 : Some of the delegates to the I.A.U. Radio Astronomy Symposium at Jodrell Bank in July 1953 (on a part of the foundations of the 250 ft Mark I radio telescope). From left to right the following can be identified: Father J.A. Koster, R. Minkowski, H.C. van de Hulst, V. Kourganov, Bart Bok, B. Lovell, G. Westerhout, and (facing away) Priscilla Bok.

I cannot discuss the researches in detail during these years but will just mention a few of the most significant developments and some of the young people associated with them, many of whom later became well known figures in the astronomical community.

The development of the Jodrell Bank interferometers

In 1948 Ryle and Smith in Cambridge had confirmed the existence of a localised source of radio emission in Cygnus and had discovered the source in Cassiopeia. By 1951 Smith had delineated the positions of these sources with sufficient accuracy to enable a search to be made with the 200 inch telescope for their optical counterparts. In September 1951 Baade and Minkowski obtained the plates which led to the identification of the peculiar galaxy in Cygnus and the supernova remnant in Cassiopeia. The question of the angular extent of these radio sources them emerged as an issue of fundamental importance. The existing radio limits were only that both sources must be less than about six minutes of arc in diameter, but this was tens of thousands of times greater than the known angular diameter of the stars. The technical situation at this time was that there seemed little hope

of using the Cambridge type of interferometer over baselines of more than a
few tens of kilometres.

In his contribution to this volume Hanbury Brown has described
how the discussions about this problem led him to the idea of the intensity
interferometer and the eventual extension of the concept to the optical
domain. At least for Jodrell Bank a sad consequence of this brilliant work
was that Hanbury Brown decided in 1962 to build a major optical intensity
interferometer at Narrabri in Australia. He has also described his development of the conventional phase correlation interferometer with H.P. Palmer
and A.R. Thompson, which, by 1956, had been extended over a baseline of
20 km (10,600λ) using a small movable array in conjunction with the 218 ft
transit telescope at Jodrell. Today it is difficult to recall the surprising
nature of their conclusion that three of the radio sources in their survey
were unresolved even at their longest baselines, implying angular diameters
of less than 12 seconds of arc and a brightness temperature at least as
great as the Cygnus source.

In 1955 Ryle had concluded that the distribution of the
intensities of the 1936 sources in the Cambridge survey implied that the
"majority of the sources are extremely rare extragalactic objects similar to
the source in Cygnus" (Ryle & Scheuer 1955). But even by 1958 there were very
few positive identifications of the large number (greater than 2000) of
sources in the Cambridge and Sydney catalogues. In fact summaries presented
in 1958 (see for example Lovell (1958)) listed identifications in the Galaxy
with 3 supernovae, 5 gaseous nebulosities (probably supernovae) and 15
emission nebulae of ionized hydrogen near hot stars. The extragalactic
identifications amounted to 16 with normal galaxies and 7 with abnormal
extragalactic objects of the Cygnus type.

Against this background the great significance of Palmer's measurements became manifest in 1960 when Minkowski announced that he had identified
one of the three unresolved sources (3C 295) with a peculiar galaxy in Bootes
at a redshift of 0.46. This was clear evidence of the correctness of Palmer's
conclusion in 1956 that the unidentified sources in his survey might be
objects analogous to Cygnus, but at much greater distances. The great
significance to cosmology of this realisation stimulated further extensions
of the baseline in the Jodrell work, and by 1958 the new 250 ft steerable
instrument (the Mark I, see later) could be used as one element of the interferometer. Between 1958 and 1961, the angular diameters of 384 radio sources
were measured using baselines up to 115.4 km (61,000λ). The 3C 295 source

itself was resolved at 4.5 seconds of arc, but 7 sources were still unresolved, indicating angular diameters less than 3 seconds of arc. One of these was 3C 48, but the direct lead from this measurement to the discovery of quasars belongs to the next decade.

Hazard and the development of the lunar occultation technique

The work of Palmer and his colleagues with the long baseline interferometer was a vital factor in the sequence of events leading to the searches which soon led to the discovery of quasars. Another important feature of the Jodrell work in the quasar story was the development of the lunar occultation technique by Hazard. His critical measurements in 1962 on 3C 273 were made in Australia, but the lead-up to this remarkable conclusion about the nature of the quasars occurred at Jodrell.

I have already mentioned Hazard's work with Hanbury Brown on the transit telescope after he joined us as a postgraduate student in 1949. In 1953 he had to leave to satisfy military service requirements, but he returned to Jodrell in late 1955 and made more measurements on the emission from normal galaxies using the transit telescope and with the Mark I steerable telescope when it came into operation in 1957. In 1960 Hazard asked me to read a memorandum in which he had worked out the high positional accuracies of radio sources which could be obtained by using the telescope to track the moon continuously while observing the time of immersion and emersion of a radio source. On 8 December 1960 he tested the scheme on the radio source 3C 212 with great success. He obtained the right ascension to ±0.3 seconds of arc and the declination to ±2 seconds of arc - improvements of 20 times and 240 times respectively over the best previous positions (Hazard 1961). A few months later Hazard left for Australia where very soon he was to make the critical measurements on 3C 273 with the Parkes 210 ft dish using this lunar occultation technique.

Lunar and planetary radar

I suppose that many of us who were associated with the wartime radar visualised the possibility of using the radar technique to obtain echoes from the moon and planets. In fact in the enthusiasm of the early post-war years and in the light of my calculations on the cosmic ray ionization, I wrote a memorandum on this matter and the project became one of our early ambitions at Jodrell. Success was reported almost simultaneously early in 1946 by the United States Army Signal Corps using a conventional radar system on a wavelength of 2.6 m, with long pulses and a large aerial array,

and from Hungary by Z. Bay using an unconventional detector consisting of a
battery of water coulometers. The third success was achieved in Australia,
using the broadcasting station 'Radio Australia' as the transmitter (see the
contribution by Kerr to this volume). At Jodrell Bank the study of lunar
radar echoes began in the autumn of 1953 using a steerable aerial to follow
the Moon. This facilitated a scientific study of the reflection process,
particularly of the puzzling variability in the strength of the reflected
signals. Both the American and Australian workers had found a short period
(seconds) fading which they attributed to the effect of lunar libration.
In addition there was a longer period (up to 30 minutes) fading for which
no satisfactory explanation·had been offered. In the Jodrell Bank observations
W.A.S. Murray and J. K. Hargreaves (1954) used a radar system on 120 MHz. The
transmitter fed a horizontally polarized broadside array of 250 m^2 area.
This was fixed, but the beamwidth was such that the moon could be observed
for 45 minutes each day at transit. The receiving aerial was a 20 ft diameter
steerable paraboloid (this reflector had been constructed originally on the
top of the Shot Tower on the South Bank of the Thames for the Festival of
Britain,1951) with a primary feed of two dipoles at right angles. By
observing the amplitude variations in the two planes of polarization of the
receiver, it was demonstrated that the long period fading of the lunar echoes
was caused by Faraday rotation in the earth's ionosphere.

          Shortly after that work, in October 1954, one of our young
Manchester graduates came to work with me for his Ph.D. degree. He was
J.V. Evans, now the director of the Lincoln Laboratories. After working in
the lunar group for one year, he had made a major contribution to the important paper (Browne et al. 1956) in which the theory of the lunar reflection
process was set out and in which it was shown that the long period fading of
the echoes could be used to measure the total electron content of the ionosphere along the line of sight to the moon. But then almost immediately his
senior collaborators left for other posts, and Evans proceeded alone to make
the first measurements of the ionospheric electron content using this method
(Evans 1956a, 1956b). In subsequent years other workers were to make substantial
use of this technique using the Mark I telescope. Evans, however, concentrated
on the study of the fading from the lunar surface and in an important paper
(Evans 1957) he proved that in the scattering of the radio waves the moon was
not behaving as a uniformly bright disc, but that it was limb-darkened.
Furthermore, he showed that on these metre wavelengths the effective scattering
area was a region at the centre of the visible disc having a radius of about

one-third that of the lunar radius. This surprising result led to the prediction that the bandwidth would be sufficient when using the moon as a reflector
in a communication circuit to make intelligible voice transmission possible.
In November 1958 I was able with the help of Evans to give a practical
demonstration of this in my BBC Reith Lectures (Lovell 1958, 1959) using the
Mark I telescope to transmit and receive. Subsequently, we established voice
communication with America and eventually with Australia using the moon
circuit. Of course the commercial and military possibilities of this system
of long distance communication were soon nullified by the use of earth
satellites.

After the success with the lunar work and having the Mark I telescope available, we proceeded to secure a pulsed 100 kw klystron working on
408 MHz which had been used in the Physics Department on a linear accelerator.
Evans was now the senior person working on this project and at the close
approach of Venus in September 1959 the first attempt to obtain radar
reflections from the planet was made (Evans & Taylor 1959). Our measurements
of the solar parallax and of the rotation rate of the planet during the next
close approach (in the spring of 1961) and subsequently are beyond the scope
of this present account (Thomson et al. 1961, Ponsonby et al. 1964). Evans was
also closely associated with the successful radar detection of the carrier
rocket of Sputnik 1 in October 1957 which is referred to below. Unfortunately
for Jodrell, Evans was enticed to America in the spring of 1960.

Other researches

In the decade to 1960 a number of other research programmes were
in progress at Jodrell Bank. A considerable number of papers was published
on physical conditions in the upper atmosphere and ionosphere, results
derived either from the study of the radio echoes from meteor trails or from
the scintillation of the radio sources. The arrival of R.D. Davies from
Australia in 1953 was the prelude to the important development of the Jodrell
researches on the 21 cm line of neutral hydrogen. With D.R.W. Williams (a
research student) he was amongst the first to draw attention to the significance of the measurement of HI absorption in the determination of the distances
of radio sources (Williams & Davies 1954). Apart from the first attempts to
detect the Zeeman splitting (Galt et al. 1960) the significant development of
these 21 cm line studies belongs to the next decade and the Mark I telescope.
In 1958 I began my own work on the detection of the radio emission from red
dwarf flare stars (Lovell et al. 1963), but the main results again belong to
a later period.

The 250 ft steerable radio telescope (the Mark I)

The research work described so far was mainly performed with equipment which had been developed and used during the war. The original 4.2 m radar equipment was soon joined by many other transmitters, receivers and aerial systems no longer required by the Army, Navy and Air Force. Apart from the £1000 grant which I obtained to procure the mast for the transit telescope, the only money available was that which Blackett allowed me from his Physics Department grant to pay for the fuel and other consumables. For many years we worked in the wartime trailers under conditions which by today's standards would be considered impossible for serious research work.

It was the immediate success of the transit telescope which quickly led me to the idea of a completely steerable instrument of at least that aperture. I have described in considerable detail elsewhere (Lovell 1968) my attempts from 1949 onwards to stimulate interest in the idea, to get financial backing, and to find an engineer who would consider the project to be feasible. The time from the autumn of 1952, when work began on the foundations (Figure 5), to October 1957 when we first used the telescope, was a period of immense anxiety exacerbated by the political and financial problems which, as I have described elsewhere, nearly destroyed Jodrell Bank as a research establishment. Nevertheless, the success of the telescope in detecting by radar the carrier rocket of the first Russian Sputnik in October 1957 (Staff of Jodrell Bank 1957), and the few years afterwards when it became a part of the American deep space network, created the conditions which enabled us to overcome those problems. We survived, and the instrument has now for over a quarter of a century been in continuous use.

REFERENCES

Anon (1948). Meteor astronomy. *The Observatory, 68,* 226-232.

Anon (1949). The Royal Astronomical Society at Manchester. *The Observatory, 69,* 121-126, 185-191 and Plate III.

Blackett, P.M.S. & Lovell, A.C.B. (1941). Radio echoes and cosmic ray showers. *Proc. Roy. Soc., A177,* 183-186.

Browne, I.C., Evans, J.V., Hargreaves, J.K. & Murray, W.A.S. (1956). Radio echoes from the moon. *Proc. Phys. Soc., B69,* 901-920.

Davies, J.G. & Gill, J.C. (1960). Radio echo measurements of the orbits of faint sporadic meteors. *Mon. Not. Roy. Astr. Soc., 121,* 437-462.

Evans, J.V. (1956a). The measurement of the electron content of the ionosphere by the lunar radio echo method. *Proc. Phys. Soc. B., 69,* 953-955.

Evans, J.V. (1956b). The electron content of the ionosphere. *J. Atmos. Terr. Phys.*, *11*, 259-271.

Evans, J.V. (1957). The scattering of radio waves by the moon. *Proc. Phys. Soc. B.*, *70*, 1105-1112.

Evans, J.V. & Taylor, G.N. (1959). Radio echo observations of Venus. *Nature*, *184*, 1358.

Galt, J.A., Slater, C.H., Shuter, W.L.H. (1960). An attempt to detect the galactic magnetic field using Zeeman splitting of the hydrogen line. *Mon. Not. Roy. Astr. Soc.*, *120*, 187-192.

Greenhow, J.S. & Lovell, A.C.B. (1960). The upper atmosphere and meteors. In *Physics of the Upper Atmosphere*, Chapter 11, 513-549. New York: Academic Press.

Hanbury Brown, R. (1953). Symposium on radio astronomy at Jodrell Bank, *The Observatory*, *73*, 185-198 and Plate V.

Hanbury Brown, R. & Hazard, C. (1950). Radio frequency radiation from the great nebula in Andromeda (M.31). *Nature*, *166*, 901-902.

Hanbury Brown, R. & Lovell, A.C.B. (1955). Large radio telescopes and their use in radio astronomy. *Vistas in Astronomy*, *1*, 542-559.

Hazard, C. (1961). Lunar occultation of a radio source. *Nature*, *191*, 58.

Jánossy, L. & Lovell, A.C.B. (1938). Nature of extensive cosmic ray showers. *Nature*, *142*, 716.

Kaiser, T.R. (1954). A symposium on meteor physics at Jodrell Bank. *The Observatory*, *74*, 195-208 and Plate VI.

Little, C.G. & Lovell, A.C.B. (1950). Origin of the fluctuations in the intensity of radio waves from galactic sources - Jodrell Bank observations. *Nature*, *165*, 422.

Lovell, A.C.B. & Wilson, J.G. (1939). Investigation of cosmic ray showers of atmospheric origin using two cloud chambers. *Nature*, *144*, 863-864.

Lovell, A.C.B. & Banwell, C.J. (1946). Abnormal solar radiation on 72 megacycles. *Nature*, *158*, 517-518.

Lovell, A.C.B., Banwell, C.J. & Clegg, J.A. (1947). Radio echo observations of the Giacobinid meteors 1946. *Mon. Not. Roy. Astr. Soc.*, *107*, 164-175.

Lovell, A.C.B. (1948). Meteoric ionization and ionospheric abnormalities. *Phys. Soc. Rep. Prog. Phys.*, *11*, 415-444.

Lovell, A.C.B. (1954), *Meteor Astronomy*. Oxford: Clarendon Press.

Lovell, A.C.B. (1957). Geophysical aspects of meteors. *Handbuch der Physik*, *48*, 427-454.

Lovell, A.C.B. (1958). Radio-astronomical observations which may give information on the structure of the Universe. In *Eleventh Solvay Conference - La Structure et L'Évolution de L'Univers*, 185-212. Brussels.

Lovell, A.C.B. (1958, 1959). *The Individual and the Universe*. Oxford: Oxford University Press.

Lovell, B., Whipple, F.L. & Solomon, L.H. (1963). Radio emission from flare stars. *Nature, 198,* 228-230.

Lovell, B. (1968). *The Story of Jodrell Bank.* Oxford: Oxford Univ. Press.

Lovell, B. (1982). J.P.M. Prentice. *Q. J. Roy. Astr. Soc., 23,* 452-460.

Murray, W.A.S. & Hargreaves, J.K. (1954). Lunar radio echoes and the Faraday effect in the ionosphere. *Nature, 173,* 944.

Ponsonby, J.E.B., Thomson, J.H. & Imrie, K.S. (1964). Radar observations of Venus and a determination of the astronomical unit, 1962. *Mon. Not. Roy. Astr. Soc., 128,* 1-17.

Prentice, J.P.M., Lovell, A.C.B. & Banwell, C.J. (1947). Radio echo observations of meteors. *Mon. Not. Roy. Astr. Soc., 107,* 155-163.

Ryle, M. (1949). Evidence for the stellar origin of cosmic rays. *Proc. Phys. Soc., A62,* 491-499.

Ryle, M. & Scheuer, P.A.G. (1955). The spatial distribution and the nature of radio stars. *Proc. Roy. Soc., A230,* 448-462.

Smith, F.G. (1950). Origin of the fluctuations in the intensity of the radio waves from galactic sources - Cambridge observations. *Nature, 165,* 422.

Staff of Jodrell Bank (1957). Radar observations of the first Russian earth satellite and carrier rocket. *Nature, 180,* 941.

Thomson, J.H., Ponsonby, J.E.B., Taylor, G.N. & Roger, R.S. (1961). A new determination of the solar parallax by means of radar echoes from Venus. *Nature, 190,* 519-520.

van de Hulst, H.C. (ed.) (1957). *IAU Symposium No. 4 on Radio Astronomy.* Cambridge: Cambridge University Press.

Williams, D.R.W. & Davies, R.D. (1954). A method for the measurement of the distance of radio stars. *Nature, 173,* 1182.

"Taffy" Bowen (left) and Hanbury Brown at the 1952 URSI General
Assembly held in Sydney (courtesy CSIRO Radiophysics Division)

PARABOLOIDS, GALAXIES AND STARS: MEMORIES OF JODRELL BANK

R. Hanbury Brown
School of Physics, University of Sydney, N.S.W. 2006, Australia.

SOME EARLY WORK WITH THE 218 FT PARABOLOID

Introduction

I shall always remember reading an article on "Cosmic Static" which Grote Reber published (Reber 1940) in the February 1940 issue of the *Proceedings of the Institute of Radio Engineers*. It was during World War II and we were working desperately hard on various "advanced" forms of airborne radar for the detection of ships, submarines and aircraft. Reber's strange article was sandwiched in between conventional papers on frequency modulation and negative feedback and, because it was so intriguingly odd, it lay open on our laboratory bench for several days until, inevitably, someone put a hot soldering iron down on it. Reber had worked at much the same frequency (160 MHz) as we had been using for airborne radar for several years, and I couldn't help thinking how nice it would be to turn some of our relatively "superior" equipment on to the sky to see if his "cosmic static" was really there. It was not until nearly a decade later that I was given the opportunity at Jodrell Bank.

In 1949 I was a partner in a firm of consulting engineers with Sir Robert Watson-Watt, the pioneer of radar. Sir Robert wanted to move our firm to Canada - which he did with disastrous results - and I did not. Instead I decided to stop advising firms on research and do some myself, perhaps in a University. As a start I went to see my old friend and war-time colleague F.C. Williams in Manchester; he suggested that I should take a look at what Bernard Lovell was doing at Jodrell Bank.

As Lovell tells us in his article, elsewhere in this volume, the original reason for his work at Jodrell Bank was to detect the ionization from cosmic rays by radar. But he and his colleagues detected meteor showers, not cosmic rays, and it was this successful use of radar which dominated the programme of Jodrell when I first visited it in 1949. But, as Lovell also relates, there was something else; in 1947 they had built, with their own hands, a fixed paraboloid (Fig. 1) with the astonishing size of 218 ft (Hanbury Brown & Lovell 1955). The original purpose of this remarkable dish had also been to detect cosmic rays by radar, but by the time of my first visit to Jodrell in 1949 they had given up that programme and realized that, by a happy accident, they had built a powerful instrument with which it should be possible to study the radio emissions from space. Lovell suggested that I should join them and take charge of a "cosmic noise" programme which had already been started by Victor Hughes. With Reber's paper on "Cosmic Static" in mind I accepted his offer with enthusiasm and went to work at Jodrell Bank in September 1949. Shortly afterwards I was joined by a research student, Cyril Hazard, with whom I was to work for several years.

As I hope to show in this brief account, our early work with this 218 ft paraboloid made a significant contribution to the development of radio astronomy, not only in observational data but in technique. All the major surveys at that time had been carried out, and were still being carried out, with interferometers. As we shall see, our observations exposed some of the limitations of those interferometers, which were ideal for measuring the precise positions of sources, but failed to see sources of large angular size. Our work with the 218 ft paraboloid demonstrated the importance of looking at the sky with a pencil-beam and so prepared the way for the 250 ft steerable paraboloid at Jodrell Bank.

Making the 218 ft Paraboloid Work at 1.89 m

As a first step we chose to work at the highest frequency at which the antenna could be expected to perform reasonably well; we wanted the narrowest possible beam. To that end we removed the original primary feed of the paraboloid, which had been designed to work on 4.2 m with a high power transmitter using open-wire feeders; in its place we substituted a primary feed designed for 1.89 m and connected it to the laboratory by 300 ft of coaxial cable with a loss of about 2.7 db. I well remember soldering this coaxial cable to the primary feed while perched on the top of the 126 ft tower at the centre of the paraboloid; it was not a job which I would care to do again.

The next step was to build a receiver with the necessary gain stability, which, in those days, was the major problem. After several experiments in which we tried rather unsuccessfully to stabilize the gain of conventional receivers, we decided to adopt the elegant "servo" system

Figure 1 An early (ca. 1949) view of Jodrell Bank showing the
218 ft fixed paraboloid in the foreground and the "Park Royal"
laboratory (right foreground) in which the receivers were
housed.

described by Ryle and Vonberg (1948). In their system the receiver
compares the signal from the antenna with that from an artificial noise
generator and, by means of a servo loop, makes them equal. The output of
the noise generator is then recorded continuously on a chart recorder.
This system has the great advantage that the gain of the receiver needs
only to be stabilized over short periods and it solved our problem. On a
single record we were able to detect a change of aerial temperature of
about 1.4 deg K which corresponded to a received flux of about
10^{-25} W m^{-2} Hz^{-1}.

Using this "servo" receiver we were able to make most
impressive records of the "cosmic noise" power received from the zenith
within a 2° beam. I shall never forget the thrill of seeing those first
records (Fig. 2). They showed clearly the broad maximum as the galactic
plane passed through the aerial beam; they also showed a few discrete
sources or, as we called them in those days, radio stars. Although we
could, and often did, observe during the day, all our best observations
were made at night. During the day the basic noise level of the system
was increased by a variety of sources, for example by radio emission from
the Sun and by ignition noise from traffic on a nearby main road. A

shallow paraboloid with an exposed and elevated primary feed is very
vulnerable to interference.

Figure 2 R. Hanbury Brown (left) and C. Hazard looking at the
output of a 1.89 m receiver attached to the 218 ft paraboloid
(1950).

Once we had stabilized the receiver and could make virtually
identical records night after night of the strip of sky in the zenith, we
decided to enlarge our field of view by swinging the beam of the paraboloid out of the zenith. The beam could, of course, only be swung by
displacing the primary feed and, in practice, that meant by tilting the
central mast. The mast itself was far from self-supporting; it was made
by slotting 6 sections of tubular steel together and supporting them by
18 guy wires. The bottom section was hinged at the ground so that it
could be tilted in a north-south plane. We set up two theodolite stations
east and west of the mast and, very gingerly, tried tilting the mast by
making successive adjustments to the 18 guys. It was a very anxious
business; the tilting had to be done in almost imperceptible stages so
as not to kink the mast and it took Hazard and myself about two hours,
running from guy to guy and shouting at each other, to tilt the beam
through one beam-width (2°). It was quite good fun on a fine day,
although it was a rather slow way of scanning the sky. The early days of

any new science are apt to be laborious. The largest tilt at which we
could persuade ourselves that the mast was safe and that the beam shape
was still respectable was 15°. This meant that we could survey a strip
of sky between declinations +38° and +68°. The final uncertainty in the
direction of the beam was about ± 10' near the zenith, increasing to about
± 25' at the limits of the field of view. This uncertainty was later
reduced when our observations could be calibrated on sources whose
positions had been found by more accurate means.

By mid-1950 we were ready to do some serious astronomy. Our
original plan had been to measure the spectrum of the galactic noise by
comparing the contours of the galactic plane at different frequencies.
But once we had a good stable receiver at 1.89 m and had learned how to
swing the beam of the paraboloid, we felt that we must first explore what
lay outside the zenithal strip. In those days very little was known about
cosmic radio noise and our paraboloid was larger and had a much narrower
beam than any other single antenna that had been used to scan the sky.
Our programme, therefore, was simple and exciting: it was to look at
everything in our field of view.

A Survey of the Andromeda Nebula (M31)

The first object which we looked at was the Andromeda Nebula.
The fact that our own Galaxy emits radio waves by some unknown process
prompted the obvious question - at least it looks obvious in retrospect -
do other similar galaxies behave in the same way? Elementary calculations
suggested that if our own Galaxy were to be placed at the distance of M31
then we should be able to detect it with our antenna.

We surveyed M31 in August and September 1950 (Hanbury Brown &
Hazard 1950, 1951a). The survey had to be carried out in the radio-quiet
period around midnight and we were driven to despair by a sequence of
thunderstorms in the middle of the night because the exposed feed of the
antenna was extremely susceptible to interference generated by electric-
ally charged rain. It took us 90 nights to make a contour map of the
region, and we just managed to complete the survey before the region moved
out of the radio-quiet period. The results (Fig. 3) established beyond
reasonable doubt that radio emission is not peculiar to our Galaxy and
that the Andromeda Nebula also emits radio waves of much the same
intensity. As a matter of interest our contours of Andromeda represent

the first measurements ever to be made of the brightness distribution across a radio source.

Following the detection of the Andromeda Nebula we went on to survey several other well-known galaxies and clusters of galaxies. This work is described later.

Figure 3   Radio map of Andromeda Nebula made in August and September 1950 with the 218 ft paraboloid (Hanbury Brown & Hazard 1950, 1951a).   ($\lambda = 1.89$ m, 1 unit = 10^{-25} W m^{-2} Hz^{-1})

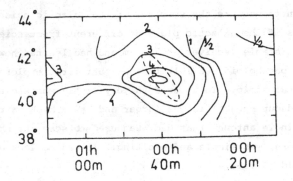

A Survey of the Galactic Radio Noise

Looking at our records the most prominent feature was the radio emission from the galactic plane and so, in 1951 and 1952, we surveyed all of the plane which lay within the field of view (Hanbury Brown & Hazard 1953a). We plotted the contours by hand; in those days there were no computers and it took ages to plot a contour map. A comparison of our contours with those measured by Reber, at almost the same wavelength but with a much wider (13°) beam, showed that our 2° beam gave much greater detail; for example, the broad maximum which Reber observed in Cygnus was seen to be a complex of sources. Thus, although this survey covered only a limited region of sky, it demonstrated the value of higher resolving power in the study of cosmic radio noise.

A Survey of the Radio Stars

The second prominent feature of our records were the discrete sources or "radio stars". We decided to find out as much about them as we could and, as a first step, we surveyed (Fig. 4) the powerful source Cygnus A (Hanbury Brown & Hazard 1951b). In addition to mapping the region around it and measuring its position, we showed that its intensity, when averaged over several minutes, was the same whether the source was fluctuating or not. Nowadays we know that these fluctuations originate in the ionosphere, but in those days there were many radio astronomers who believed that the fluctuations were intrinsic to the source and were, perhaps, analogous to the radio emission from the active Sun. Our measurements confirmed the conclusion, derived from experiments with spaced receivers, that the fluctuations were ionospheric in origin.

Figure 4   A record of the transit of the source Cygnus A through the beam of the 218 ft paraboloid. The maximum intensity is 5×10^{-23} W m^{-2} Hz^{-1}. The record was made on 15 September 1950.

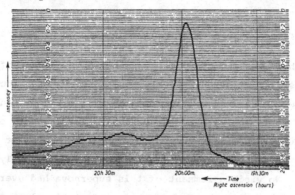

By the end of 1952 we had completed a survey of 23 sources within our field of view (Hanbury Brown & Hazard 1953b). Our results were significant for two reasons. Firstly they showed, clearly, the concentration of the more intense sources into the galactic plane and thereby confirmed the observations of Mills (1952) who had suggested that they should be called Class I sources (see his contribution to this volume). Our survey helped to establish beyond doubt the existence of these Class I sources as members of the Galaxy.

Our survey was also valuable because it drew attention to a significant defect in at least one of the major existing surveys of radio stars. At Cambridge Ryle, Smith and Elsmore (1950) in a survey of 50 sources made at a wavelength of 3.7 m with an interferometer with a base-line of 110 wavelengths, had reached the conclusion that the distribution of the radio stars showed no detectable concentration into the galactic plane (see the article by Smith elsewhere in this volume). Our survey

showed precisely the opposite. A comparison of the two surveys showed
that of the 13 most intense sources detected with our pencil-beam (10 of
which lay within $5°$ of the galactic plane) only 4 appeared in their survey.
I shall never forget the strenuous arguments we had with those dedicated
interferometrophiles in our efforts to convince them that our sources in
the galactic plane were real and not side-lobes of our pencil-beam. It
was part of the conventional wisdom at Jodrell Bank that Cambridge had
only three standard reactions to our work: (1) "it is wrong", (2) "we
have done it before", or (3) "it is irrelevant". Indeed, as we later
showed, at least 6 of the 10 sources in the galactic plane had angular
sizes exceeding $1°$ and were either partially or totally resolved, and
therefore largely undetected, by their interferometer.

Tycho Brahe's Supernova of A.D. 1572 and the Great Loop in Cygnus

Although it was commonly believed in 1952 that many of the
discrete sources must lie within our Galaxy, only one had been identified
with a galactic object. In 1949 Bolton, Stanley & Slee (1949) had
identified a source in Taurus with the Crab Nebula, the remnants of the
supernova of A.D. 1054. In mid-1952 we decided to take a look at the
remnants of another supernova, the one reported so carefully by Tycho
Brahe in A.D. 1572. No visible remnant of this supernova had ever been
observed and the position given by Tycho was uncomfortably close to the
intense source Cassiopeia A. We had, therefore, to take special measures
to ensure that any source which we might detect was not a side-lobe
response to Cassiopeia A. We solved this problem by constructing an
auxiliary interferometer with which we could test the mutual coherence
with Cassiopeia A of any source detected by the 218 ft paraboloid. Our
survey showed a weak source close to the position reported by Tycho and
we identified it, provisionally, with the remnants of his supernova
(Hanbury Brown & Hazard 1952). This identification has since been
confirmed.

As far as I can remember it was Zedenek Kopal who first
suggested that we should look for Tycho's supernova. He was Professor of
Astronomy in the University of Manchester and used to come out to Jodrell
once a week to teach us some astronomy. In those days most radio astronomers were either physicists or engineers and most of us knew very little
conventional astronomy. Kopal was encyclopaedic and enthusiastic; he

did quite a lot to bridge the local gap between radio and optical astronomy and to make us feel that we were part of a wider community of astronomers.

There was one other obvious object in our field of view which looked as though it might be the remains of a supernova, the Great Loop in Cygnus. Dennis Walsh and I surveyed it late in 1954 and provisionally identified it with a radio source of large angular size (Walsh & Hanbury Brown 1955). This identification has also since been confirmed.

The Angular Size of Radio Stars

One of the most interesting and valuable programmes carried out with the 218 ft paraboloid was the measurement of the angular size of those radio sources which lay within its field of view. In 1953 Cyril Hazard left to do his National Service and I was joined by Henry Palmer and Richard Thompson. Together we built a small mobile antenna and connected it by a cable to make a simple interferometer with the large dish. In this way we showed that the majority of the Class I sources which we had detected close to the galactic plane had large angular diameters (Hanbury Brown et al. 1954); 6 out of 19 had diameters in the range of $1\frac{1}{2}$ to $3°$. One of these sources was subsequently identified by Minkowski with an extended filamentary nebulosity in Auriga and, as four other sources had already been identified with nebulosities in the Galaxy, it began to look as though the Class I sources were radio nebulosities, not radio stars.

As a next step we set out to measure a sample of the Class II sources, that is to say the majority of sources which appeared to show no concentration towards the galactic plane. We chose five sources with galactic latitudes above $10°$. Technically this was a much more difficult job because it turned out that our simple interferometer failed to resolve them and, of necessity, we had to extend the baseline. This was the start of a long and arduous programme in which, for several years, we progressively increased the length of our baseline, solving new technical problems at each state. For example, to compare the "fringe visibility" of sources at widely different baselines we had to develop the "rotating lobe interferometer" (Hanbury Brown et al. 1955); to operate at very long baselines we had to develop rather complex radio links, signal delay circuits and so on (Elgaroy et al. 1962). By 1955 we had reached a baseline of 6720λ (13 km) and to our surprise found that 3 of our sources were still unresolved and must therefore have angular sizes less than 25 seconds of arc. We were not expecting this. At that time most of us had come to suspect that the Class II sources were extragalactic, but we expected to find that their angular sizes were typical of galaxies; we had expected minutes, not seconds, of arc. And so we pressed on and by 1956 had successfully established a remote antenna at the 'Cat and Fiddle' in Derbyshire, the highest pub in England. The baseline was then 10600λ (20 km) and to our even greater surprise our 3 sources were still unresolved; we now knew that their angular size must be less than 12 seconds of arc!

In their account of this experiment (Morris <u>et al.</u> 1957) Henry Palmer and his colleagues pointed out that the apparent brightness temperatures of these three sources, at least 2×10^7 K at 158 MHz, were comparable with that of the intense source in Cygnus A which had been identified with a collision between two galaxies. They made the important suggestion that these 3 sources, and perhaps the Class II sources in general, might be extragalactic objects of the same type.

In summary, this programme with the 218 ft paraboloid yielded two truly remarkable results. Firstly, it showed that the Class I sources had large angular sizes and pointed to their association with extended filamentary nebulosities. Secondly, it yielded the astonishing result that 3 out of 5 Class II sources had angular sizes much smaller (< 12") than anything which had been measured previously. It showed that they had brightness temperatures comparable with that of Cygnus A and pointed to their association with objects of similar type. This was a highly signifi- cant new piece of evidence about the Class II sources and, in due course, proved to be an important step towards understanding their origin.

I should add that the programme of measuring the angular size of the radio sources did not stop there. To an onlooker it was an excit- ing chase to find how small the smallest sources really were. To the people who did the actual work, this excitement was, I suspect, tempered by the very hard work involved. It involved an awful lot of plodding about in muddy fields; in my mind's eye I always see Henry Palmer in wellington boots. Nevertheless he and his colleagues soldiered on and by 1961 had achieved a baseline of 61,000λ (115.4 km) and were talking about sources less than 1 or 2 seconds of arc. Some of the sources which he was then studying, for example 3C 48, never were resolved, but it was

the efforts to identify them with optical objects which led, eventually,
to the discovery of quasars.

THE RADIO EMISSION FROM NORMAL GALAXIES

Observations of Normal Galaxies with the 218 ft Paraboloid

When we started work in 1949 one of the few established facts
of radio astronomy was that our own Galaxy emits radio waves, albeit by
some unknown process. It was this which prompted us to survey the Andromeda
Nebula and, having found the same mysterious radiation, to survey other
bright galaxies in the field of view. Our aim was simple, it was to find
out how the radio emission from *normal* galaxies depends upon their type
and apparent photographic magnitude. We defined a *normal* galaxy as one
which was completely normal when examined photographically. At that time
two intense radio sources had been identified with galaxies in Virgo (M87)
and Centaurus (NGC 5128) both of which showed distinctly odd features
(Bolton et al. 1949). In those days they were known as *abnormal* galaxies;
later they were to be called *radio-galaxies*.

By mid-1953 we had identified radio sources with 6 of the
brightest galaxies in our antenna's field of view, most of which were
spirals of type Sb or Sc, and had reached the tentative conclusion that
the ratio of radio flux to light flux is roughly constant for *normal* late-
type spirals (Hanbury Brown & Hazard 1953b). No radio emission had, as
yet, been detected from a *normal* elliptical galaxy, and our failure to
detect the two bright elliptical companions (NGC 205,221) of the Andromeda
Nebula further suggested that this type of galaxy is a comparatively weak
radio emitter.

Observations of Normal Galaxies with the 250 ft Steerable Paraboloid

Our programme of looking at normal galaxies with the 218 ft
reflector was finished in 1953 and no more of this work was done until
1957 when we were able to use its successor, the 250 ft "Mark I" steerable
paraboloid. After having worked with the old 218 ft dish, the Mark I was
ridiculously easy to use. Instead of spending two or three hours every
morning running about outside in the rain adjusting guy wires, we simply
telephoned the "controller" and asked him to scan the telescope in whatever coordinates we wanted. In two or three hours we could collect as
much data as we had previously gathered in two or three *weeks*.

As a start Hazard, who had returned to Jodrell in 1955, and I surveyed the Andromeda Nebula at wavelengths of 1.89 m and 1.26 m (Hanbury Brown & Hazard 1959); it was later also surveyed by Michael Large and his colleagues at 73 cm (Large et al. 1959). From our surveys, with beamwidths in the range 2° to $\frac{3}{4}^\circ$, we reached the conclusion that the radio emission from M31 had two major components, a *disk* component roughly coextensive with the visible nebula and a *corona* which was much broader and roughly spherical. Our analysis suggested that 90 per cent of the radio emission originated in this corona with an emissivity comparable with that which Baldwin (1955) had found for a coronal distribution in our own Galaxy.

Altogether, Hazard and I surveyed 28 normal galaxies with the 250 ft paraboloid (Hanbury Brown & Hazard 1959; 1961a, b). To compare the radio and optical emission from these galaxies we defined a *radio index* R, where

$$R = m_r - m_{pg}$$

and m_r and m_{pg} are the integrated radio and photographic magnitudes of the galaxy, respectively. In our earlier work with the 218 ft paraboloid we had defined the radio magnitude as

$$m_r = -53.45 - 2.5 \log S_{158}$$

where S_{158} (W m^{-2} Hz^{-1}) was the radio flux observed at 1.89 m (158 MHz). This curious definition, which has persisted in the literature for many years, was chosen to make the radio and optical magnitudes of the bright galaxy M51 (NGC 5194/5) equal at our wavelength of 1.89 m. It was a convenient definition for us, but I have often wished that we had chosen something which looked less peculiar, for example 53 instead of 53.45! The results on 20 spirals gave a mean value R = +1.3 with an r.m.s. dispersion of ± 0.7. There was no significant difference between Sb and Sc galaxies; as far as radio emission was concerned, they all behaved the same.

One of the principal aims of our work was to establish the differences in radio emission from different types of normal galaxies in the hope that this would be a useful clue to the origin of the radio emission. We were however unable to detect radio emission from any of

the 4 early-type galaxies which we surveyed; also we failed to detect 3
of the irregulars, although in the position of one (IC 1613) we found a
radio source. These results, taken together with the observations of
irregular galaxies by Mills (1959), suggested that the radio index for
irregular galaxies was greater than that of spirals, having a value,
perhaps, of about +3.0; in other words all the available evidence
suggested that, as radio emitters, irregulars were about 6 times weaker
than spirals with the same optical brightness.

The results for early-type galaxies were even less satisfactory.
According to the literature at that time, 18 had been surveyed and only 1
had been reported as a radio source. Our analysis of these results
suggested that the mean radio index for early-type galaxies must be greater
than +2.0 and that they also must be weaker radio emitters, in relation to
their light, than spirals.

In summary, our work on normal galaxies showed that the ratio
of radio emission to light is fairly constant for spirals of type Sb and
Sc; and taken together with other surveys, it suggested there is less
radio emission from early-type and irregular galaxies.

The Radio Emission from Clusters of Galaxies

One of the many things which we did in our early efforts to
establish the intensity of radio emission from normal galaxies was to
survey clusters at 1.89 m with the 218 ft paraboloid. In 1951/52 we
detected radio sources in the position of the Perseus and Ursa Major II
clusters (Hanbury Brown & Hazard 1952a; 1953b). We then compared the radio
intensity of these sources with their integrated light and found that it
was greater, by roughly 6 times, than we had expected on the assumption
that all galaxies in the clusters had the same ratio of radio to light as
we had found for normal spirals. Subsequent work showed that this
discrepancy, at least in the case of the Perseus cluster, was due to the
presence of a radio-galaxy (NGC 1275).

Another attempt which we made to measure the aggregate radiation from galaxies was to survey with the 218 ft dish the well-known
bright band of galaxies which stretches across the sky, almost at right
angles to the galactic plane and centred roughly on R.A. 12^h. We reached
the tentative conclusion that some of the minor irregularities which we
detected in the general background radiation from that region were associated with this band of galaxies. Some years later, when we had access to

the Mark I, Hazard and I debated whether or not to continue with this work
which we had found so interesting. However at that time we were busy
mapping the radio emission from the North Polar Spur (Hanbury Brown et al.
1960), and we decided that any attempt to map irregularities in the extragalactic emission must wait until we knew more about the irregularities
which we were finding in the emission from our own Galaxy.

THE INVENTION AND EARLY DEVELOPMENT OF THE INTENSITY INTERFEROMETER

The First Radio Intensity Interferometer

If you read the "Report on the Progress of Astronomy" written
by J.S. Hey in 1949 (Hey 1949), you will see how little was then known
about the "radio stars". To be sure, Bolton, Stanley and Slee (1949)
had tentatively identified three of these "stars" with the Crab Nebula
and the galaxies M87 and NGC 5128, but the two most intense, in Cygnus and
Cassiopeia, were a complete mystery; in their positions there was nothing
obvious to be seen. Many people thought that these radio sources might
originate in some unrecognized type of star which, like the Sun, emitted
intense bursts of radio waves, but which was very much more powerful.
This idea was supported by the curious fact that the sources fluctuated
rapidly, a phenomenon which we now know to be due to the ionosphere. No
wonder we were keen to solve this mystery; it is not often that one gets
the chance to discover a new type of object in the sky!

One obvious way of finding out more about the sources was to
measure their angular size. If they were nebulae or galaxies, then we
expected to measure minutes of arc; if they were stars, then we expected
fractions of a second of arc. To measure minutes of arc looked fairly
easy; all we had to do was to build an interferometer with a baseline of
a kilometre or so. But if, as seemed more likely, we had to measure
fractions of a second of arc, then we should need very long baselines
indeed. For example, we knew that to measure the largest known stellar
diameter (Betelgeuse, 0.047") would require a baseline of about 4000 km
at a wavelength of 1 m and, in 1949, we could see no practicable way in
which this could be done with a conventional interferometer. For one
thing, we failed to see how to preserve the relative phase of the signals
received at two such widely separated places. It was while trying to find
some way of comparing two remote signals that I thought of the basic idea
of an intensity interferometer. Late one night in 1949 I was wondering

whether, if I were to take "snapshots" of the noise received from a radio source on oscilloscopes at the outputs of two spaced receivers, I could then compare these snapshots. The answer to that question led me directly to the idea of an interferometer in which the *intensities* of two noise-like signals are compared instead of their amplitude and phase. To put this proposal on a sound mathematical basis I sought the help of Richard Twiss, who made a remarkably thorough analysis and confirmed that the proposal was sound (Hanbury Brown & Twiss 1954).

Our next step was to join with Roger Jennison in building a radio intensity interferometer to measure the angular sizes of the two intense sources in Cygnus and Cassiopeia. This instrument consisted of two *completely independent* receivers tuned to 125 MHz with a bandwidth of 200 kHz. The intermediate-frequency output of each receiver was rectified in a square-law detector (to simplify the theory!) and then passed through a low-frequency filter with a pass-band extending from 1 to 2.5 kHz. The two low-frequency signals were then brought together and their product or *correlation* measured in a linear multiplier. We had shown theoretically that this correlation, when suitably normalized, was proportional to the square of the fringe visibility in a conventional interferometer with the same baseline.

The reason for building this novel instrument was that it could be made to work with very long baselines. The two receivers, being completely independent, were simple to build, and the technical problem of correlating two low-frequency signals without upsetting their relative phase looked easy to solve. For short baselines we proposed to modulate one of them on to a radio-link, or to send it by land-line; for very long baselines, should they prove essential, we intended to record the signals on tape and then correlate the tapes. I had in mind from the beginning that we might need to work even across the Atlantic.

We built our first model of the interferometer in 1950 and tested it by measuring the angular size of the quiet Sun. It worked perfectly well and we decided to press on with the measurements of Cygnus and Cassiopeia. It took some time to build the final instrument and the measurements were done by Roger Jennison and M.K. Das Gupta in 1952. The antenna of the remote station was loaded on to a lorry and set up in a farmyard about a mile away. All went well, and in subsequent measurements the remote station worked its way slowly across Cheshire, farm by farm. But to our surprise both sources were resolved with baselines of only a few kilometres. Cassiopeia proved to be roughly circular in outline with an angular size of about 3.5 minutes of arc, while Cygnus was elongated and measured roughly 2 x 0.5 minutes of arc (Hanbury Brown et al. 1952). These radio stars were clearly not stars. Although our results were valuable, there was really no need to have developed the intensity interferometer; we could have done the same job with a conventional interferometer in half the time and with half the effort. We had built a steam-roller to crack a nut.

But, as it turned out later, building this novel instrument was not a waste of time. While Richard Twiss and I were watching it one day, we noticed that it worked perfectly well even when the ionosphere was making the signals fluctuate violently. Richard looked into the theory and it dawned on us that we had overlooked a most important feature of an intensity interferometer: it can easily be made to work through a turbulent medium. We realized that, if we could make an optical interferometer working on the same principle, we could overcome one of the main obstacles which had so far prevented the development of Michelson's stellar interferometer, namely the effects of turbulence in the atmosphere.

Figure 5   R.C. Jennison (at rear) and M.K. Das Gupta with the radio intensity interferometer (ca 1951).

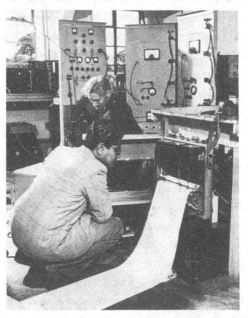

Jennison and Das Gupta then completed a more detailed survey of the two radio sources, in the process discovering that the "elongation" of Cygnus A was (Fig. 6) actually two, distinct sources (Jennison & Das Gupta 1953; 1956). It was while trying to interpret the measurements on this first double source that Jennison developed his method of phase closure (Jennison 1958) which is so important today. Richard Twiss and I then turned our attention to the problems of making an optical interferometer.

Figure 6   The approximate distribution of intensity across the radio source in Cygnus found by Jennison and Das Gupta (1953).

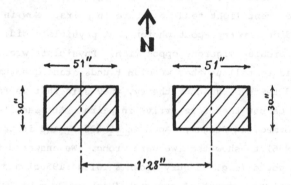

The First Intensity Interferometer for Light Waves

It took us some time to work out a quantitative theory of an intensity interferometer for light waves; the trouble was that we had to start thinking about photons, which don't normally worry radio engineers. For example, at radio wavelengths the signal-to-noise ratio of our intensity interferometer was limited by "wave noise", and "photon noise" was negligible; but at optical wavelengths it would be limited by "photon noise" because the energy of the photon is so much greater. To our disappointment our theory showed that to work on bright stars we would need very large light collectors which looked to be absurdly expensive. It took us several months, however, to realize that all these collectors were required to do was to direct the starlight on to a photo-electric detector and that, compared with an optical telescope, they could be relatively crude and inexpensive.

We then decided to check our theory by experiment. Early in 1955 we set up an optical system in a room at Jodrell Bank which had been built to house a spectro-heliograph. A beam of light from an intense

"point source" was viewed by two photo-electric detectors whose separation
could be varied. The outputs of these two detectors were taken to an
electronic correlator with a bandwidth of 27 MHz. With this equipment we
demonstrated that the fluctuations in the two spaced detectors were, in
fact, correlated and that this correlation decreased with their separation
as we had predicted theoretically.

In publishing the results of the experiment (Hanbury Brown &
Twiss 1956a) we drew attention to the fact that if, as we had shown, the
intensity fluctuations in a light wave arriving at two spaced receivers
are correlated, then it followed that the time of arrival of the photons
at two spaced receivers must also be correlated. In other words, the
photons of mutually coherent light tend to arrive in pairs. Now in those
days radio engineers didn't worry about photons; but physicists did, and to
our surprise our work aroused vigorous opposition. Physicists wrote to us
from all over the world to tell us that we didn't understand quantum theory;
brandishing the sacred texts of quantum theory, they said that photons are
emitted at random and cannot therefore arrive in pairs. At least two
experiments were performed, in Hungary (Ádám et al. 1955) and in Canada
(Brannen & Ferguson 1956), to show that we were wrong. We answered these
objections as best we could (e.g. Hanbury Brown & Twiss 1956c) and in due
course published a complete account of our work (Hanbury Brown & Twiss
1957a, b; 1958a, b). In settling the controversy we were helped by
Edward Purcell who published a short theoretical paper giving welcome
support to our conclusions (Purcell 1956). The most effective reply would
have been to repeat our experiment with equipment which could count the
coincidences between individual photons. Unfortunately we couldn't do this
without a mercury isotope lamp which, at that time, was temporarily
unobtainable in England. In fact it was not until 1957 that Richard Twiss
managed to get hold of a suitable lamp and, with Alec Little, repeated
our original experiment in Australia with equipment which could count
individual photons (Twiss et al. 1957).

Encouraged by our laboratory experiment we decided in 1955 to
build a pilot model of a stellar interferometer (Fig. 7) to measure the
angular size of the brightest star in the sky, Sirius. To save time and
money I persuaded the Army to lend us two of their largest (156 cm) searchlights; we removed the arc lamps, substituted photomultipliers, and mounted
the searchlights on railway sleepers in a field at Jodrell Bank. The

electronic correlator was installed in one of the laboratories of the
control building of the 250 ft antenna (under construction at that time);
in the absence of the control desk and computer, the equipment which
directed the two searchlights was mounted in the control room so that it
had a good view of the sky.

The weather in that part of the world is far from ideal for
optical astronomy and, to make matters worse, Sirius reaches a maximum

Figure 7 Pilot model of stellar intensity interferometer
used in 1955 to measure the angular diameter of Sirius.

elevation of only 20° above the horizon. Indeed the experiment was the
most tedious that I have ever undertaken. I had to keep the equipment
running for 60 nights over a period of five months (November 1955 – March
1956) in order to get a total of 18 hours of measurements on Sirius.
Nevertheless, the work was a success (Hanbury Brown & Twiss 1956b; 1958b);
although the equipment was crude, it worked well enough. Since most of
the time Sirius was twinkling violently, it certainly demonstrated that an
intensity interferometer would work through a turbulent atmosphere. The
measurements were made at four different baselines, the longest being
about 9 m, and gave an angular diameter for Sirius of 0.0071" ± 0.00015".

This was, by the way, the first angular diameter of a main sequence star ever to be measured.

From our pilot model we learned a number of valuable practical lessons; for example, we appreciated the importance of screening the whole electronic installation, including the photomultipliers, against radio interference. We also realized that large exposed mirrors must be heated at night to avoid the formation of dew. But the most important result of this test was, of course, that it showed that a stellar intensity interferometer was a practicable proposition.

In 1958 we put forward a proposal that a full-scale stellar intensity interferometer capable of measuring all stars brighter than $+2^m.5$ should be built. To cut a long story (Hanbury Brown 1974) short we built the instrument as a joint venture of the Universities of Manchester and Sydney; it arrived at the Narrabri Observatory of the School of Physics (University of Sydney) in January 1962. Originally I had intended to spend one or two years in Australia and then return to Jodrell Bank, leaving someone else to run the Observatory. But things didn't work out that way. So much money and work had gone into building the interferometer, and it was so very difficult to install and run such a complicated instrument in the bush, that there was really no choice. I had to stay and look after it.

RETROSPECT

I have always regretted leaving radio astronomy in 1962 — there was so much to be done and the new large instruments were coming into use. Nevertheless we only live once and, looking back, I am glad that I had the rare priviledge of working on a "young" science before it grew up into a "Big Science".

REFERENCES

Ádám, A., Jánossy, L. & Varga, P. (1955). Observations on coherent light beams by means of photomultipliers. Acta Phys. Hung., 4, 301.

Baldwin, J.E. (1955). The distribution of the galactic radio emission. Mon. Not. R. Astr. Soc., 115, 691-700.

Bolton, J.G., Stanley, G.J. & Slee, O.B. (1949). Positions of three discrete sources of galactic radio-frequency radiation. Nature, 164, 101-102.

Brannen, E. & Ferguson, H.I.S. (1956). Photon correlation in coherent
    light beams. Nature, 178, 481.

Elgaroy, O., Morris, D. & Rowson, B. (1962). A radio interferometer for
    use with very long baselines. Mon. Not. R. Astr. Soc., 124,
    395-403.

Hanbury Brown, R. (1974). The Intensity Interferometer. London: Taylor
    & Francis.

Hanbury Brown, R., Davies, R.D. & Hazard, C. (1960). A curious feature of
    the radio sky. Observatory, 80, 191-198.

Hanbury Brown, R. & Hazard, C. (1950). Radio frequency radiation from the
    Great Nebula in Andromeda (M31). Nature, 166, 901-903.

—   (1951a). Radio emission from the Andromeda Nebula. Mon. Not.
        R. Astr. Soc., 111, 357-367.

—   (1951b). A radio survey of the Cygnus region. Mon. Not. R.
        Astr. Soc., 111, 576-584.

—   (1952a). Extragalactic radio frequency radiation. Phil. Mag.,
        43, 137-152.

—   (1952b). Radiofrequency radiation from Tycho Brahe's super-
        nova (A.D. 1572). Nature, 170, 364.

—   (1953a). A radio survey of the Milky Way in Cygnus,
        Cassiopeia and Perseus. Mon. Not. R. Astr. Soc., 113, 109-122.

—   (1953b). A survey of 23 localized sources in the northern
        hemisphere. Mon. Not. R. Astr. Soc., 113, 123-133.

—   (1953c). Radio frequency radiation from the spiral nebula
        Messier 81. Nature, 172, 853.

—   (1953d). An extended radio frequency source of extragalactic
        origin. Nature, 172, 997.

—   (1959). The radio emission from normal galaxies. 1.
        Observations of M31 and M33 at 158 MHz and 237 MHz. Mon. Not.
        R. Astr. Soc., 119, 297-308.

—   (1961a). The radio emission from normal galaxies. 2. A study
        of 20 spirals at 158 MHz. Mon. Not. R. Astr. Soc., 122,
        479-490.

—   (1961b). The radio emission from normal galaxies. 3.
        Observations of irregular and early-type galaxies at 158 MHz
        and a general discussion of the results. Mon. Not. R. Astr.
        Soc., 123, 279-283.

Hanbury Brown, R., Jennison, R.C. & Das Gupta, M.K. (1952). Apparent
    angular sizes of discrete radio sources. Nature, 170, 1061.

Hanbury Brown, R. & Lovell, A.C.B. (1955). Large radio telescopes and
        their use in radio astronomy. In Vistas in Astronomy, 1,
        ed. A. Beer, pp.542-560. London: Pergamon Press.

Hanbury Brown, R., Palmer, H.P. & Thompson, A.R. (1954). Galactic sources
        of large angular diameter. Nature, 173, 945.

—   (1955). A rotating-lobe interferometer and its application
        to radio astronomy. Phil. Mag., 46, 857-866.

Hanbury Brown, R. & Twiss, R.Q. (1954). A new type of interferometer for
        use in radio astronomy. Phil. Mag., 45, 663-682.

—   (1956a). Correlation between photons in two coherent beams
        of light. Nature, 177, 27-29.

—   (1956b). A test of a new type of stellar interferometer on
        Sirius. Nature, 178, 1046-1048.

—   (1956c). The question of correlation between photons in
        coherent light rays. Nature, 178, 1447-1448.

—   (1957a). Interferometry of the intensity fluctuations in
        light. 1. Basic theory: the correlation between photons in
        coherent beams of radiation. Proc. Roy. Soc. A., 242, 300-324.

—   (1957b). 2. An experimental test of the theory for partially
        coherent light. Proc. Roy. Soc. A., 243, 291-319.

—   (1958a). 3. Applications to astronomy. Proc. Roy. Soc. A.,
        248, 199-221.

—   (1958b). 4. A test of an intensity interferometer on Sirius.
        Proc. Roy. Soc. A., 248, 222-237.

Hey, J.S. (1949). Reports on the progress of astronomy - radio astronomy.
        Mon. Not. R. Astr. Soc., 109, 179-214.

Jennison, R.C. (1958). A phase sensitive interferometer technique for the
        measurement of Fourier Transforms of spatial brightness
        distributions. Mon. Not. R. Astr. Soc., 118, 276-284.

Jennison, R.C. & Das Gupta, M.K. (1953). Fine structure of the extra-
        terrestrial radio source Cygnus I. Nature, 172, 996-997.

—   (1956). The measurement of the angular diameters of two
        intense radio sources. Phil. Mag., Ser. (8), 1, 66.

Large, M.I., Mathewson, D.S. & Haslam, C.G.T. (1959). A high resolution
        survey of the Andromeda nebula at 408 MHz. Nature, 183,
        1250-1251.

Mills, B.Y. (1952). The distribution of the discrete sources of cosmic
        radio radiation. Aust. J. Sci. Res. A., 5, 266-287.

Mills, B.Y. (1959). Radio frequency radiation from external galaxies. In Handbuch der Physik, 53, ed. S. Flügge, pp.239-274. Berlin: Springer-Verlag.

Morris, D., Palmer, H.P. & Thompson, A.R. (1957). Five radio sources of small angular diameter. Observatory, 77, 103-110.

Purcell, E.M. (1956). The question of correlation between photons in coherent light rays. Nature, 178, 1449-1450.

Reber, G. (1940). Cosmic static. Proc. I.R.E., 28, 68-70.

Ryle, M. & Vonberg, D.D. (1948). An investigation of radio frequency radiation from the Sun. Proc. Roy. Soc. A., 193, 98.

Ryle, M., Smith, F.G. & Elsmore, B. (1950). A preliminary survey of the radio stars in the northern hemisphere. Mon. Not. R. Astr. Soc., 110, 508-523.

Twiss, R.Q., Little, A.G. & Hanbury Brown, R. (1957). Correlation between photons in coherent beams of light detected by a coincidence counting technique. Nature, 180, 324-326.

Walsh, D. & Hanbury Brown, R. (1955). A radio survey of the Great Loop in Cygnus. Nature, 175, 808.

(l. to r.)   Don Mathewson, Michael Large, and Glynn Haslam working on 408 MHz data taken in 1958 with the new 250 ft reflector at Jodrell Bank.   It took roughly one week for one man to reduce one night's good observations.   (courtesy Mathewson)

The radio astronomy group at the Cavendish Laboratory, Cambridge in 1954: (back row, l. to r.) John Thomson, John Baldwin, George Whitfield, Pat O'Brien, Peter Scheuer, Robin Conway, R.L. Adgie, a technician; (middle) Graham Smith, Martin Ryle, Tony Hewish; (front) M.B. Turner, John Shakeshaft, Johnny Blythe (photo courtesy Bruce Elsmore)

EARLY WORK ON RADIO STARS AT CAMBRIDGE

F.G. Smith
Nuffield Radio Astronomy Laboratories
Jodrell Bank

Two powerful influences acted on the early development of radio astronomy at Cambridge. The first was the existing radio research under J.A. Ratcliffe, directed primarily at the ionosphere. J.W. Findlay, who is known amongst radio astronomers as the inspiration for the 300 ft transit telescope at Green Bank and, later, as one of the designers of the Very Large Array, was a member of this research group before and after the 1939-45 war. The second influence was wartime experience in radar, when Martin Ryle in particular developed his genius for experimental methods which were at once bold, original and economical. His most important wartime work was in airborne counter-measures, involving the analysis of enemy radar and the desperate scramble to provide aircraft with warnings of radar-directed fighter attack. Ratcliffe would maintain that his own contribution was to attract Martin Ryle to the Cavendish, and to encourage him to develop his own techniques in investigating radio waves from the sun. Nevertheless Ratcliffe's influence in our understanding of radio, and even more of Fourier analysis, were other vital ingredients.

Radio research re-started in the Cavendish in 1945. Ryle, who was an Imperial Chemical Industries Research Fellow, was joined by Derek Vonberg; both were registered for Ph.D.s, although neither ever wrote a thesis. Their first approach to the measurement of solar radio waves was to build a radio version of the Michelson interferometer which would distinguish the sun from other extraneous sources of noise. They built a switched receiver known as the Cosmic Radio Pyrometer (Ryle & Vonberg 1948) in which a controllable noise diode was switched against the input noise signal. The interferometer aerials, working at 175 MHz, were simple broadside arrays of full-wave dipoles over a wire mesh reflector.

I joined the research group in 1946, having spent my first year back at Cambridge in completing my first degree in Physics. My wife, who had also been involved in wartime radar work, was already working with Ryle as

an assistant. There are many of these personal connections which can be
traced to radar work in the Telecommunications Research Establishment: this
one led ultimately to the coincidence that the wives of the 12th and 13th
holders of the office of Astronomer Royal (Ryle 1972-82; Smith 1982-) were
sisters.

My first experience in radio astronomy was to assist in setting
up the pair of interferometer aerials at larger and larger spacings in an
attempt to resolve the angular diameter of the giant sunspot of July 1946
(Ryle & Vonberg 1946). The baseline soon spread outside the allotted area
and into the University rugger ground, reaching a length of 140 wavelengths.

The resolution of the diameter of the July 1946 sunspot, as well
as the detection of circular polarisation, opened the way for rapid developments in the subsequent work on "radio stars". Hey, Parsons and Phillips
(1946) published their paper on fluctuations in radio noise from the direction of Cygnus in the same month as our sunspot work. J.S. Hey had been working both on the sunspot radiation and on cosmic radio noise; he pointed out
an analogy between the discrete source in Cygnus and sunspots, arguing that
the fluctuations in the Cygnus source would not be observed in a distributed
interstellar source. There must be some sort of star, emitting radio waves
more powerfully than the sun. We therefore employed interferometric
techniques to follow up Hey's discovery, as did Bolton and Stanley in
Australia (see the contribution of Bowen in this volume).

In 1948 we made our first observations of Cygnus A using a
Michelson interferometer at 80 MHz. The servo receiver was connected to two
groups of 4 Yagi aerials spaced 500 m apart. The first night's run showed,
to our surprise, not one but two interference patterns, three hours apart.
The second set of interference fringes obviously had a lower periodicity -
another surprise. We knew little or nothing about spherical astronomy, and
it was the next day before I could tell Ratcliffe that the new source was "in
a constellation called Cassiopeia". We published a reasonably good position
and showed that the source was unpolarised (Ryle & Smith 1948). We had, of
course, rather hoped that it would be circularly polarised, like sunspot
radiation.

A larger pair of aerials was obviously needed, as there were
other weak sources to be seen in these first interferometer recordings. The
so-called "Long Michelson" interferometer (Ryle, Smith & Elsmore 1950)
consisted of two broadside arrays 110λ apart, each 20λ by 1λ at 3.7 m wavelength, using a wire reflector stretched over a horizontal angle-iron frame
(Figures 1, 2 and 3). Our wartime experience of radio engineering at

Figure 1. One of the elements of the "Long Michelson" interferometer, a broadside array operated at 3.7 m wavelength on the Old Rifle Range in Cambridge (ca. 1950). This instrument produced the 1C survey of radio stars. Martin Ryle's house can be seen just to the right of the tower of the University Library in the distance.

Figure 2. A close-up view of one element of the "Long Michelson" interferometer, with one of the 7.5 m diameter Würzburg paraboloids visible in the background. Research student J.H. Thomson stands next to the aerial in this 1955 photograph by B. Elsmore.

Figure 3. Graham Smith (left) and Martin Ryle soldering part of the "Long
         Michelson" interferometer in 1949.

Telecommunications Research Establishment was a great help. We used a very
long slotted line for impedance measurements, and stub tuning on the
parallel wire transmission lines. Coaxial cables and insulating dipole
supports were supplied from captured German radar equipment by the Royal
Aircraft Establishment (RAE) at Farnborough.

A new phase-switching receiver (Ryle 1952) was used for the
recordings. Ryle's invention of this system is supposed to have derived from
his advisory work in underwater sonar: for a time we called the system the
"mermaid-hunter". Phase-switching allows the radio receiver to discriminate
between sources with large angular size, which are rejected, and smaller
sources which can then be recorded at a much greater receiver gain. It is
exactly suited to a search for star-like radio sources, and to the accurate
measurement of their positions.

The phase-switched records, first obtained in 1949, were marvellous
we turned up the gain and sources showed up through the whole 24 hours. The
resultant catalogue contained 50 sources and is now known as the 1C survey
(Ryle, Smith & Elsmore 1950). Although the survey was complete in 24 hours,
as the aerials were fixed to look only at the zenith, we ran the system

nevertheless for some months and established that the sources were constant
in strength and position (Ryle & Elsmore 1951). I subsequently continued
recording the four strongest sources, exploiting the accuracy inherent in
the phase-switched interferometer to look for parallax or proper motion
(Smith 1951a). This was the start of the position-finding work that
eventually led to the identifications of Cyg A and Cas A (see below).

Although the 1C catalogue, like its successor 2C (see below),
suffered hopelessly from confusion, it did show clearly that there were
lots of radio sources distributed fairly evenly over the sky. As expected,
the five optically brightest extragalactic nebulae, including M31, were
detectable, based on positional coincidence good to $\sim 2^{\circ}$, with radio
strengths similar to our own Galaxy. But this gave no clue about the
generality of the sources. What were they? The only clue, before Bolton,
Stanley and Slee (1949) published the probable identifications of Taurus A,
Centaurus A and Virgo A, lay in the fluctuations. Were these intrinsic,
indicating a small star-like source, or were they imposed by the ionosphere?
I undertook some spaced-receiver observations at 45 MHz to find out. The
results were published in 1950, in collaboration with Little and Lovell at
Jodrell Bank (Smith 1950; Little & Lovell 1950). Most of the fluctuations
were definitely due to ionospheric scintillation, but some odd 10 sec bursts
of radiation were received simultaneously over baselines of up to 160 km.
We were never able to explain these bursts. Incidentally, these observations
included the first use of Defford airfield for radio astronomy: it was later
used by Hey for an interferometer, and one of his 25 m telescopes is now
used in the 6-station "Merlin" interferometer network run by Jodrell Bank.

The observations for parallax and proper motion, mentioned above,
also led to an incidental achievement in measuring the total electron
content of the ionosphere through the East-West component of ionospheric
refraction (Smith 1952d). This component, which was detected in the series
of transit measurements at 80 MHz, was correlated with ionosonde measurements
of electron content below the maximum of the F-region. It was found that the
total electron content of the ionosphere was about three times the content
which could be measured by the ionosonde. This is of some historic interest
as the first observation of the upper ionosphere.

The search for correlated fluctuations was extended to even
longer baselines when I went to the USA in 1952 on a Carnegie Fellowship. I
took with me a 45 MHz interferometer receiver, which was set up with a simple
pair of dipoles first at the Department of Terrestrial Magnetism, Washington,

and then for a short while on Palomar Mountain. No correlated fluctuations
were found, and no results were published: the subject was closed until
Hewish re-opened it with studies of interplanetary and interstellar
scintillation.

Identifications of the discrete sources were still elusive, and
a substantial improvement in positions was needed. An interferometer used
as a transit telescope was obviously a good instrument, but we had no
calibrating sources to measure collimation error. We therefore set up two
7.5 m Würzburg parabolic radio reflectors, captured German radar antennas
(Figure 4) provided by RAE, at 280 m spacing, using feed systems, preamplifiers and cables which could all be interchanged, just as one reverses
the telescope in its bearings in a classical transit instrument. We also had

Figure 4. The 7.5 m Würzburg paraboloids in 1950. With this interferometer
          Graham Smith (pictured) provided the accurate positions which
          led in 1951 to the identifications of Cas A and Cyg A by W. Baade
          and R. Minkowski (photo by B. Elsmore).

to measure the length and direction of the baseline, which was nominally East-West but actually ran along the edge of a convenient field with an azimuth more than 5° off true East-West. I learned spherical trigonometry the hard way, and only later found that Bessel had done it all more than a century earlier.

These interferometer recordings were made on paper charts, run at fairly high speed, with 1 second time marks. The time was obtained from a Shortt free pendulum clock on loan from the Royal Greenwich Observatory, and checked by Rugby radio time signals. Most of the accurate recordings were made at 214 MHz; the times of transit were measured to an accuracy of about 1 second of time, corresponding to about ten arc seconds. Angular rates of movement were obtained by inserting measured lengths of coaxial cable in the two arms of the interferometer; this was more accurate than relying on a precise knowledge of the centre frequency of the receiver, which was not well defined by the various tuned circuits.

Again I attempted to measure parallax by recording through the year. This meant that I visited the observatory four times a day at least twice a week for a whole year, setting the Würzburgs to the correct elevation and running the chart recorder at high speed for the few crucial minutes near transit. We now know, of course, that parallax measurements were quite hope- less, but with no certain identifications and with what appeared to be an isotropic distribution of radio sources over the sky it was not unreasonable to surmise that they might be very local.

At last we had positions to better than one minute of arc (Smith 1951b, 1952a). We went to Professor R.O. Redman, Director of the Cambridge Observatories, and he started Dr. David Dewhirst on the identification process for Cyg A and Cas A. Dewhirst (1951) did in fact find part of the Cas A supernova remnant on a Cambridge 40" plate, but the poor observing condi- tions made it impossible to follow up these identifications. At Redman's suggestion, and encouraged by J.H. Oort, I sent the new positions to W. Baade and R. Minkowski. They were sufficiently encouraged by the excellent positional coincidence between Tau A and the Crab, and turned the Palomar 200 inch telescope on to the two positions, with the well-known result (Baade & Minkowski 1954; Smith 1955; see also the article by Greenstein elsewhere in this volume) announced in the following excerpts of letters from Baade to Smith:

1951 October 23. "I would like to let you know that my search for the Cassiopeia radio source at the 200 inch has turned up an exceedingly

interesting object close to your measured position. It is an emission
nebulosity ...."

1952 April 29. "Regarding the Cygnus source the situation is
this: I photographed the field at the 200 inch last fall after you had sent
me your accurate position. The result was very puzzling. At the place you
gave me there is a rich cluster of galaxies and the radio position coincides
closely with one of the brightest members of the cluster as you may see from
the following two lines:

radio source in Cygnus 1950.0 $19^h57^m45^s.3 \pm 1^s$ $+ 40^\circ35!0 \pm 1'$

galaxy   " $19^h57^m44^s.6 \pm 0^s.3 + 40^\circ35!7 \pm 0!1$

This galaxy, which has a diameter of about 0'.6 of arc, is a queer object.
In fact, the 200 inch picture suggests strongly that we are dealing with two
galaxies which are in actual collision..."

1952 May 26. "Last week Minkowski obtained the spectrum of the
nebula with the new grating spectrograph at the 100 inch. Its outstanding
feature is a strong emission spectrum with [NeV] present!! That we are dealing with an extragalactic object is convincingly shown by the large redshifts
of the emission lines which amount to about 15000 km/sec."

It may seem strange that we were still speaking of "radio stars"
at this time. In my letter to Baade giving him the accurate positions (1951
August 22) I referred to an anlysis of the population of the galaxy, which
was the first application of population statistics in radio astronomy,
including the now familiar 3/2 power law. This analysis is mentioned in a
survey paper by Ryle (1950) which shows clearly that we were still thinking
in terms of radio stars rather than galaxies. We suggested that radio stars
might be as common as visual stars, without knowing at all what sort of stars
they might be.

The identification of the four strongest radio sources did not
exactly clarify the basic problem of the nature and distance of the general
run of sources. There were two supernova remnants, clearly belonging to our
Galaxy: and there were two extragalactic nebulae, Cyg A being very distant,
and Cen A being at best understood as "peculiar". At the same time there were
obviously at least some larger diameter sources, seen by Hanbury Brown and
Hazard at Jodrell Bank, and by Mills in Australia: these we regarded mainly
as large-scale galactic structure . It required a leap of the imagination to
place the majority of the radio sources into a single category which populated
the universe on a truly cosmic scale. This step was made by Ryle and Scheuer
(1955) by putting together the number-flux relation of radio sources (the

"log N-log S" relation) and the integrated background intensity (for a full discussion, see the article by McCrea elsewhere in this volume). Although the details of the log N-log S relation were the subject of fierce contro-versy for several years, this paper represents the entry of radio astronomy into cosmology. The origin of the Galactic component remained unknown until we later came to understand the Soviet work on synchrotron radiation. The statistics used by Ryle and Scheuer were based on the 2C survey, to which I now turn.

The 2C and 3C surveys were made using a new interferometer, with four parabolic cylinder aerials (Ryle & Hewish 1955). The total collecting area was over an acre, roughly equal that of the 250 ft Mark I telescope at Jodrell Bank. The four units formed a rectangle, with long axis East-West. A complete survey involved four different phasings of the North-South pairs, and 25 different elevation settings. In addition, the system could be re-arranged as a North-South phase-switched interferometer; in this configura-tion it was sensitive to sources with large angular diameter, which were of special importance at that time in discussions of population statistics.

The four parabolic cylinders of the new interferometer were designed by Ryle and built with the help of a local engineer, D. Mackay, subsequently Mayor of Cambridge. The stringing of the reflector wires and the assembly of the line dipole feed were undertaken by a growing team of staff and students, with very little effort from any staff of technicians. Changing the elevation meant climbing each of 35 towers, heaving on the steel cantilevers, and pushing a pin into another locating hole. This was unpleasant in icy conditions, but on the whole the research team enjoyed field work, even including a weekly maintenance day with grass cutting and hedge trimming.

The 2C survey (Shakeshaft et al. 1955) was embarrassingly success-ful. The large collecting area coupled with the phase-switching receiver gave beautiful records filled with a profusion of radio sources. But as we now see in retrospect, the 2C survey is the classic example of confusion. This was a problem that could not have been anticipated, since there was no reason to expect so many radio sources that their individual traces overlapped all along the chart recordings. Even so, it seemed possible to disentangle them: we did not appreciate that the source statistics were such that individuals could never be sorted out below a certain definite intensity. The catalogue of 1936 sources is too near one source per beam area to be at all reliable. But it again showed that the sky is full of radio sources, and although the individual catalogued sources could not be relied on, we soon had Scheuer's

statistical "P(D)" analysis (Scheuer 1957), which allowed a meaningful
investigation of the population of radio sources. A better catalogue, 3C
(Edge et al. 1959), came from the next stage, which was to halve the wavelength. Even here there was something to be learnt, and it required a
completely new analysis with rigorous criteria before the revised version,
3CR, was published (Bennett 1962). It is some reward for years in the firing
line to see that 3CR catalogue numbers (although still cited as "3C") are
still widely and even affectionately in use.

The first measurements of source diameters and brightness
distributions, undertaken in the early 1950s, could be regarded as the
beginning of the story of aperture synthesis, which is the subject of a
separate essay in this volume by Scheuer. But the idea of synthesis grew
slowly, and our initial approach was essentially the more analytic approach
due to Michelson. We were very familiar with the laboratory technique of
analysing the width and shape of spectral lines using a two-element interferometer. The concept of fringe visibility was in fact applied to the sunspot measurements in 1946. From this derived the concept of combining measurements at a number of interferometer spacings to produce a brightness distribution; the interest here was to look for limb-brightening on the sun (Stanier
1950). The same idea was used in Australia, with McCready, Pawsey and Payne-
Scott (1947) pointing out more clearly the relation between interferometry
and Fourier transformation.

The interferometers which we built for the "radio stars" were
intended to distinguish between point radio sources and the diffuse background. The idea of using a variable spacing interferometer to explore the
totally unknown field of radio star diameters apparently occurred independently
to the groups at Jodrell Bank, Sydney, and Cambridge, who published their
results together in Nature in 1952 (Mills 1952; Hanbury Brown et al. 1952;
Smith 1952b). We were fortunate in that the large spacings which could only
be tackled by radio links were not needed for Cyg A and Cas A; we were able
to resolve the sources using simple cable-linked interferometers. The only
technical innovation of any later importance was the use of three antennas,
which allowed a self-calibration of the system through the now familiar use
of ratios of Fourier components (Smith 1952c).

These measurements came just at the time when Baade was exploring
the optical nebulosity of Cas A, whose diameter coincided exactly with the
radio measurements of 5 arc minutes. Cyg A was more of a puzzle, as our
diameter showed the radio source to be larger than the visible "colliding

galaxies". The Jodrell Bank measurements soon also showed that there was more structure to be resolved than could be expressed as a single measured diameter (see the contribution by Hanbury Brown in this volume). This was the first indication of the complex structure of the extragalactic radio sources, a topic which subsequently engaged a major part of the research at Jodrell Bank and elsewhere.

One might expect that a retrospect of this kind would above all bring up such memories as opening a letter from Baade telling us about Cygnus A and Cassiopeia A. Not so. I remember most the slogging out to our observatory, four times every 24 hours through most of a year to reset those Würzburg antennas and the chart recorder, in the experiment to look for parallax in radio sources that turned out to be, after all, at cosmological distances. And learning to use a sledgehammer, a slotted line, and the Nautical Almanac, and sitting down to analyse the sources of noise in a triode amplifier and the action of diode switches in hybrid networks. Above all, perhaps, days of honest toil in good company. It was a long time before we even realised that we were astronomers.

References

Baade, W. & Minkowski, R.   1954.   Identification of the radio sources in Cassiopeia, Cygnus A, and Puppis A. *Ap.J., 119,* 206.
Bennett, A.S.   1962.   The revised 3C Catalogue of radio sources. *Mem. R. Astr. Soc., 58,* 163.
Bolton, J.G., Stanley, G.J. & Slee, O.B.   1949.   Positions of three discrete sources of galactic radio-frequency radiation. *Nature, 164,* 101.
Dewhirst, D.W.   1951.   (Report of a discussion during 12 October 1951 meeting of Royal Astronomical Society). *Observatory, 71,* 209.
Edge, D.O., Shakeshaft, J.R., McAdam, W.B., Baldwin, J.E. & Archer, S. 1959.   A survey of radio sources at a frequency of 159 Mc/s. *Mem. R. Astr. Soc., 68,* 37.
Hanbury Brown, R., Jennison, R.C. & Das Gupta, M.K.   1952.   Apparent angular sizes of discrete radio sources. *Nature, 170,* 1061.
Hey, J.S., Parsons, S.J. & Phillips, J.W.   1946.   Fluctuations in cosmic radiation at radio-frequencies. *Nature, 158,* 234.
Little, C.G. & Lovell, A.C.B.   1950.   Origin of the fluctuations in the intensity of radio waves from galactic sources. *Nature, 165,* 423.
McCready, L.L., Pawsey, J.L. & Payne-Scott, R.   1947.   Solar radiation at radio frequencies and its relation to sunspots. *Proc. Roy. Soc., A 190,* 357.
Mills, B.Y.   1952.   Apparent angular sizes of discrete radio sources. *Nature, 170,* 1063.
Ryle, M.   1950.   Radio astronomy. *Rep. Prog. Phys., 13,* 184.
Ryle, M.   1952.   A new radio interferometer and its application to the observation of weak radio stars. *Proc. Roy. Soc., A 211,* 351.
Ryle, M. & Elsmore, B.   1951.   A search for long-period variations in the intensity of radio stars. *Nature, 168,* 555.

Ryle, M. & Hewish, A.   1955.   The Cambridge radio telescope. *Mem. Roy. Astr. Soc., 67,* 97.

Ryle, M. & Scheuer, P.A.G.   1955.   The spatial distribution and the nature of radio stars. *Proc. Roy. Soc., A 230,* 448.

Ryle, M., Smith, F.G. & Elsmore, B.   1950.   A preliminary survey of the radio stars in the northern hemisphere. *Mon. Not. R. Astr. Soc., 110,* 508.

Ryle, M. & Smith, F.G.   1948.   A new intense source of radio-frequency radiation in the constellation of Cassiopeia. *Nature, 162,* 462.

Ryle, M. & Vonberg, D.D.   1946.   Solar radiation on 175 Mc/s. *Nature, 158,* 339.

Ryle, M. & Vonberg, D.D.   1948.   An investigation of radio frequency radiation from the Sun. *Proc. Roy. Soc., A 193,* 98.

Scheuer, P.A.G.   1957.   A statistical method for analysing observations of faint radio stars. *Proc. Camb. Phil. Soc., 53,* 764.

Shakeshaft, J.R., Ryle, M., Baldwin, J.E., Elsmore, B. & Thomson, J.H.   1955. A survey of radio sources between declinations $-38°$ and $+83°$. *Mem. Roy. Astr. Soc., 67,* 106.

Smith, F.G.   1950.   Origin of the fluctuations in the intensity of radio waves from galactic sources. *Nature, 165,* 422.

Smith, F.G.   1951a.   An attempt to measure the annual parallax or proper motion of four radio stars. *Nature, 168,* 962.

Smith, F.G.   1951b.   An accurate determination of the positions of four radio stars. *Nature, 168,* 555.

Smith, F.G.   1952a.   The determination of the position of a radio star . *Mon. Not. R. Astr. Soc., 112,* 497.

Smith, F.G.   1952b.   Apparent angular sizes of discrete radio sources. *Nature, 170,* 1065.

Smith, F.G.   1952c.   The measurement of the angular diameter of radio stars. *Proc. Phys. Soc. B., 65,* 971.

Smith, F.G.   1952d.   Ionospheric refraction of 81.5 Mc/s radio waves from radio stars. *J. Atmos. & Terr. Phys., 2,* 350.

Smith, F.G.   1955.   Positional radio-astronomy and the identification of some of the radio stars. *Vistas in Astronomy, 1,* 560.

Stanier, H.M.   1950.   Distribution of radiation from the undisturbed Sun at a wavelength of 60 cm. *Nature, 165,* 354.

THE DEVELOPMENT OF APERTURE SYNTHESIS AT CAMBRIDGE

P.A.G. Scheuer
Mullard Radio Astronomy Observatory, Cavendish Laboratory
Madingley Road, Cambridge CB3 OHE, England

By the beginning of 1954 the principles of aperture synthesis were fully understood all over the world, but the world of radio astronomy was then very small, and the world I mean, in which radio astronomy was controlled by radio engineers who were learning astronomy, was smaller still. In the Netherlands and in the United States radio astronomy was in the hands of real astronomers, to whom a telescope meant a paraboloidal mirror and nothing else; their contributions were of a different kind. So the little world that understood aperture synthesis consisted of CSIRO Radiophysics Division in Sydney, the English radio astronomers at Cambridge and Manchester, and the French group at Nançay. Furthermore, the Manchester group were preoccupied with building a very large paraboloid indeed, so they were not very interested.

The underlying principles were neatly written down by Bracewell and Roberts (1954). I wrote them down too, but only in my thesis (Scheuer 1954), because Martin Ryle took a severe line, that on engineering topics you shouldn't write mere theory, you should jolly well build the thing first. It took ten years to get from the principles to the first full-scale working instrument of the modern kind: the Cambridge one-mile telescope. I don't think that was because we couldn't make the next mental step to earth rotation synthesis. Pat O'Brien had observed the sun with interferometers along various position angles at various times of day (O'Brien 1953), and I had observed the profile of the Milky Way at such time as it makes a vertical arch over Cambridge, so the fact that the earth turns the interferometer had been forced into our consciousness quite hard enough. But if earth rotation synthesis was discussed at all, it would have been in the jocular way in which one might have discussed the benefits of a large space telescope. Not one, but several technologies that are now taken for granted were at that time not up to the job.

The most obvious is computing.   Electronic digital computers were being developed, one of them in the Cambridge University Mathematical Laboratory, but up to that time we still did our modest Fourier transforms by hand, using Lipson-Beevers strips borrowed from the Crystallography section of the Cavendish.   (These were sets of cardboard strips printed with A cos nx and A sin nx, which one lays neatly side by side before adding up columns.)   Any computing had to be kept simple, and that was a severe constraint on design.   Indeed, there was much to be said for doing much or all of the Fourier inversion in hardware;   that is why Mills Crosses were such a good option, and several groups built them.

Rather less obvious are the many implications of the fact that there were no low-noise amplifiers for decimetre and centimetre wavelengths.   At metre wavelengths any antenna with large forward gain is extremely large and unwieldy, not the sort of thing one wants to put on an equatorial mounting;   so if one wanted large gain, a transit telescope was implied.   At the same time, the alternative of many small individual elements had grave drawbacks, not only because of the increase in the sheer quantity of computing, but even more because each element would then admit radiation from the intense sources in Cassiopeia and Cygnus, sources 10,000 times brighter than the sort of level one might hope to reach.   To reduce the sidelobe contributions from these monsters to an acceptable level would have required a phase and amplitude stability that one seemed unlikely to be able to reach.   I suspect that Cas A and Cyg A are to some extent responsible for differences in the early development of radio astronomy in the Northern and Southern Hemispheres.   With the short wavelengths and quite large paraboloids now in use, the forward gain of each antenna is so great that Cas A and Cyg A are no longer bugbears, and probably many younger radio astronomers are not even aware of the problem. (John Baldwin's group has now built a 151 MHz telescope of many small elements, of the kind that might have been contemplated long ago, but the software for eliminating residual rings due to strong sources, as well as the basic Fourier synthesis, would have been unthinkable on 1960 vintage computers.)   One more comment in the same vein: surely a key factor in the early successes of Caltech's Owens Valley observatory was John Bolton's confidence that he would be able to build good receivers at the extremely high frequency of 960 MHz by the time his interferometer was finished.

Through the latter half of the 1950s we find many possible telescope designs mulled over in Ryle's laboratory notebooks.   Most of these thoughts are directed towards optimising transit-type synthesis instruments; I have indicated above what I think are the technical reasons for this.   Besides, one must bear in mind that at the time one was thinking chiefly in terms of better surveys, for which a transit instrument is no great disadvantage.   Thus the earliest aperture synthesis instruments (and the Mills Crosses too) were transit instruments only.

The first synthesis of a substantial two-dimensional aperture was made by a research student, John Blythe (1957a, b, c).   He built a line of 48 full-wave dipoles (Figure 1) running roughly East-West, plus a single dipole over a reflecting screen as the element to be carried, step by step, down the North-South axis.

Figure 1.   Blythe's East-West aerial.   "By judicious choice of the line of the aerials to avoid huts, hedges, paths, a stream, telephone lines and other aerials it was found possible to construct an array 1200 feet long" (Blythe 1957c), part of which is shown here.

This 'moving T' synthesised the equivalent of a Mills Cross, with a 2° beam at 38 MHz.   Why did he not build a Mills Cross, then?   A cynic might say that the furious row then raging over source counts made that emotionally impossible.   In fact, there were very clear practical reasons. Constructing a large phased array was a fairly formidable proposition, and was seen as a problem that would rapidly become worse in later, larger

versions;   for example larger individual elements would shadow each other
as soon as one observed away from the zenith.   On the other hand, we did
by then have a computer (EDSAC 1) in the University, which Blythe was able
to use for constructing his map of the sky (Figure 2).   Blythe's system

Figure 2.   Blythe's map of the sky.   The missing portions
are those corrupted by low-level sidelobes of Cas A and Cyg A.
Apart from the Milky Way, the most prominent feature is the
North Polar Spur.

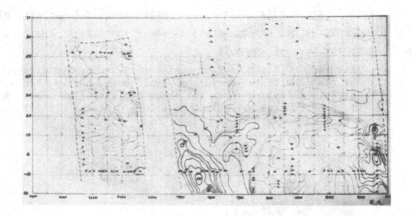

was the prototype for the 178 MHz interferometer (Figure 3) that produced
the 4C catalogue (Scott et al. 1961; Pilkington & Scott 1965; Gower et al.
1967), and the '4CT' (Crowther & Clarke 1966; Caswell et al. 1967) and
38 MHz pencil beam surveys (Costain & Smith 1960; Williams et al. 1966).
These aerials were built on the new site at Lords Bridge, with funds
provided by the Department of Scientific and Industrial Research (DSIR)
and by Mullard Ltd;   the opening of the observatory on July 25th 1957 was
celebrated with a dinner at Gonville and Caius College, and a photograph
of that occasion shows the faces of most of those concerned (Figure 4).

Figure 3.   The fixed aerial of the interferometer system
which produced the 4C survey.   It is 442 m long and 20 m
wide.   The moving aerial is a 58 m section of exactly
similar parabolic cylinder, mounted on 305 m of North-South
rails, 783 m East of the fixed aerial.   Both were fed by
lines of dipoles at 178 MHz, and the synthesised beam was
25' x 35' (at the zenith), crossed by interference fringes.

While the 3C aerials (also parabolic cylinders) had to be
tilted in declination by shoulder power, one tower at a
time, the 4C aerials incorporated a system of counterweights,
trip switches, and electric motors that enabled them to be
tilted (or driven along the rails) in unison.

Figure 4 (on following page). The dinner on 25 July 1957 celebrating the opening of the Mullard Radio Astronomy Observatory, in the hall of Gonville and Caius College, Cambridge. This formal occasion is not at all typical of the Cambridge radio astronomers at that time, but shows the faces of most of them, and of many other astronomers besides. Martin Ryle is at the top table (second from right, near side); also Sir Edward Appleton (who performed the opening ceremony), the Vice-Chancellor of the University, the Astronomer Royal (Sir Richard Woolley, second from left, far side), and representatives of Mullard Ltd. and D.S.I.R. Mr. J.A. Ratcliffe, then head of the radio group at the Cavendish, is not at the top table; he said "this is Martin's day", and placed himself at the foot of the table on the left (just out of view).

Labelling the rows A to F from left to right and counting from the ends of the tables nearest the camera, the members of the radio astronomy group were:

A3 Jan Högbom, A9 Graham Smith, A12 John Baldwin.

B3 Mike Turner, B7 Ken Machin, B10 Tony Hewish.

C2 Peter Scheuer, C6 Carman Costain, C9 Bruce McAdam, C10 Bill Hackett.

D6 Harriet Tummer, D8 John Blythe, D12 John Shakeshaft.

E4 George Whitfield, E9 David Edge.

F2 Bruce Elsmore.

Derek Vonberg (B12?) was a research student in the earliest years of the group. Mrs. Ryle (B9) is one of the very few ladies present. Readers with keen eyes will identify many other British astronomers, and a few well known physicists too. Note that Fred Hoyle (D3) came to this celebration; although the source-count controversy was already three years old, it was still at the stage of the retort courteous. (Photograph courtesy of Cambridge Daily News).

[The contingent from Jodrell Bank included: Bernard Lovell (second to left of Ryle), A6 Robert Hanbury Brown, D7 J. G. Davies, and D9 R. C. Jennison.-Ed.]

The first written mention of earth rotation synthesis is an entry in Ryle's notebook dated July 22nd 1954 (Figure 5), together with diagrams on the opposite page.

Figure 5.   Entry in Ryle's laboratory notebook, 22nd July 1954.

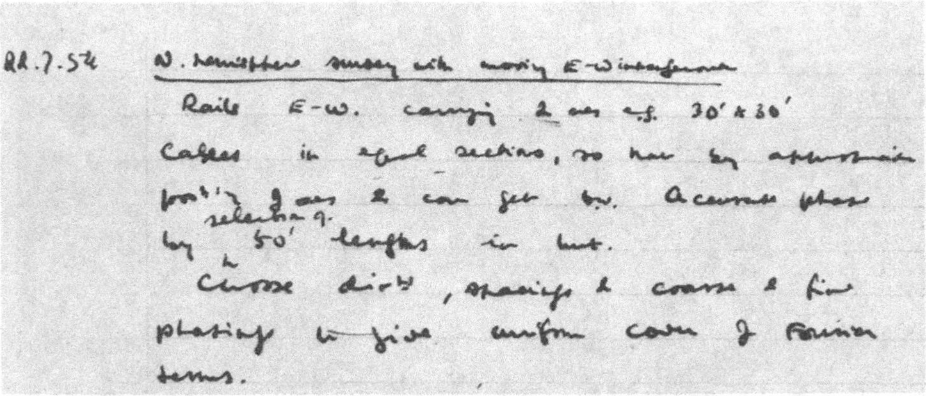

"N hemisphere survey with moving E-W interferometer.   Rails East-West carrying 2 aerials e.g. 30 ft x 30 ft.   Cables in equal sections, so that by appropriate pos[itionin]g of ae[rial]s and can get b.[and] w.[idth]. Accurate phase by selection of 50 ft lengths in hut.

Choose dir[ectio]ns, spacings and coarse and fine phasings to give uniform cover of Fourier terms."   Four sides and a month earlier there is a note: "22.6.54 High-brow computers.   Preliminary analysis done at time of recording:- e.g. clock driven sine pot[entiometer] fed from set to give appropriate no. of revs each minute (say) ...   Final analysis requires add[itio]n of all amplitudes at some time.

Ideally should fill up 100 x (24 x 60) integrators (e.g. P.O. counters) each giving intensity in a unit direction.

Some means for 2 dimensional array of stores seems necessary - Copper plate drum (or rows of wires)."

The references to band width and lengths of cable in hut concerns path compensation.   The total paths from source to receiver via the two elements of an interferometer must be kept nearly equal, to ensure that the phase difference between the two incoming signals is almost the same throughout the band of frequencies received.   Thus when observing a source away from meridian transit with an East-West interferometer, the

extra air path in front of one of the aerials must be compensated by
inserting extra cables in the path to the other.   The extra cables to be
inserted, as the source moves across the sky, must be in precise multiples
of a wavelength (or at any rate half-wavelengths), not only to preserve
the phase, but in order to ensure that the impedances seen by the receiver
are unchanged.   It was clear that the cable insertion should be done by
electronic switching, but even in the late 1950s that was not a routine
matter (cf. 15.12.1959 entry).

The devices half-seriously described as "high-brow computers"
could have been used merely to "unwind" the fringes - and thus reduce the
sampling rate required - but the wording strongly suggests that they were
intended to perform a Fourier synthesis.   With appropriate binning of the
outputs this is clearly possible, and many years later the software of some
two-dimensional mapping programs worked in just this way.   However, in
1954 the progress of digital computing could not be predicted with enough
confidence, hence these thoughts about analogue devices.   One way of
accumulating the final map that was considered at the time involved the
electrolytic deposition of copper - hence the "copper plate drum".

Many other types of telescope were considered, including
twisted open-wire transmission lines, a lunar occultation aerial, and a
horrendous cylindrical parabola mounted on circular rails (20.7.58).
Perhaps (or perhaps not) it was the latter that brought his thoughts back
to the biggest turntable of all.   The next relevant entry of any size is
not till 15.12.59:   "Ext. of idea proposed 22.7.54.   2 30'-40' paraboloids - rail E-W with polar mounts + ? electronic b.w. control (? maybe at
IF where stability better).   12 hour recording of 7.5° circle of sky."
A page of numerical estimates of signal:noise ratios, beamwidths, etc.
follows, and the entry ends: "Also be very suitable for simple study for
complexity in bright sources.
Find out abt. - 1.   Paraboloids + param. ampl.)   Hence opt.
          2.   Rails   )
                                            }   length of rail
)
          3.   ? Xtal controlled L.O. ...
          4.   Delay cables."
This entry indicates that the essential decision had already been made,
for it describes the broad features of what became the One-mile telescope,
and, as we shall see, the resolution to "find out about paraboloids" was
acted upon immediately after Christmas.   It was a big decision.   The cost
of the operation was an order of magnitude greater than any earlier

Cambridge radio telescope.   It also represented a leap upwards in frequency.   Nearly all Cambridge radio astronomy before 1960 was done at
frequencies $\lesssim$ 210 MHz, with fairly robust front ends that lived in stout
metal boxes out in the field, kept warm and dry only by their power
supplies.   The "opt[imum] length of rail" is largely a matter of
economics;   an array of given total length costs least when the total cost
of paraboloids equals the cost of the rails.

On the other hand, the idea of earth rotation synthesis, which
we called 'supersynthesis', had certainly been around for some time, and
required no explanation;   thus there is an entry in February or March 1958
that reads simply: "Uses of Steerable Wz.   [referring to the ex-World War
II German Würzburg dishes]   1. Pol^n - done in 1 day instead of 1 year.
2. Supersynthesis.   3. With moving ae. solar distribn. all axes etc.
4. Maser testing."

In the meantime a research student from Sweden, Jan Högbom,
had made an experimental test of 'supersynthesis'.   It was a very modest
test, using the two arrays of 4 half-wave dipoles that he had used in his
main project of locating solar bursts, and he detected just two sources,
M87 and the Crab nebula.   The last chapter of his thesis (August 1959)
also contains the clearest theoretical account of earth rotation synthesis
to that time, though it did not reach the full simplicity of Ryle's 1962
paper - perhaps because Högbom preferred to use a North-South baseline,
which complicates the geometry.   Högbom noted that earth rotation had been
used much earlier: "The periodicity of an interferometer recording is
often used to determine the declination of a source (Smith 1952), and it
was noted in the first Cambridge survey (Ryle, Smith & Elsmore 1950) that
it was possible to sort out some confused regions because sources at
different declinations had different periodicities.   McAdam (1958) treated
aerials as filters and relied essentially on the periodicity for his declination resolving power in a high declination survey.   This can now be
looked upon as the synthesis of extra aperture by the rotation of the
earth."   Högbom's work and thoughts seem to have been independent and
rather private.   Ryle was apparently scarcely aware of the experiment
and cannot recall it now;   conversely, Chapter 9 of Högbom's thesis reads
as though he had invented earth rotation synthesis all by himself.

The next trial of earth rotation synthesis was made by Ryle
and Ann Neville in May 1960 and June 1961, and is one of the most beautiful
experiments in radio astronomy.   By dividing the long fixed antenna and

the moving antenna of the 178 MHz "4C" interferometer into 32 and 4 short
units respectively, and adding just one small antenna placed between them,
they could get all spacings from 1 to 75 units, and thus map an 8° patch
of sky round the North celestial pole with a 4!5 beam.   This enabled them
to measure sources down to 250 mJy, eight times fainter than the limit of
the 4C survey.   Not until the advent of the Very Large Array (VLA) in
recent years has there been another such single step in sensitivity and
resolution.   The map (Figure 6) was exhibited at a Herstmonceux conference

Figure 6.   The first earth rotation synthesis map, a
North pole survey at 4!5 angular resolution made by
Ryle & Neville (1962) using the elements of the 4C aerial.

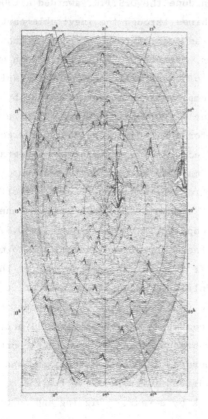

and some who were there reported that Prof. Oort had looked at it thought-
fully for a very long time.   Whether or not that was the moment of
decision, the Benelux Cross Project soon metamorphosed into plans for the
Westerbork Synthesis Radio Telescope.

It would be quite wrong to suppose that these tests led up to the design of the One-Mile telescope; they went on in parallel with it. The design of the One-Mile telescope must have begun in earnest at about the time of the December 1959 laboratory notebook entry, for on January 20th 1960 enquiries about the manufacture of 30 to 50 foot paraboloids, and equatorial mounts for them, went out to at least 8 firms. The correspondence files of that time (at the Cavendish Laboratory) also contain an itemised estimate of costs in Ryle's handwriting, adding up (with 10% contingencies) to £368,000, and this estimate went into an application to the D.S.I.R. in April 1960. In April there was a preliminary meeting with possible building contractors. In June the D.S.I.R. awarded a £3000 grant for a design study, and it shines through the inevitable caution of official letters that they wanted the project to go ahead. As the construction phase approached, however, everything had to go through government channels (e.g., "the Ministry of Works will have to be the sole channel for instructions to firms employed" in August 1960) and by the spring of 1961 there were increasingly frustrating delays. There was still no decision about the grant for the project, so no tenders could be issued, and the correspondence reflects Ryle's increasingly desperate state of mind about such practical matters as getting foundations in before the return of winter. The reason for the apparent dragging of official feet became clear in mid-July 1961 with a letter from D.S.I.R. announcing that the grant for the radio telescope would be delayed. The D.S.I.R. did not have enough money for all the major grants they wanted to make, so to be fair they froze them all: "It is with the greatest regret that I have to tell you this, realising full well that this means that it will, in all probability, not be possible for the outside work on your project to be started until the spring of next year. I hope you will accept my personal assurance that it is only because I can see no other course open to me." The grant was finally approved in December 1961, and was followed by congratulations from various influential well-wishers, but the project had clearly been pushed back by a year. Digging and concreting went on through 1962, and the later stages of construction were not helped by the exceptionally severe winter of 1962-3. Aerial erection was in progress in April 1963. The first trial survey at 408 MHz was made in

Figure 7.   The completed One-mile telescope.   The central
dish and the dish on rails are visible in this photograph;
the second fixed dish is behind the reader.   A small section
of the 38 MHz corner reflector (Williams et al. 1966) is also
visible.

April-June 1964 (Neville, Ph.D. thesis; it was not published) and the
first observations at 1407 MHz were made later in the same year: the
first real maps of Cyg A and Cas A (Ryle, Elsmore & Neville 1965). There
were many later improvements, but clearly it all worked   (Figs. 7 and 8).

Figure 8.   The first maps of Cygnus A (left) and Cassiopeia A,
made with the One-mile telescope (Ryle et al. 1965).   These
were 1.4 GHz observations, at 23" resolution.

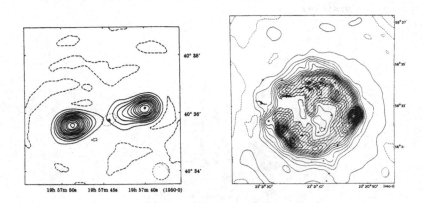

By the end of 1964, John Shakeshaft and John Baldwin had firm
plans for the 'half-mile telescope', principally for observing the 21 cm
hydrogen line.   Four small paraboloids were bought, essentially at scrap
value, and mounted on the rails of the One-mile telescope.   This imposed
inevitable restrictions on the scheduling of the Half-mile telescope, but
made it remarkably cheap to build.

By the autumn of 1966, Martin Ryle was already planning the
next major instrument, confident in the knowledge that the performance of
the One-mile telescope was close to theoretical.   It was to have a baseline nearly three miles long, to operate at higher frequencies, and have
more dishes so as to allow faster operation.   In November 1966 the layout
was not yet settled but by March 1967 it was fixed in its present form of
42 foot diameter dishes, 4 fixed and 4 on a rail track.   Superficially it
might appear to be just a bigger version of the One-mile with some improved
technology.   Certainly the 5 km telescope (Britain went metric) incorporated new techniques, and its construction involves some interesting bits of
history.   The closing of the Cambridge to Bletchley railway line made a
narrow East-West strip of land adjoining the Lords Bridge site available
and the timing of this event in relation to the timetable of government
spending on science was crucial and at times worrying.   There was also a
battle over the course of the M11 motorway, which produced much negotia-
tion, much anxiety, and even an undergraduate examination question:

Discuss the use of the Cornu spiral in solving diffraction problems.

Signals from a transmitter of 0.1 m wavelength situated at a height of 2 m above the ground are picked up by a receiver in a radio telescope 22 m above the ground at a distance of 1 km. It is desired to screen the receiver from these signals by erecting an absorbing screen 50 m from the transmitter. How high must it be if the received power is to be reduced by a factor of 100? (Natural Sciences Tripos Pt IB 1972, Adv. Physics (2)).

But the really fundamental change this time was in the astronomical aims. The One-mile was still designed with survey work in mind as one of its primary purposes. It was to extend the source counts to the faintest flux levels, without ambiguities of confusion or unknown angular sizes, and to provide precise positions for optical identifications. By 1966 we knew the essential answers: the source counts absolutely required an evolving universe, they were isotropic, and they converged strongly (in the sense that $d(\ln N)/d(\ln S) < 1.5$) below about 0.1 Jy. Anyone who was not convinced by then would not be convinced by more evidence of the same kind (and might not even be convinced by the discovery of the microwave background in the previous year). We now wanted to find out how the radio source population evolved, and for that we needed to understand much more about the inner workings of sources. Besides, we wanted to get to grips with radio source physics for its own sake, and the One-mile had shown enough of source structure to make this an exciting prospect. Thus the new instrument was designed to get the best possible angular resolution on individual sources. The ability to map patches of sky completely was therefore sacrificed to economy and high-frequency operation: the mobile antennas can only be set down and attached to cables at a limited number of fixed stations, for phase stability at short wavelengths forbids the use of long flexible cables above ground. Some sensitivity was also sacrificed to high-frequency operation, by buying dishes that are actually smaller than those of the One-mile, but have a solid surface.

There have, of course, been further technical developments in aperture synthesis: not only the construction of the VLA, the biggest instrument of all, but also new techniques of Very Long Baseline Interferometry, of the practical utilization of closure phase and closure amplitude, of various forms of hybrid mapping and maximum entropy analysis. But one has to stop the story somewhere, and I stop at this point because so far it is very much the story of one remarkable man, who not only provided the inspiration and the driving force, but actually designed most

of the bits and pieces, charmed or savaged official persons according to
their deserts, wielded shovels and sledgehammers, mended breakdowns, and
kept the rest of us on our toes.   Sir Martin Ryle retired in 1982, and I
dedicate this piece to him, confidently expecting that he will tell me
that it's all wrong, but also that he will not let my appalling errors
affect his friendship.

Figure 9.   Sir Martin Ryle.

REFERENCES

Blythe, J.H. (1957). A new type of pencil beam aerial for radio astronomy.
        Mon. Not. R. Astr. Soc., 117, 644.
Blythe, J.H. (1957). Results of a survey of galactic radiation at 38 Mc/s.
        Mon. Not. R. Astr. Soc., 117, 652.
Blythe, J.H. (1957). A Survey of galactic radio emission. Ph.D. Thesis,
        Cambridge Univ.
Bracewell, R.N. & Roberts, J.A. (1954). Aerial smoothing in radio
        astronomy. Austral. J. Physics, 7, 615.
Caswell, J.L., Crowther, J.H. & Holden, D.J. (1967). A survey of radio
        emission at 178 MHz. Mem. R. Astr. Soc., 72, 1.
Costain, C.H. & Smith, F.G. (1960). The radio telescope for 7.9 m wave-
        length at the Mullard Observatory. Mon. Not. R. Astr. Soc.,
        121, 405.

Crowther, J.H. & Clarke, R.W. (1966). A pencil-beam radio telescope operating at 178 Mc/s. Mon. Not. R. Astr. Soc., 132, 405.

Gower, J.F.R., Scott, P.F. & Wills, D. (1967). A survey of radio sources in the declination ranges -07° to 20° and 40° to 80° (4C). Mem. R. Astr. Soc., 71, 49.

Högbom, J. (1959). The structure and magnetic field of the solar corona. Ph.D. Thesis, Cambridge Univ.

McAdam, W.B. (1958). The observation of discrete radio sources. Ph.D. Thesis, Cambridge Univ.

Neville, A.C. (1964). A new method of synthesising large radiotelescopes. Ph.D. Thesis, Cambridge Univ.

O'Brien, P.A. (1953). The distribution of radiation across the solar disk at metre wavelengths. Mon. Not. R. Astr. Soc., 113, 597.

Pilkington, J.D.H. & Scott, P.F. (1965). A survey of radio stars between declinations 20° and 40°. Mem. R. Astr. Soc., 69, 183.

Ryle, M., Elsmore, B. & Neville, A.C. (1965). High resolution observations of the radio sources in Cygnus and Cassiopeia. Nature, 205, 1259.

Ryle, M., Smith, F.G. & Elsmore, B. (1950). A preliminary survey of the radio stars in the northern hemisphere. Mon. Not. R. Astr. Soc., 110, 508.

Ryle, M. & Neville, A.C. (1962). A radio survey of the North polar region with a 4.5 minute of arc pencil-beam system. Mon. Not. R. Astr. Soc., 125, 39.

Ryle, M. (1962). The new Cambridge radio telescope. Nature, 194, 517.

Scheuer, P.A.G. (1954). Radio emission from the Galaxy. Ph.D. Thesis, Cambridge Univ.

Scott, P.F., Ryle, M. & Hewish, A. (1961). First results of radio star observations using the method of aperture synthesis. Mon. Not. R. Astr. Soc., 122, 95.

Smith, F.G. (1952). The determination of the position of a radio star. Mon. Not. R. Astr. Soc., 112, 497.

Williams, P.J.S., Kenderdine, S. & Baldwin, J.E. (1966). A survey of radio sources and background radiation at 38 Mc/s. Mem. R. Astr. Soc., 70, 53.

Solar eclipse expedition of the U.S. Naval Research Laboratory to Khartoum, Sudan in 1952. With the 10 cm wavelength system are (l. to r.) Russell Sloanaker, C.F. White, and Fred Haddock (courtesy NRL)

SECTION FOUR: The Rest of the World

   While there can be little argument that England and Australia dominated most aspects of the fledgling radio discipline, noteworthy contributions also originated elsewhere before 1960. In this section we find accounts of early developments in several of these nations.[*] First, Alexander Salomonovich recounts the important work of several observational groups early active in the Soviet Union. But it was the Soviet *theorists*, specifically Vitaly Ginzburg and Iosef Shklovsky, who wielded the greater influence on radio astronomy as a whole -- over the years neither of them shirked from proposing radical ideas to match the radical observations. Their successes include explanations for the solar emission (hot corona) and the radiation from the galactic background and discrete sources (synchrotron mechanism), as well as predictions regarding spectral lines (CH) and variable radio intensity from discrete sources (Cassiopeia A).

   Also after the war, major groups in France, Canada, and Japan began studying the sun and throughout the 1950s maintained this specialty. The story of the radio sun and the men who perceived it from Paris, Ottawa, and Nagoya is laid out in the final three articles in this section.

----

[*] Notably missing here are contributions describing the early work done in the Netherlands and in the United States (see Preface).

At the base of the 22 meter reflector of the Lebedev Physical Insitute at Serpukhov, near Moscow (1960): (leftmost) A.E. Salomonovich, A.D. Kuz'min, and R.N. Bracewell (courtesy George Swenson)

THE FIRST STEPS OF SOVIET RADIO ASTRONOMY

A. E. Salomonovich
P. N. Lebedev Physical Institute, Academy of Sciences of the
USSR, Moscow

INTRODUCTION

In this volume we celebrate the fiftieth anniversary of the beginning of radio astronomy, when Karl Jansky observed cosmic radio emission for the first time. In the Soviet Union, radio astronomy began in 1946 with theoretical papers by V. L. Ginzburg and I. S. Shklovsky. Then in the following year Soviet scientists carried out their first observations of an extraterrestrial radio source, the Sun.

I must say some words about favourable conditions for the development of radio astronomy in the 1940s in our country. There existed a high level of theoretical physics, in particular of electrodynamics, combined with a deep interest in the problems of radio wave propagation, ionosphere and plasma physics, statistical physics, and radio engineering. In particular, investigations made by the scientific school of L. I. Mandel'shtam and N. D. Papaleksi paved the way.

Mandel'shtam and Papaleksi addressed the idea of radar measurements of the distance to the Moon at least twice -- in 1925 and in 1943. On the first occasion they had to admit the impossibility of such an experiment with existing radio equipment. Their estimates made in 1943, however, were more reassuring. (By the way, they also analyzed the possibilities of trying an *optical* reflection experiment; the advantage of monochromatic pulses was especially noted (Papaleksi 1946).) The first successful lunar radar experiments, however, were carried out in 1946 in the U.S.A. and in Hungary; the first Soviet radar astronomy was also in 1946, but used for the study of meteors (Levin 1946).

At the end of World War II Papaleksi carefully considered radar measurements not only of the Moon, but also of the Sun. In this connection he suggested to a young theoretician, V. L. Ginzburg, a researcher at the P. N. Lebedev Physical Institute (FIAN), that the problem of radio wave absorption and emission in the solar atmosphere should be considered. The

resultant study of Ginzburg's (1946) is discussed fully in his own contribution to this volume. Simultaneously and independently, a researcher at
the Sternberg Astronomical Institute of Moscow State University, I. S.
Shklovsky (who is of the same age as Ginzburg), published an article (1946)
where he showed in particular that thermal solar radiation in the meter
wavelength range could not be emitted by the photosphere and chromosphere,
but was due to the solar corona. This article also contains the theory of
solar bursts and the first call to experimenters to organize observations
of solar radio activity (see also Shklovsky (1982)).

Thus did it become necessary to test the hypothesis of the
coronal origin of the meter-wavelength radio emission of the Sun. The
method which was soon found consisted in the observation of solar radio
emission during an eclipse. The next suitable solar eclipse was that in
Brazil in 1947. Papaleksi, head of the Oscillations Laboratory of FIAN,
had earlier partaken in observations of the behaviour of the ionosphere
at the time of an eclipse. This time, as head of the large Soviet expedition, he also included in the program observations of solar radio emission.
He recognized that a sufficiently experienced specialist was required to
cope with the radio observations, and so B. M. Chikhachev, a post-graduate
of FIAN, was entrusted with the task. He was a student and collaborator of
Papaleksi, as well as one of the pioneers of the Soviet radio metal tube
industry. When preparation for the expedition was at its height, however,
Papaleksi suddenly died in February, 1947.

Professor S. E. Khaykin (Figure 1), one of Mandel'shtam's
pupils and a colleague at the same laboratory, then took personal control
of all radio observations on the expedition. The job of deputy chief was
taken by the well-known Arctic explorer G. A. Ushakov. Not only (optical)
astronomers, but future prominent radio astronomers such as Ginzburg and
Shklovsky also took part in the expedition (Figure 2). But at that time
Ginzburg was interested in the ionosphere and Shklovsky in *optical* observations during the eclipse.

After a complicated navigation across the Atlantic Ocean, the
motor ship "Griboyedov" reached the Bay of Baía. The optical astronomers
went inland to the most sunny place in Brazil, Araxá, while the radio
astronomers stayed aboard ship. On the day of the eclipse, it rained from
morning till evening in Araxá, but in the Bay of Baía it was absolutely
clear weather all day! The radio observations were carried out on the deck
at a wavelength of 1.5 m with a phased array of dipoles having its

Figure 1.   Professor S. E. Khaykin (1901-1968).

Figure 2. Participants in the expedition for the eclipse of
May 20, 1947. In the front row, the first from the right is
S. E. Khaykin and the fourth is G. A. Ushakov; in the middle
row, the sixth from the right is B. M. Chikhachev and the
eleventh is V. L. Ginzburg; in the back row, the second from
the right is I. S. Shklovsky.

electrical axis directed close to the zenith (Figure 3). But how could the
azimuthal tracking of the Sun be done at eclipse time? Khaykin found a
characteristically ingenious solution, namely use the ship itself as a
turning device. Thus, for the first time in the history of the Navy, the
crew slackened and tightened the chain cables at the command of a professor
of physics, and the anchored ship obediently slewed. (Of course, these
commands were actually given through the captain.) In order to obtain reliable records of the changes in the solar signal, several observers,
placed in different rooms, read indications on meters at set times determined by the ship's chronometer.

The observational results, published by Khaykin and Chikhachev
in 1947, became classical. While the optical eclipse had been total, the
radio eclipse was only annular (see Figure 4). The theoretically predicted
coronal nature of meter-wavelength solar radio emission was experiment-
ally verified. A quarter of a century later, this discovery was registered
in the USSR with a commendation, or "diploma", to Papaleksi, Khaykin and
Chihkachev.

While returning from Brazil, Khaykin worked out a comprehensive
plan to develop observational radio astronomy in our country. Note that
there were little technical means for such observations at that time. The
problem was to measure the intensity of cosmic radio noise at a level well
below the noise of the receivers themselves. Scientists had to develop
unusual receiving equipment and radio telescopes (although note that there
was not then such a term). Furthermore, they had to train personnel who
would possess a deep knowledge in both the fields of radiophysics and
astrophysics.

Khaykin performed great services for Soviet radio astronomy,
managing to draw into the field a wide circle of specialists. He connected
the first steps of radio astronomy with the solution of an important radiophysical problem, the investigation of conditions for radio wave propagation through the entire extent of the earth's atmosphere. This was done
using the methods of radio astronomy to determine atmospheric refraction,
scattering, and absorption, at wavelengths ranging from meters to centimeters, of waves from natural transmitters such as the Sun, the Moon, the
planets, and discrete sources.

The intensive development of radio astronomy in the USSR was
also aided by the existence of strong radiophysical schools outside of
Moscow. At Gorky, radio engineering and radiophysics had been successfully

Figure 3. Motor ship "Griboyedov" in the bay of Baiá (Brazil).
In 1947 Soviet radio scientists observed a total eclipse of
the Sun, using the phased-dipole-array antenna on the deck,
and first measured the large extent of the solar corona at
1.5 m wavelength.

Figure 4. The record of the radio eclipse (versus GMT) obtained at a wavelength of 1.5 m (1); the variation in the
solar optical disk area (2); the "eclipse curve" calculated
for hydrogen prominences and filaments by E. I. Mogilevsky
(3) (from Khaykin and Chikhachev 1947).

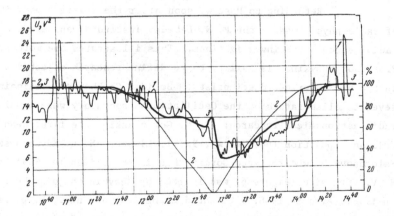

developed as far back as the 1920s. Such leaders at Gorky University as
A. A. Andronov, G. S. Gorelik, and M. T. Grehova in fact formed a "branch"
of the Mandel'shtam-Papaleksi school. As long ago as 1947 the Department
of Statistical Radiophysics of the Institute of Technical Physics of Gorky
University, headed by Gorelik, began experimental work on radiometric
equipment. Meanwhile, the Department of Radio Wave Propagation of the
radiophysical faculty started theoretical investigations into the mechanism of generation of the solar radio emission; this work was initiated and
headed by Ginzburg. In 1947-1948 in Gorelik's group, a young radiophysi-
cist, V. S. Troitsky, built their first chopping type radiometer, for
4 meter wavelength, and I. L. Bershtein developed a Dicke-type radiometer
for 10 cm wavelength. In 1948 Troitsky also developed a 1.5 meter radio
telescope and worked out the theory of chopper radiometers. This theory
formed the basis for the creation of centimeter- and meter-wavelength
radiometers, the latter being completely "state-of-the-art" for those days.
These investigations in Gorky were under the leadership of Gorelik and
Troitsky, but were at Khaykin's initiative and carried out in close conjunction with the research at FIAN.

A radio astronomy field station for Gorky University was established in 1949 near the village of Zimenki, on the higher bank of the
Volga river. Only a single radio telescope was initially installed there
(Figure 5), but after a while two more telescopes were set up. Besides
those mentioned above, others such as S. V. Zhevakin, the late G. G. Getmantsev, and A. G. Kislyakov also participated in the early work and
observations as radio astronomy began in Gorky.

Returning to Moscow, soon after the Brazil expedition a group
of radio-physicists in the P. N. Lebedev Institute was selected to conduct
radio astronomical investigations. This illustrates one more trait of
Professor Khaykin, his ability to re-define research objectives and to
enlist new experienced scientists. Besides Chikhachev, beginning in 1948
several collaborators of the Oscillations Laboratory were involved in the
radio astronomical research: B. N. Gorozhankin, the first head of the
Crimean expedition (see below), Ya. I. Likhter, N. L. Kaydanovsky, A. E.
Salomonovich (who replaced Gorozhankin in the Crimea in 1949), and others.
V. V. Vitkevich (Figure 6) was invited by Khaykin to join the FIAN group
in 1948. Also in 1948, a talented mechanical engineer, P. D. Kalachev,
started working at FIAN; over the subsequent years Soviet radio astronomy
greatly benefited from his scientific and engineering skill.

Figure 5. One of the first Gorky radio telescopes, installed at Zimenki on the bank of the Volga river.

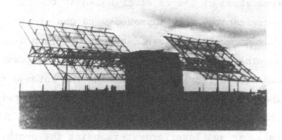

Figure 6.   Professor V. V. Vitkevich (1917-1972) in 1948.

The first FIAN radio astronomical station was organized on the basis of a Crimean expedition in 1948. The expedition used two observation posts on the southern shore of the Crimean peninsula, overlooking the Black Sea at Alupka and Alushta. The posts had been founded by Papaleksi after World War II for the purpose of measurements of distances by radio inter-ferometric methods. Following an agreement with the director of the Crimean Astrophysical Observatory, G. A. Shayn, Mt. Koshka, where the optical observatory was located, became a center for the new expedition. Installation of the first FIAN radio telescope was begun there in 1948 and

and completed in early 1949; the same phased array of dipoles used on ship-
board during the 1947 eclipse served as its basis. Within a comparatively
short period the FIAN group then built a number of large radio telescopes
of meter-, decimeter- and centimeter-wavelength ranges, all with receivers
allowing for compensation of the receiver noise by the chopping method. At
first these radio telescopes were improved versions of radar antennas. An
example of this was a "Würzburg-Riese" antenna of 7.5 m diameter captured
from the Germans. The mesh surface of the antenna was covered with aluminum sheets formed from a template, permitting the antenna to be used at
wavelengths as short as 10 cm. Later on, numerous radio telescopes of the
phased-dipole-array type and paraboloidal-section type were created. They
operated, as a rule, as sea interferometers using the nearby Black Sea.

One of the programs in the Crimea was the development of a
radiometer for 50 cm wavelength by Vitkevich. In the beginning of 1949,
he constructed an original capacitative switch and carried out a successful
test of a two-antenna radio interferometer at this wavelength. After the
test, Vitkevich and his group settled down in Alushta and began their
fruitful work in the field of radio astronomy. Many noted scientists of
today began their work at these Crimean stations: F. V. Bunkin, V. G.
Veselago, N. V. Karlov, B. D. Osipov, R. L. Sorochenko, and others.

In 1951-1952 the Lebedev Institute organized one further observational station, in the Kaluga region about 150 km southwest of Moscow.
This group, headed by Kaydanovsky, was concerned with radio observations
of the Moon, as well as the first successful Soviet observations of discrete sources in the centimeter range. Well-known scientists, such as
N. S. Kardashev and the late E. G. Mirzabekyan began their observational
work there (Kaydanovsky & Kardashev 1956; Kaydanovsky et al. 1956).

Gradually more and more interest was taken in the astrophysical
problems. Thus, in 1949 Chikhachev carried out a very interesting investigation of the active regions on the solar disk in the meter wavelength
range (Chikhachev 1950, 1956). In the same year Vitkevich began observations of radio emission of the quiet and slightly active Sun (Vitkevich
1956). In 1951 he suggested the idea of investigating the solar corona
using the radio emission of a background, discrete cosmic radio source -
the Crab Nebula (Taurus A) (Vitkevich 1951). He discovered by this method
the extended outer solar corona, the famous "supercorona". These observations were possible thanks to his development of the technique of radio
interferometric measurements. The discovery of the supercorona is a

remarkable achievement of Soviet radio astronomy; accordingly, Vitkevich was awarded a State Prize of the USSR for this work. This work entailed a series of investigations with his collaborators and pupils (V. A. Udal'tsov, B. N. Panovkin, N. A. Lotova, V. I. Shishov, R. D. Dagkesamansky, and others) which were carried out right up to his premature death in 1972. Examination with different interferometric base lines and different wavelengths in the meter range allowed the detection and study of outer coronal inhomogeneities, polarization characteristics, peculiarities of corpuscular solar radiation and, later on, scintillation, the solar wind, etc.

The first works of the Gorky radio astronomers (V. D. Krotikov, A. G. Kislyakov, N. M. Tseytlin, K. S. Stankevich, M. R. Zelinskaya, and others) headed by Troitsky date from the beginning of the 1950s. These were concentrated in the first place with investigations of the physical properties of the lunar surface layer and, later on, of the planetary surfaces. This work was done at a high theoretical and experimental level and covered the whole radio frequency region available to terrestrial observations, from millimeter to meter wavelengths.

This research is the pride of Soviet radio astronomy. The theory of thermal radio emission of the Moon developed by Troitsky (1954) has become classical, with all subsequent investigators referring to this theory. The Gorky radio astronomers outstripped the corresponding foreign investigations for several years. In particular, an unexpected phenomenon was discovered, viz., the mean brightness temperature of the moon (that part which does *not* depend on lunar phase) increases with wavelength. This was interpreted (Krotikov & Troitsky 1963) as evidence for the existence of a constant heat flow from the lunar interior, taken to be caused by radioactive processes. This discovery was recognized by the State with an official "registration under diploma".

Of course the modern means of space investigations have by now permitted us to take samples of lunar soil and return them to the Earth for laboratory analysis. Even whole laboratories have landed on Mars and Venus. One should remember, however, that terrestrial radio measurements of the Moon and planets gave important initial data for the design of such wonderful apparatus. Only a few people now remember the former dust model for the lunar surface. However, it was just these radio observations, and in the first place those by Troitsky, which allowed the invalidity of this model to be established.

Other important results were also obtained in Gorky in the
1950s. For instance in 1956 V. A. Razin (1956, 1958) discovered linear
polarization in the Galaxy's background radio emission. Later he was re-
warded with a diploma for this discovery.

At the beginning of the 1950s the experimental basis of Soviet
radio astronomy was considerably extended. Vitkevich, the head of the
Crimean expedition of FIAN since 1952, proposed to shift its center from
Mt. Koshka to nearby Golubov ("Blue") Bay, still on the Black Sea. There
Kalachev headed the design of a series of phased-dipole-array and paraboloidal antennas for use in interferometers of meter and decimeter wavelengths (Figure 7). New solar spectrographs and other equipment for radio
measurements of the Sun and Galaxy were also developed. New young collaborators such as Yu. I. Alekseyev, Yu. L. Kokurin, A. D. Kuz'min, L. I.
Matveyenko, and V. A. Udal'tsov participated in this research. In particular, a unique, sloping, parabolic mirror 31 m in diameter was built
(Figure 8). It consisted of a fixed, concrete surface coated with a metal
film, and a movable platform which permitted scanning of the receivers and
horns in the vertical plane near the focus. The first two-dimensional
radio images of the solar disk at 3 cm wavelength were obtained with this
telescope in 1957 (Vitkevich *et al*. 1958). Also in 1957, Kuz'min and
Udal'tsov (1959; also Udal'tsov 1962) managed to detect polarization of
the Crab Nebula emission at 10 cm wavelength. Polarization was detected at
first using a ferrite polarizer and later using a correlation method.
This discovery, as well as the discovery by M. A. Vashakidze and V. A.
Dombrovsky of the polarization of visible light from the Crab Nebula,
contributed a great deal to the proof that nonthermal synchrotron radiation came from space. With all this activity the Crimean expedition had
become by the end of the 1950s one of the largest radio astronomical
observatories in the world.

In 1952-1954 Khaykin, who considered that radio astronomy had
to develop in close practical contact with optical astronomy, transferred
his scientific activities to the Main Astronomical Observatory of the
USSR Academy of Sciences in Pulkovo, near Leningrad. In 1954 he created
there a new Department of Radio Astronomy. Until that time radio astronomical research in Leningrad was carried out only by A. P. Molchanov and
his graduates in the Department of Radiophysics of Leningrad University.
Already in 1948-1949 he had suggested using solar radio emission for
antenna calibration measurements and for navigation, and he became one of

Figure 7. A radio telescope of the Lebedev Institute with a truncated parabolic mirror 18 x 8 m^2, installed in Katsiveli, Crimea, in 1953. This telescope was used for solar and 21 cm spectral line observations.

Figure 8. The large fixed radio telescope operated by the Lebedev Institute in Katsiveli, Crimea. The mirror is 31 m in diameter. With this radio telescope was discovered in late 1957 the polarization of the Crab Nebula radio emission at decimeter wavelengths. In 1957 it also produced the first radio images of the Sun at a wavelength of 3 cm (Vitkevitch *et al.* 1958).

the first collaborators in the new department at Pulkovo. The main body
of the department was formed by local researchers already in Leningrad
(D. V. Korol'kov and V. N. Ikhsanova); by Kaydanovsky, N. F. Ryzhkov, and
T. M. Egorova, who had transferred there from FIAN together with Khaykin;
and by graduating students of the Sternberg Astronomical Institute of
Moscow University (Yu. N. Parysky and N. S. Soboleva). The department
quickly acquired other able young people such as G. B. Gel'freykh, A. F.
Dravskikh, and N. A. Esepkina.

The main purpose of the new department was the creation of a
large microwave radio telescope of a new type, a Variable Profile Antenna
(VPA), as proposed by Khaykin and Kaydanovsky. Such an antenna was developed in a very short time and put into operation in 1956 (Figure 9). The
main lobe width of the VPA in one coordinate was excellent, eventually
being reduced, at a wavelength of 8 mm, to 15 seconds of arc. Many first-
class investigations were carried out with the VPA, e.g., precise measurements of coordinates of a number of discrete sources, investigations of
their structure and polarization, detailed study of the spectra of the Sun
and of planetary atmospheres, etc. The honor of the discovery (in 1955)
of polarized radiation from local sources connected with sunspots, as well
as of bursts in the centimeter wavelength range, also belongs to the radio
astronomers of Pulkovo (Khaykin 1966).

Shortly before his move to Leningrad, Khaykin turned the attention of the radio astronomers at FIAN to the promising centimeter and
millimeter wavelengths. There were considerable difficulties in mastering
these ranges, for example, low sensitivity of receivers and the lack of
large antennas with the necessary surface accuracy. Over the period
1953–1958 the FIAN group developed a number of centimeter- and millimeter-
wavelength range radiometers using non-mechanical ferrite modulators and
polarizers, while also improving the theory of the threshold sensitivity of
such radiometers. In 1954 the first radio telescope for 8 mm wavelength,
2 m in diameter, was built. In 1959 a 22 m radio telescope, RT-22,
developed under the guidance of Salomonovich and Kalachev and with the
participation of various industrial organizations, came into operation
(Figure 10). The width of the main lobe of its antenna pattern, 2 x 2
minutes of arc at a wavelength of 8 mm, was at that time the best in the
world for steerable instruments.

Later on, this radio telescope was equipped with updated millimeter and centimeter-wavelength receivers such as masers and parametric

Figure 9. The "VPA" Pulkovo microwave radio telescope, a
strip reflector of variable profile built in 1956. It consists
of 90 tiltable plates, each 1.5 x 3 m.

Figure 10. The 22 meter RT-22 radio telescope of the Lebedev
Institute at Pushchino, near Serpukhov, in operation since
1959 at millimeter and centimeter wavelengths.

amplifiers which were used for the first time there. Since 1959 the
scientists of FIAN have been continuously observing the Sun, the Moon and
discrete sources with the RT-22 radio telescope. They have obtained milli-
meter-wave images of these sources (Salomonovich 1962), compiled the first
Soviet catalogues of discrete sources (Kuz'min 1962a,b), obtained the
radiation spectrum of Venus, and completed a number of other investigations.

          It became necessary at this time to increase sharply the angu-
lar resolution of the radio observations in the meter-wavelength range.
In order to create a broad-band telescope with a flat gain over its entire
usable wavelength range, Vitkevich and Chikhachev advanced in 1955-1956 the
idea of creating a giant (1 km x 1 km) radio telescope of the Mills Cross
type, the DKR-1000 telescope. The important differences from the Mills
Cross itself were the steerability of the electrical axis and also the
rather wide (1:4) range of operating frequency.

          In connection with the creation of the RT-22 and DKR-1000
radio telescopes, the question of founding a new FIAN radio astronomical
station also arose. In 1956 the place for this was selected, in Pushchino
on the river Oka, near the city of Serpukhov. At the beginning of the
1960s the FIAN laboratory of radio astronomy headed by Vitkevich separated
from the Oscillations Laboratory and moved to Pushchino.

          In 1956, after the creation of the Radiophysical Research In-
stitute, radio astronomy in Gorky was also further developing. Besides
the investigations of the Moon already mentioned, the department headed by
Troitsky carried out measurements of the absolute intensities of the
discrete sources, and of the Galaxy background polarization. Meanwhile
another department, headed by Getmantsev, began the first radio astronomical measurements using artificial Earth satellites (analogous investigations were simultaneously begun by V. I. Slysh in Moscow at the Sternberg
Astronomical Institute (Getmantsev *et al.* 1969)). Another department,
headed by M. M. Kobrin, was engaged in systematic observations of radio
frequency radiation of the Sun over a wide wavelength range.

          A number of other scientific institutions also began radio
astronomical research in the 1950s. For example, in 1951 at the Byurakan
Astrophysical Observatory of the Armenian Academy of Sciences V. A. Sanam-
yan, G. M. Tovmasyan and others began, on V. A. Ambartsumyan's initiative,
construction of several radio interferometers of meter and decimeter wavelengths. These were for observations of the Sun and bright discrete
sources at the Byurakan Observatory.

In 1954 the Crimean Astrophysical Observatory commenced observation of solar radio emission. A series of radio spectrographs and radio
telescopes were built and operated in the department of radio astronomy,
headed by I. G. Moiseyev. The creation of a second, improved version of
the RT-22 radio telescope, located at Simeiz in the Crimea, should be
credited to A. B. Severny and Moiseyev. Kalachev designed a better steer-
ing device, improved the reflector accuracy, and modernized the control
system for this new model. The second RT-22 radio telescope was finished
in 1966 and has since carried out numerous interesting investigations at
very short wavelengths.

A department headed by E. I. Mogilevsky in the Institute of
Terrestrial Magnetism, Ionosphere and Radio Wave Propagation (Moscow) also
joined in the 1950s in the development of solar radio astronomy. This was
one of the few laboratories in the country that carried out systematic and
laborious, but very necessary, measurements of solar activity.

At the end of the 1950s one more large radio astronomical center was founded, at Kharkhov in the Institute of Radio Engineering and
Electronics of the Academy of Sciences of the Ukraine. The head of the
center, S. Ya. Braude, began a well-thought-out "attack" upon the decameter
wavelength region -- the longest waves available to terrestrial observations. In 1960-62 his group began developing and building the decameter
interferometers ID-1 and ID-2; later on, the T-shaped radio telescopes,
UTR-1 and UTR-2 (1800 x 900 m), were built. The UTR-2 radio telescope,
with its computerized pencil beam, has given unique information on the
spectra of discrete sources in the range of 10 to 25 MHz.

Also in the 1950s a large solar radio telescope was developed
and partly built in East Siberia, in the Siberian Institute of Terrestrial
Magnetism, Ionosphere and Radio Wave Propagation, near Irkutsk, due to
the energy of G. Ya. Smol'kov and his collaborators. One of the most
notable authorities in antenna theory, A. A. Pistol'kors, took an active
part in the creation of this instrument.

Talking about the first steps of radio astronomy in the Soviet
Union, I cannot but mention its most romantic branch, radio observations
of the solar eclipses. Having begun with the eclipse of 1947 in Brazil,
Soviet radio astronomers have continued to observe almost all total and
annular eclipses. Following the whims of the motions of the celestial
bodies, they moved to Ashkhabad, where they trembled with cold in February,
1952, and then to Novomoskovsk on the Dnepr river in the Ukraine, where

they were tormented by heat and mosquitoes in the summer of 1954. Over
the years eclipses were observed not only on the territory of the Soviet
Union, but also in Africa (Mali), on the island of Hainan and in the Cook
Islands, and in Mexico and Cuba. At first all of these observations were
organized by Molchanov.

The beginning of the 1960s was also marked by rapid development in Soviet radar astronomy. Although the first radar observations of
meteors go back to 1946, for some time this type of research was at a
standstill, excluding occasional work by Kobrin in 1954-1957. His work
was devoted to radar observations of the Moon at cm wavelengths by a phase
method (1956). Then, thanks to the tenacious efforts of the group headed
by V. A. Kotel'nikov in the Institute of Radio Engineering and Electronics
of the USSR Academy of Sciences, Moscow, the capabilities of new resources
connected with space exploration were brilliantly exploited (Figure 11).

These scientists obtained first-class results on radar measurements of the distances to Venus (1961), to Mercury (1962), and then to
Mars and Jupiter. In particular, they accurately determined the astronomical unit, solved the complicated problem of measuring the velocity and
direction of the rotation of Venus, and obtained important data on the
nature of the planetary surfaces (Kotel'nikov *et al*. 1964). For this work
Kotel'nikov and his collaborators were awarded the Lenin Prize in 1964.

A weak point in the initial stages of development of Soviet
radio astronomy was, probably as in other countries, insufficient relations
between the radio astronomers and the optical astronomers and astrophysicists. Only a few astrophysicists managed to see the radio astronomical
techniques in proper perspective and therefore were motivated to discuss
and provide problems relevant to radio astronomy. On the other hand, as
often happens in such matters, the radiophysicists took a great interest
in the new technique, but were little interested in the participation of
"ordinary" astronomers in their work. Gradually these relations grew
stronger, however, and a new generation of radio astronomers, well prepared both in radiophysical and astrophysical aspects, came into being.
Such astrophysicists of this new type were trained in Shklovsky's and
S. B. Pikel'ner's department in the Sternberg Astronomical Institute in
Moscow.

I can only briefly touch upon the dramatic and striking history of the development of theoretical radio astronomy in our country. As
a rule, the theoretical understanding of radio astronomical data, as well

Figure 11. The array of eight 16 m dishes in the Crimea with
which the first radar signals reflected from Venus and other
planets were obtained at a wavelength of about 40 cm by
Academician V. A. Kotel'nikov and his collaborators.

Figure 12. B. M. Chikhachev (1910-1971) (left) and M. Ryle at
the I.A.U. General Assembly in Dublin (1955).

as the development of astrophysical and later of cosmological problems, outstripped the process of creating new radio telescopes and radiometers. Interesting information on these topics may be found in the article in this volume by Ginzburg, as well as in another recent article by Shklovsky (1982).

The history of the development of spectral-line radio astronomy in the Soviet Union illustrates the process of establishing relations between radiophysicists and astrophysicists. As far back as 1947, Shklovsky became acquainted with a brief report of a communication given in 1944 by the young Dutch astrophysicist, H. C. Van de Hulst. The latter pointed out the possibility of observing the spectral line of interstellar neutral hydrogen at a wavelength of 21 cm. Shklovsky (1949) made all the necessary calculations and came to the conclusion that such an experiment could be set up with then-available equipment. He appreciated the importance of the experiment and suggested to the radiophysicists that they build a spectral-line radiometer and make these measurements. His suggestion, however, for some reason did not meet with a response from the experimentalists. As is generally known, the first observations of monochromatic cosmic radiation were carried out in 1951 in the U.S.A., Holland, and Australia. Only in 1952 did Chikhachev (Figure 12) and Sorochenko conduct similar experiments, obtaining in 1955 their first results (1956). These were continued by Sorochenko and his collaborators in Puschino, and by Ryzhkov, Egorova, I. V. Gosachinsky, and others at Pulkovo.

On the other hand, monochromatic radio frequency radiation from the recombination lines of interstellar ionized hydrogen was predicted by Kardashev (1959) in spite of earlier evaluations made by such an authority as J. P. Wild. Later on, such lines were discovered in the Soviet Union through the efforts of Sorochenko and E. V. Borodzich of the P. N. Lebedev Institute (1964), and also of A. F. Dravskikh, Z. V. Dravskikh, and V. A. Kolbasov (1964) from Pulkovo. They observed the recombination lines in emission from the Orion Nebula at wavelengths of 3 and 5 cm. The observations were carried out in close contact with Kardashev. All of the above mentioned scientists were recognized with a commendation for their discovery.

Overall, the first steps of Soviet radio astronomy, as described above, provided a solid foundation for the many areas of modern progress in the field.

REFERENCES

Chikhachev, B. M. (1950). The investigation of local sources of the solar radio emission. Dissertation for Cand. of Sci., P. N. Lebedev Physical Institute, Moscow.

Chikhachev, B. M. (1956). On sunspot radio emission. *In* Proceedings of the Fifth Conference on Cosmogony: Radio Astronomy, p. 246, Moscow: Acad. Sci. of USSR.

Chikhachev, B. M. & Sorochenko, R. L. (1956). A communication on preliminary observations of radio emission of hydrogen at the wavelength 21 cm. Radiotechnika i Electronika, $\underline{1}$, 886.

Dravskikh, A. F., Dravskikh, Z. V. & Kolbasov, V. A. (1964). Detection of radio emission in a line of excited hydrogen. Paper delivered to 12th General Assembly of IAU, Hamburg.

Getmantsev, G. G., Salomonovich, A. E. & Slysh, V. I. (1969). Extraterrestrial radio astronomical investigations. Vestnik Akad. Nauk SSSR, No. 4, 55.

Ginzburg, V. L. (1946). On solar radiation in the radio spectrum. Dokl. Akad. Nauk SSSR, $\underline{52}$, 491 = C. R. (Doklady) Acad. Sci. USSR, $\underline{52}$, 487.

Kardashev, N. S. (1959). On the possibility of detection of allowed atomic hydrogen lines in the radio range. Astron. Zh., $\underline{36}$, 838.

Kaydanovsky, N. L., Mirzabekyan, E. G. & Khaykin, S. E. (1956). The polarization radiometer at wavelength of 3.2 cm and its applications. *In* Proceedings of the Fifth Conference on Cosmogony: Radio Astronomy, p. 113, Moscow: Acad. Sci. of USSR.

Kaydanovsky, N. L. & Kardashev, N. S. (1956). The results of observations of radio emission of discrete cosmic sources at a wavelength of 3.2 cm. *In* Proceedings of the Fifth Conference on Cosmogony: Radio Astronomy, p. 436, Moscow: Acad. Sci. of USSR.

Khaykin, S. E. & Chikhachev, B. M. (1947). An investigation of solar radio emission during the solar eclipse on May 20, 1947 by the Brazil expedition of the Academy of Sciences of the USSR. Dokl. Akad. Nauk SSSR, $\underline{58}$, 1923.

Khaykin, S. E. (1966). Radio astronomical investigations at Pulkovo. *In* Pulkovskoy Observatory 125 Lyet, p. 57, Moscow: Nauka.

Kobrin, M. M. (1956). On the phase radar method for Earth-Moon distance measurements. *In* Proceedings of the Fifth Conference on Cosmogony: Radio Astronomy, p. 146, Moscow: Acad. Sci. of USSR.

Kotel'nikov, V. A., Dubrovin, V. M., Kuznetsov, B. I., Petrov, G. M., Rzhiga, O. N. & Shakhovskoy, A. M. (1964). Successes with radar investigations of the planets. Priroda, No. 9, 2.

Krotikov, V. D. & Troitsky, V. S. (1963). On the heat flow from the Moon interior. Izvest. Vuzov Radiofiz., $\underline{6}$, 333.

Kuz'min, A. D. & Udal'tsov, V. A. (1959). An investigation of the polarization of radio emission of the Crab nebula at 10 cm wavelength. Astron. Zh., $\underline{36}$, 33 = Sov. Astron.-A.J., $\underline{3}$, 39.

Kuz'min, A. D. (1962). Results of investigations of some discrete sources at a wavelength of 9.6 cm. Trudy Lebedev Inst., $\underline{17}$, 84; Spectra of discrete sources of radio emission observed with the 22 meter radio telescope of FIAN. Astron. Zh., $\underline{39}$, 22 = Sov. Astron.-A.J., $\underline{6}$, 15.

Levin, B. Yu. (1947). The flux of the Draconids in 1946. Astron. Calendar, Izd. of Gorly region, p. 121.

Papaleksi, N. D. (1946). On the measurement of the Earth-Moon distance by electromagnetic waves. Uspekhi Fiz. Nauk, $\underline{29}$, 250.

Razin, V. A. (1956). Preliminary results of the polarization of cosmic
        radio emission measurements at a wavelength of 1.45 m. Radio-
        technika i Elektronika, 1, 846.
Razin, V. A. (1958). The polarization of cosmic radio emission at wave-
        lengths of 1.45 and 3.3 m. Astron. Zh., 35, 241 = Sov.
        Astron.-A.J., 2, 216.
Salomonovich, A. E. (1962). Some results of investigations carried out
        with the radio telescope RT-22. Trudy Lebedev Inst., 17, 42.
Shklovsky, I. S. (1946). On the emission of radio waves by the Galaxy and
        by upper layers of the solar atmosphere. Astron. Zh., 23,
        333; The current status of the problem of the nature of the so-
        lar corona. Uspekhi Fiz. Nauk, 30, 63.
Shklovsky, I. S. (1949). Monochromatic radio emission from the Galaxy and
        the possibility of its observation. Astron. Zh., 26, 10.
Shklovsky, I. S. (1982). On the history of the development of radio astron-
        omy in the USSR. News on Life, Science, and Technology,
        No. 11, Moscow: Izd. Znanie.
Sorochenko, R. L. & Borodzich, E. V. (1964). The detection of radio emis-
        sion from NGC 6618 (Omega Nebula) in a line of excited hydro-
        gen. Paper delivered (by V. V. Vitkevich) to 12th General
        Assembly of IAU, Hamburg.
Troitsky, V. S. (1954). On the theory of the Moon radio emission. Astron.
        Zh., 31, 511.
Udal'tsov, V. A. (1962). On the polarization of radio emission from the
        Crab nebula at a wavelength of 21 cm. Astron. Zh., 39, 849 =
        Sov. Astron.-A.J., 6, 665.
Vitkevich, V. V. (1951). A new method for investigation of the solar cor-
        ona. Doklady Akad. Nauk SSSR, 77, 585.
Vitkevich, V. V. (1956). On radio emission of the quiet and faintly dis-
        turbed Sun. In Proceedings of the Fifth Conference on Cosmog-
        ony: Radio Astronomy, p. 149, Moscow: Acad. Sci. of USSR.
Vitkevich, V. V., Kuz'min, A. D., Salomonovich, A. E. & Udal'tsov, V. A.
        (1958). A radio image of the Sun at a wavelength of 3.2 cm.
        Doklady Akad. Nauk SSSR, 118, 1091 = Sov. Phys.-Doklady, 3, 12.

REMARKS ON MY WORK IN RADIO ASTRONOMY

V. L. Ginzburg
P. N. Lebedev Physical Institute, Academy of Sciences of the
USSR, Moscow

INTRODUCTION

An attempt to elucidate the history of the development of
radio astronomy in different aspects and using different sources seems to
be quite relevant and useful. But unfortunately, in spite of sincere
aspiration, I have not been able to write a paper which would correspond
to the requirements of the editor of the present book. The main reason
is that I am evidently not an astronomer, but a physicist both by education and by the experience of many years. Astronomy became for me a
"part-time job". While in my personal life it happened rather accidentally, from a more general point of view it happened on the wave of a truly
great process of transformation of astronomy from optical to "all-wave".
In the course of this process many physicists, radio engineers and other
professionals took interest in new astronomical methods and problems and
"came" to astronomy (often irritating professional astronomers by their
poor knowledge of classical astronomy and even of proper terminology).
Some of the neophytes became real astronomers, while others did not give
up their previous specialities and spent only part of their time in
astronomy. It is actually not even a question of *time*, but rather the
style of the work. A physicist, say, may devote all his efforts to
astrophysics (which is now rather difficult to separate from physics
proper), but if he insufficiently commands purely astronomical material,
he nevertheless does not become a real astronomer, who must know well the
astronomical classics and literature, observational methods and results,
etc. In a word, I am not a professional astronomer and therefore my
astronomical works are somewhat fragmentary and episodic, except, possibly, those connected with cosmic ray astrophysics.

In such a situation I can here only outline and briefly comment on my radio astronomical papers in the hope that it will be useful for
the history of radio astronomy.

EARLY SOLAR WORK

When war with Germany broke out on June 22, 1941, I made up my mind to devote myself to something which would be at least potentially useful for defense; before then, I had been mainly engaged in the theory of elementary particles, or as it now would be called, high-energy physics. Taking rather accidental advice, I turned to radio-wave propagation in the ionosphere. I shall not dwell here upon this activity, which lasted many years and was eventually reflected in a monograph (Ginzburg 1967). But it was in fact the ionosphere which was the starting point for my passing over to solar and then to extra-solar radio astronomy.

Well-known Soviet physicists and radio specialists L. I. Mandel'shtam and N. D. Papaleksi contemplated the problem of Moon radio-location [radar] long before the war. In 1944, stimulated by progress in radio-location (for more details see Ginzburg (1948) and Migulin *et al.* (1981)), Papaleksi turned back to this idea, also considering radio-location of the planets and the Sun. In this connection, at the end of 1945 or beginning of 1946, he asked me to clarify the conditions of radio-wave reflection from the Sun. As a matter of fact, it was a typical ionosphere problem and all the formulae were at hand. The results of the calculations did not seem very optimistic since for a large set of parameters such as electron concentration and temperature in the corona and chromosphere, which then remained unknown in many respects, radio waves would be strongly absorbed in the corona or chromosphere and not even reach the level of reflection. (The question of reflection due to inhomogeneities was not considered; the "point" $n(\omega) = \sqrt{1-\omega_p^2/\omega^2} = 0$ was, roughly speaking, taken for the level of reflection.) But from this a more interesting conclusion immediately followed: the source of solar radio emission must not be the photosphere, but rather the chromosphere and, for longer waves, the corona. At that time the corona was already assumed to be heated to hundreds of thousands or even to a million degrees. Thus, even under equilibrium conditions, i.e., in the absence of perturbations and sporadic processes, the temperature of emission from the corona at waves longer than about a meter must reach approximately a million degrees in spite of the photospheric temperature of only T ∿ 6000 K. This is what was presented in my first astronomical paper (1946). In that same year D. F. Martyn and I. S. Shklovsky published similar conclusions. My paper was

submitted for publication on March 27, 1946, while the dates of submission
of the other papers are not indicated; they appeared respectively in
Nature for November 2, 1946, and in *Astronomicheskii Zhurnal* for November-
December, 1946. In the calculations I used absolutely clear and reliable
formulae known from the theory of ionospheric propagation (Ginzburg 1949,
1967). Martyn did not present formulae, but he evidently acted in the
same way. Shklovsky believed that one should separately take into account,
and then sum up, the absorption connected with "free-free" transitions as
well as the absorption due to electron-proton collisions. As a matter of
fact, these are one and the same mechanism (Ginzburg 1967), and therefore Shklovsky's quantitative results were incorrect. But this circumstance was of no importance since the parameters of the corona were not
exactly known at that time.

The existence of thermal radiation from the corona with
$T \sim 10^6$ K was confirmed in a paper by Pawsey (1946), published immediately
after Martyn's paper, in which it was shown that such radiation plays the
role of a lower limit to the received signal, reached under the conditions
when the sporadic component of solar radio emission becomes sufficiently
weak.

AN ECLIPSE EXPEDITION IN 1947

A weak point in radio astronomy during that period was its low
angular resolution, preventing investigation of the Sun even for regions
with sizes of minutes of arc -- it is difficult to believe this today when
radio interferometers are far ahead of the best optical telescopes in
angular resolution power. In this connection Papaleksi suggested measurement of solar radio emission during the total eclipse of May 20, 1947,
with the help of a 1.5-m wavelength antenna which was installed on board
a ship and had a wide directivity pattern (several degrees). These measurements were successful (Ginzburg 1948a,b; Khaykin and Chikhachev
1947, 1948; Migulin *et al.* 1981) and turned out to be the first of their
kind. Although the intensity of optical emission from the Sun during a
total eclipse diminishes by several orders of magnitudes, at 1.5-m wavelength the intensity during the eclipse diminished no more than by 60%.
Thus meter-wave radio emission was proved to come from the corona, which
remains uncovered by the Moon even under a total optical eclipse. They
also succeeded in observing some details concerning the distribution of

active radio-emitting regions on the solar disk.

I took part in this Brazil expedition of the U.S.S.R. Academy of Sciences, on board the ship "Griboyedov". It seems that I was included on the staff of the expedition as recognition for my early work in the development of Soviet radio astronomy. I did not participate, however, in the measurements of solar radio emission, which were carried out on board the ship. Rather I was on the main part of the expedition which made its way inside Brazil to perform optical measurements, which unfortunately failed because of bad weather. This main part of the expedition included also a small ionospheric group, headed by Ya. L. Al'pert, of which I was a part; weather could not of course prevent ionospheric measurements. (See Salomonovich's article for more about this expedition.)

As a result of the expedition and related activities, I became for a while almost a professional radio astronomer. I tried to get acquainted with all the available material, methods of measurements, etc. As a result I wrote two reviews of radio astronomy (1947, 1948b), amongst the first in the world literature.[1] Now that 35 years have passed it is difficult to judge the value of these papers and I do not want to analyze them in detail. I shall only mention the proposal made in the 1948 review to use radio-wave diffraction on the Moon edge in order to improve angular resolution of details on the Sun during eclipses. This question was further considered in more detail in my 1950 paper with Getmantsev where we were already thinking more of discrete sources of cosmic radio emission, rather than of the Sun. The method of radio-wave diffraction on the Moon edge has since been widely used, and therefore I wish only to add here that in these papers we discussed the possibility to enhance the angular resolution still more if the source under observation happens to be located on the line connecting the Moon center with the point of observation; here we are evidently dealing with an analogue to the Arago-Poisson light spot.[2] Such observations are of course strongly hampered by the non-spherical shape of the Moon and the necessity to have a source on or very near to the above-mentioned line, and I do not know of any attempts to exploit this method. But perhaps one should nevertheless analyze in more detail such a possibility? This could work not only for the Moon, but also planets and their satellites, as well as artificial screens (both flat and spherical).

MY MAIN AREAS OF RESEARCH RELATED TO RADIO ASTRONOMY

If I dwelt on the entirety of my subsequent papers in so much detail, it would take too much space. As has already been mentioned, I worked in the field of astrophysics in a rather sporadic and chaotic way; that portion closer to radio astronomy can perhaps be divided into three main trends:

a) Twinkling of radio sources in and beyond the ionosphere, oscillations of the intensity of solar radio emission, the use of polarization measurements, and earth satellite measurements (Ginzburg 1952; 1956a, 1956b, 1960; Gershman and Ginzburg 1955; Ginzburg and Pisareva 1956, 1963; Benediktov *et al.* 1962).

b) The theory of sporadic radio emission from the Sun. V. V. Zheleznyakov and I engaged ourselves in this range of questions beginning in 1958. A number of other papers followed, but it would be out of place to refer to them here since the corresponding results (with references) are fully given in the 1964 book *Radio Emission of the Sun and Planets* by Zheleznyakov.

c) The theory of cosmic synchrotron radio emission and its connection with the problem of the origin of cosmic rays and with high-energy astrophysics as a whole. This was the subject of my main astrophysical interest and I still work in this field, although less than before. From the point of view of historical information which I can offer, these problems also play the most important role.

Radio galaxies

The origin of radio galaxies and, specifically, the question of the energy source of their radio emission was not immediately clear. For instance one can mention the prevalent hypothesis of colliding galaxies, which as a rule proved incapable of explaining the mechanism of energy output in radio galaxies. Another idea suggested was a sharp increase in the number of supernova flares in radio galaxies, an idea which has always been rather groundless. Therefore in my opinion a 1961 note of mine was not without value since it stated that the required energy output and cosmic-ray acceleration in radio galaxies is in principle easily provided by gravitational energy, in particular in the process of star formation. It was pointed out that "it seems more attractive to associate the galactic flares not with supernova flares, but with another large-scale mechanism, for example, gravitational instability of a galaxy or of its central part, as schematically outlined above." This trend of thought continued to a certain extent in later papers devoted to quasars (see Ginzburg and Ozernoy (1977) and the literature cited therein, as well as Ginzburg (1964)).

Pulsars

The later discovery of pulsars gave rise to the temptation to clarify the mechanism of their radio emission. Zheleznyakov and I (and partially V. V. Zaytsev) published several articles on this subject, the last being a 1975 review. But the problem proved to be much more complicated than it at first seemed (it happens sometimes that a problem turns out to be particularly sophisticated, and that this is not suspected in advance). For this reason I decided long ago to leave this field and, from a recent review (Michel 1982), I see that I was right. I think that only representatives of the younger generation (or even generations) will be able to gain a real understanding of this very interesting but complicated and many-sided range of questions, including the mechanisms of pulsar radiation and the theory of pulsar magnetospheres.

Synchrotron theory

Now I shall turn finally to the synchrotron theory of cosmic radiation, as well as cosmic ray astrophysics (often traditionally called the problem of the origin of cosmic rays). Fortunately (at least for me) I can restrict myself mainly to brief remarks, and here refer the reader to my contribution in *Early Cosmic Ray Studies* (1982), where I have already touched upon the history of cosmic ray astrophysics.

Approximately in 1947-1949 (as much as I remember) it became quite clear that comparatively long-wave, non-solar cosmic radio emission (including, in particular, the very first radio astronomical results, namely, the measurements by Jansky carried out on a wavelength of about 15 m) possesses an effective temperature T_{eff} exceeding 10^4 K. Therefore it was impossible to interpret such radio emission as thermal radiation from interstellar gas since this gas, generally speaking, has a temperature $T \lesssim 10^4$ K, and in any case radiation with $T_{eff} \gg 10^4$ cannot be explained by thermal radiation of gas. Thus, one had to assume the existence of some non-thermal source analogous, for example, to the

sporadic sources of non-thermal solar radio emission. This was how the
"radio-star hypothesis" arose in a quite natural way. According to this
idea, some stars were anomalously powerful radio sources responsible for
the non-thermal, cosmic radio emission with its continuous spectrum and
diffuse directional distribution (Ryle 1949; Unsöld 1949; Shklovsky
1951). The radio star hypothesis, however, faced a lot of difficulties,
mainly involving assumptions, sometimes arbitrary and unrealistic, concerning the hypothetical radio stars. Besides, an alternative soon
appeared and with time became stronger and stronger -- the synchrotron
hypothesis of the origin of non-thermal radio emission, which in the end
proved to be valid.

Competition or struggle between these two hypotheses took
several years. From the physical point of view synchrotron radiation had
been known and quite clear for many years (Schott 1912), and in the
1940's it was especially widely discussed in the literature of physics in
connection with the analysis of synchrotrons (e.g., see references cited
in Ginzburg (1982)). But it was only in 1950 that the first papers
appeared in which the synchrotron mechanism was considered as related to
cosmic radio emission; Alfvén and Herlofson (1950) discussed radiation
from radio stars and Kiepenheuer (1950) that from interstellar space. It
seems rather strange to me that these papers appeared in a journal of
physics and, besides, only in the form of short letters. In any case, as
far as I know, they did not attract attention from astronomers. But on
the contrary, I at once believed that the synchrotron mechanism was responsible for non-thermal cosmic radio emission. I ascribe this not to
any keen insight, but to the above-mentioned fact that I was closer to
physics and rather far from classical astronomy. In this situation the
synchrotron mechanism seemed clear and realistic, whereas hypothetic
strange "radio stars" remained purely speculative. I immediately verified the calculations in these two papers, but, if I am not mistaken, I
did not add anything essential. (I have not now compared all the express-
ions and estimates from these 1950 papers and from my own 1951 paper (submitted for publication on October 31, 1950) because it does not seem
essential here, the more so as I have never claimed priority.) But the
fact is that my 1951 paper was the first, and remained for some time the
only one, that responded to the proposals of Kiepenheuer and of Alfvén and
Herlofson to use the synchrotron mechanism in astronomy. The point is

probably that the reaction of astronomers was quite the opposite, i.e.,
the synchrotron mechanism seemed mysterious and speculative, whereas
"radio stars", although posing riddles, were more acceptable -- for what
kinds of stars cannot exist? In this respect Shklovsky was not an exception. He not only developed the radio star hypothesis (Shklovsky 1951,
1952), but also positively denied the synchrotron hypothesis ("which for
a number of reasons does not seem to us to be acceptable"[3]). I dare make
this remark here only because it is just Shklovsky's 1952 paper which has
been repeatedly cited in the world literature as the first principal one
discussing the application of the synchrotron hypothesis. A similar error
is often spread concerning the question of exploiting polarization as a
criterion for establishing the validity of a synchrotron origin of cosmic
radiation. Polarization measurements were proposed by I. M. Gordon
(1954a[4], 1954b) and supported by me (1953, 1954), whereas Shklovsky (1954)
felt such measurements would be of insufficient sensitivity.[5]

        Concerning the papers by Shklovsky, I shall make only two more
remarks. After the appearance in 1953 of a very important work by S. B.
Pikel'ner, who emphasized that the interstellar magnetic field exists in
the entire galactic volume, Shklovsky (1953a) realized that his earlier
(1952) objection to the efficiency of the synchrotron mechanism was
groundless. In this same 1953 paper he came to the conclusion that the
radio star hypothesis, which he had supported not long before (1951,
1952), was "a complete failure". Furthermore, the "interpretation of
'radio stars' as a very numerous category of galactic objects" was described in this paper as "quite groundless". It is also relevant to note
here that Shklovsky's paper (1953a) did not in any respect develop the
theory of synchrotron radiation (it contains not a single formula concerning this theory). In a subsequent paper published in the same year
(1953b), however, Shklovsky proposed a synchrotron interpretation of optical radiation from the Crab Nebula, which was undoubtedly an achievement
and played an important role. Polarization measurements were not, however, mentioned in this second paper, and in any case the possible efficiency of the synchrotron mechanism in astrophysics not only in the radio,
but also in the optical and even the X-ray bands, had been emphasized
before by Gordon (1954a). Finally, I would like to note that the only
source of information about the synchrotron mechanism cited in Shklovsky
(1953b) was his previous paper (1953a).

Here and in my 1982 paper I have made some remarks concerning
the papers by Shklovsky which were devoted to the origin of cosmic radio
emission. I emphasize an obvious discrepancy between their actual content
and what is often ascribed to them in the literature. I have already
tried to do this in a brief and somewhat implicit form at the beginning of
a review (Ginzburg and Syrovatsky 1965), but without any success. That
is why I came to the conclusion that I must either not write about the
history of radio astronomy at all, or tell the truth as I see it. It is
the concern of others, especially the historians of science, to verify the
facts; the more impartially and thoroughly this is done, the better.

Judging from the proceedings of the 1955 Manchester Symposium
on radio astronomy (van de Hulst 1957), which included a paper on the
"radio star" hypothesis while my paper sent to the symposium was not even
published, the "astronomical public opinion" in 1955 was still on the side
of the "radio star" model. But already at the next radio symposium, in
Paris in 1958 (Bracewell 1959), the synchrotron mechanism was uncondi-
tionally accepted as dominating in the production of non-thermal cosmic
radio emission (we do not mean, of course, the radiation coming from the
solar atmosphere and generally from relatively dense regions). This time
my report "Radio Astronomy and the Origin of Cosmic Rays" was included in
the proceedings of the Symposium (page 589), although I myself was not
able to participate.[6]

The establishment of a relation between radio astronomy and
cosmic rays has led, as a matter of fact, to the appearance of a new field
of research in astronomy, namely, cosmic ray astrophysics, now usually
included in the term *high-energy astrophysics*.[7] This has been just the
main road I have taken (since 1950-1953) in my own astronomical activities. Further historical details can be found in Ginzburg (1956, 1958,
1980, 1982), Ginzburg and Syrovatsky (1963), and Ginzburg and Ptuskin
(1976).

A COMPARISON OF THE PHYSICS AND ASTRONOMY COMMUNITIES

As is clear from what has been said at the beginning of the
present note, I am able to compare the situation in physics and in astronomy, at least in some respects and aspects. It seems to me that it would
not be out of place here to touch upon such a comparison.

A person, if compared with a measuring device, reacts mainly
not to a function (i.e., to the actual value of the quantity or to the
"state"), but to its derivative. As a result some astronomers do not seem
sufficiently to appreciate the very favorable situation which is theirs
as a "scientific community." The principal circumstance is the very existence of such a community united by the International Astronomical Union
(IAU). The IAU and its services (half a hundred commissions, continuously
held symposia and colloquia, exchange of information, a well organized
system of publication of journals, various proceedings, books, etc.) contribute much to the development of astronomy. Physics as a whole does not
experience anything of the kind at the present time. This can of course
be explained by a much larger number of physicists as compared with astronomers, by the multiplicity of scientific fields of research, by the
ties of physics with industry and applied fields, etc. But the fact is,
I think, that at the present time one cannot speak of a united scientific
community of physicists. The fact that it is rather isolated or more
narrow communities, unions and commissions which hold their congresses and
symposia is on the one hand of course useful, but on the other hand it
leads to a still greater separation of physicists of different specialities. All this very much hampers work both from the psychological and
organizational points of view. I, for example, am 75-85% a physicist,
judging by the number of my papers and results. But by the amount of
literature received (reprints, preprints, journals), by the number of invitations to symposia, etc., and by the number and "strength" of contacts
with colleagues, I am rather an astronomer (though I include here in
astronomy the range of questions particularly close to physics, namely
high-energy and cosmic ray astrophysics). Specifically, considering that
the most important part of my physics research is in the field of superconductivity and superfluidity, I feel myself in bad need of contacts
(even indirect ones, to say nothing of personal contacts) with corresponding specialists outside the USSR. My work suffers much from this
circumstance. Were these astronomical problems, the situation would be
quite different.

It would be out of place to dwell here longer on these
thoughts, which are aimed only at emphasizing how valuable the activity
of the IAU is. Therefore, in spite of all difficulties, for progress in
the further development of astronomy it seems necessary to maintain a
strong, united IAU as long as possible.

NOTES

1. How far I still remained from astronomy as a whole, in spite of what has been said, is seen from the incident I related in the 1980 volume dedicated to J. Oort (Ginzburg 1980). On the way back from Brazil the participants of the expedition were lucky to visit Leiden, although quite accidentally. And there, instead of getting acquainted with Oort and generally taking part in the discussion of astronomical problems, I rushed to the Kamerling Onnes Cryogenic Laboratory, since I was then interested most of all in low-temperature physics.

2. As an objection to the wave theory of light, Poisson pointed out a consequence of this theory which seemed to him quite absurd: on the axis of the geometrical shadow from a round opaque screen a light spot must be observed (the source is considered to be point-like and is located behind the screen on the axis perpendicular to the screen plane). The presence of a light spot was confirmed by an experiment conducted immediately after this by Arago. For an opaque sphere as screen, the conditions of observation of a central peak are facilitated (Getmantsev and Ginzburg 1950).

3. This quotation from Shklovsky (1952) is presented in more detail in Ginzburg (1982).

4. Publication of this paper was delayed for technical reasons. Its main content was reported and known in the Soviet Union at least since late 1952.

5. An English translation of the relevant discussion at the 1953 conference on "Origin of Cosmic Rays" is given in Ginzburg (1982).

6. In the proceedings, the Section "Mechanisms of Solar and Cosmic Emission" opens with a long report, including a "historical review" by G. R. Burbidge. This review, as well as other sections of the same paper, however, contains statements quite opposite to those presented above (and in my 1982 paper) on the basis of the available published materials. It would be of interest to know which sources were used by Burbidge.

7. *Cosmic rays* is the term used at the present time only in application to charged particles. Therefore gamma ray astronomy, X-ray astronomy, astronomy of high-energy neutrinos, and cosmic ray astrophysics are more and more often collectively referred to as high-energy astrophysics. This term is used here in this sense.

REFERENCES

Alfvén, H. A. & Herlofson, N. (1950). Cosmic radiation and radio stars.
     Phys. Rev., 78, 616.
Benediktov, B. A., Getmantsev, G. G., & Ginzburg, V. L. (1962). Radio
     astronomical investigations employing artificial satellites
     and space rockets. Planet. Space Sci., 9, 109.
Bracewell, R. N. (ed.) (1959). Paris Symposium on Radio Astronomy (IAU
     Symp. No. 9; URSI Symp. No. 1) Stanford: Stanford Univ. Press.
Gershman, B. N. & Ginzburg, V. L. (1955). On the mechanism of the appear-
     ance of ionospheric inhomogeneities. Doklady Akad. Nauk SSSR,
     100, 647. See also: Izvestiya Vuzov. Radiofyzika, 2 (No. 1),
     8 (1959).
Getmantsev, G. G. & Ginzburg, V. L. (1950). On Solar and cosmic radio
     emission diffraction on the Moon. Zh. Exper. Teor. Fiz., 20,
     347.
Ginzburg, V. L. (1946). On solar radiation in the radio spectrum. Doklady
     Akad. Nauk SSSR, 52, 491 = C. R. (Doklady) Acad. Sci. URSS,
     52, 487.
Ginzburg, V. L. (1947). Solar and galactic radio emission. Uspekhi Fiz.
     Nauk, 32, 26.
Ginzburg, V. L. (1948a). N. D. Papaleksi and radio astronomy. Izvestiya
     Akad. Nauk SSSR: Ser. Fiz., 12, 34.
Ginzburg, V. L. (1948b). New data on Solar and galactic radio emission.
     Uspekhi Fiz. Nauk, 34, 13.
Ginzburg, V. L. (1949). On the absorption of radio waves in the Solar
     corona. Astron. Zh., 26, 84.
Ginzburg, V. L. (1951). Cosmic rays as a source of galactic radio emission.
     Doklady Akad. Nauk SSSR, 76, 377.
Ginzburg, V. L. (1952). Interstellar matter and ionospheric perturbations
     leading to radio star twinkling. Doklady Akad. Nauk SSSR,
     84, 245.
Ginzburg, V. L. (1953). Supernova and nova stars as sources of cosmic
     and radio emission. Doklady Akad. Nauk SSSR, 92, 1133.
Ginzburg, V. L. (1954). In Proceedings of the Third Conference on Cosmog-
     ony: The Origin of Cosmic Rays, p.260, Moscow: Acad. Sci. of
     USSR.
Ginzburg, V. L. (1956a). On the mechanisms of formation of ionospheric
     inhomogeneities leading to "radio star" twinkling. In Proceed-
     ings of the Fifth Conference on Cosmogony: Radio Astronomy,
     p.512, Moscow: Acad. Sci. of USSR.
Ginzburg, V. L. (1956b). On non-ionospheric intensity oscillations of
     radio emission from nebulae. Doklady Akad. Nauk SSSR, 109,
     61.
Ginzburg, V. L. (1956c). The nature of cosmic radio emission and the
     origin of cosmic rays. Nuovo Cim. (Ser. 10), Suppl. 3, 38.
Ginzburg, V. L. (1958). The origin of cosmic radiation. In Progress in
     Elementary Particle and Cosmic Ray Physics (Vol. 4), p.339,
     Amsterdam: North Holland.
Ginzburg, V. L. (1960). On the possibility of determining the magnetic
     field strength in the external Solar corona by transmission
     through it of polarized radio emission from discrete sources.
     Izvestiya Vuzov. Radiofizika, 3, 341.
Ginzburg, V. L. (1961). On the nature of radio galaxies. Astron. Zh.,
     38, 380 = Sov. Astron.-A.J., 5, 282.
Ginzburg, V. L. (1964). On magnetic fields of collapsing masses and the
     origin of superstars. Doklady Akad. Nauk SSSR, 156, 43 =
     Soviet Phys.-Doklady, 9, 329.

Ginzburg, V. L. (1967). The Propagation of Electromagnetic Waves in Plasmas. [English edition: Pergamon Press (1970)].

Ginzburg, V. L. (1980). On high-energy astrophysics. *In* Oort and the Universe, ed. H. Van Woerden, W. N. Brouw and H. C. Van de Hulst, p.129, Dordrecht: D. Reidel.

Ginzburg, V. L. (1982). On the birth and development of cosmic ray astrophysics. *In* Early History of Cosmic Ray Studies: Some Personal Reminiscences, ed. Y. Sekido and H. Elliot, Dordrecht: D. Reidel.

Ginzburg, V. L. & Ozernoy, L. M. (1977). On the nature of quasars and active galactic nuclei. Astrophys. and Space Sci., 50, 23.

Ginzburg, V. L. & Pisareva, V. V. (1956). On the nature of Solar radio emission intensity oscillations and inhomogeneities in the Solar corona. *In* Proceedings of the Fifth Conference on Cosmogony: Radioastronomy, p.229, Moscow: Acad. Sci. of USSR.

Ginzburg, V. L. & Pisareva, V. V. (1963). Polarization of radio emission from discrete sources and the study of metagalactic, galactic and near-solar space. Izvestiya Vuzov Radiofizika, 6, 877.

Ginzburg, V. L. & Ptuskin, V. S. (1976). On the origin of cosmic rays: some problems in high energy astrophysics. Rev. Mod. Phys., 48, 161.

Ginzburg, V. L. & Syrovatsky, S. I. (1963). The Origin of Cosmic Rays. [English edition: Pergamon Press (1964)].

Ginzburg, V. L. & Syrovatsky, S. I. (1965). Cosmic magnetobremsstrahlung (synchrotron radiation). Ann. Rev. Astron. Astrophys., 3, 297.

Ginzburg, V. L. & Zheleznyakov, V. V. (1958). On possible mechanisms of sporadic Solar radio emission (radiation in an isotropic plasma). Astron. Zh., 35, 694 = Sov. Astron. - A.J., 2, 653.

Ginzburg, V. L. & Zhelezhnyakov, V. V. (1975). On the pulsar emission mechanisms. Ann. Rev. Astr. Astrophys., 13, 511.

Gordon, I. M. (1954a). On the problem of the physical nature of chromospheric eruptions. Doklady Akad. Nauk SSSR, 94, 813.

Gordon, I. M. (1954b). *In* Proceedings of the Third Conference on Cosmogony: The Origin of Cosmic Rays, p.253 and p.268, Moscow: Acad. Sci. of USSR.

Khaykin, S. E. & Chikhachev, B. M. (1947). Investigation of Solar radio emission during the Solar eclipse on May 20, 1947 by the Brazil expedition of the Academy of Sciences of the USSR. Doklady Akad. Nauk SSSR, 58, 1923; (1948) Izv. Akad. Nauk SSSR: Ser. Fiz., 12, 38.

Kiepenheuer, K. O. (1950). Cosmic rays as the source of general galactic radio emission. Phys. Rev., 79, 738.

Martyn, D. F. (1946). Temperature radiation from the quiet Sun in the radio spectrum. Nature, 158, 632.

Michel, F. K. (1982). Theory of pulsar magnetospheres. Rev. Modern Phys., 54, 1.

Migulin, V. V. *et al.* (1981). Uspekhi Fiz. Nauk., 134, 519-550 = Sov. Phys.-Uspekhi, 24, 614-632.

Pawsey, J. L. (1946). Observation of million degree thermal radiation from the Sun at a wave-length of 1.5 metres. Nature, 158, 633.

Pikel'ner, S. B. (1953). Kinematic properties of interstellar gas in connection with cosmic ray isotropy. Doklady Akad. Nauk SSSR, 88, 229.

Ryle, M. (1949). Evidence for the stellar origin of cosmic rays. Proc. Phys. Soc., A62, 491; (1950) Radio astronomy. Rep. Progress Phys., 13, 184.

Schott, G. A. (1912). Electromagnetic Radiation. Cambridge: Cambridge
    Univ. Press.
Shklovsky, I. S. (1946). On the radiation of radio waves by the Galaxy
    and by upper layers of the Solar atmosphere. Astron. Zh.,
    23, 333.
Shklovsky, I. S. (1951). Radio stars. Doklady Akad. Nauk SSSR, 79, 423.
Shklovsky, I. S. (1952). On the nature of galactic radio emission.
    Astron. Zh., 29, 418.
Shklovsky, I. S. (1953a). The problem of cosmic radio emission. Astron.
    Zh., 30, 15.
Shklovsky, I. S. (1953b). On the nature of the Crab Nebula's optical
    emission. Doklady Akad. Nauk SSSR, 90, 983.
Shklovsky, I. S. (1954). In Proceedings of the Third Conference on Cosmog-
    ony: The Origin of Cosmic Rays, p.254 and p.276, Moscow: Acad.
    Sci. of USSR.
Unsöld, A. (1949). On the origin of radio frequency and ultra shortwave
    radiation from the Milky Way. Zs. f. Astrophys., 26, 176;
    (1951) Cosmic radiation and cosmic magnetic fields. Phys.
    Rev., 82, 857; (1955) Astrophysical comments on the origin of
    the cosmic radiation, Z. Phys., 141, 70.
Van de Hulst, H. C. (ed.) (1957). Radio Astronomy (IAU Symp. No. 4).
    Cambridge: Cambridge Univ. Press.
Zheleznyakov, V. V. (1964). Radio Emission of the Sun and Planets.
    [English edition: Pergamon Press (1970)].

At the Nançay station circa 1960: (l. to r.) Emile Blum, a technician, Don McLean, and André Boischot (courtesy Blum)

THE EARLY YEARS OF RADIO ASTRONOMY IN FRANCE

J.F. Denisse
Astronome titulaire de l'Observatoire de Paris, France.

ORIGIN

It was Yves Rocard, appointed Director of the Physics Laboratory of the "Ecole Normale Supérieure" at the end of the Second World War, who introduced radio astronomy in France. A member of the Free French Forces, Rocard had very closely followed the technical developments that the allied war efforts stimulated, and he knew the importance of the first observations by British radars.

In 1946, Rocard formed a group of physicists around J.F. Denisse and J.L. Steinberg, including J. Arsac, E.J. Blum, A. Boischot, E. Le Roux, P. Simon, among others. This group initially had at its disposal two Würzburg radar antennas retrieved in Germany: a 7.5 m reflector with full steerability and usable for decimetric waves, which was installed about 20 km from Paris, and a second 3 m reflector mounted on the roof of the Paris laboratory.

By 1953 it became evident that there was a need for systems of antennas with large dimensions and also for protecting these from the radio interference of the urban environment. This led to creation of a specialized observatory in Nançay, about 200 km south of Paris, where the first important instruments were installed.

In 1954, on the initiative of A. Danjon, Director of the Paris Observatory, the team moved to its new laboratories at the Meudon Observatory. At that time, I had just been appointed to the observatory, after several years in the USA and Canada, working with such scientists as G. Reber, F. Haddock, and A.E. Covington. I was to lead the group until the beginning of the 1960's, by which time it had grown to about 40 scientists and students, engineers and technicians. Among them were Y. Avignon, F. Biraud, B. Clavelier, J. Delannoy, M. Ginat, J. Heidmann, I. Kazès, M. Kundu, J. Lequeux, A.M. Le Squeren, M. Parise, M. Pick, M. Vinokur.

At this same time, at the Institute of Astrophysics, another group of radio astronomers was formed as part of the CNRS (French National Center for Scientific Research), headed by M. Laffineur and P. Coupiac. At the beginning, M. Laffineur used a Würzburg reflector installed at the Meudon Observatory; later, he began the construction of an antenna of large dimensions at the Observatory at Haute Provence for the observation of radio sources.

During the war years, French research had been isolated from the scientific and technical advances that ensured the rapid development of radio astronomy in the Anglo-Saxon world. The first objective of the groups was therefore to make up for the delays in the French program. It might be pointed out that the first discoveries in radio astronomy had aroused but little interest among the astronomers active at the end of the war: the observations of K.G. Jansky, as well as those of G. Reber, were hardly even noticed. With little help from "professional" astronomers, the young radio astronomy physicists oriented their research according to the instrumental possibilities, mainly towards solar studies, because they were easier to carry out with the unsophisticated equipment then available. Although the astronomical community did not directly aid the radio astronomers, it welcomed them heartily into the field. In France, for example, André Danjon, who was the "Astronomer Royal" of French astronomy, showed throughout his life a total confidence and real support which did much for the development of research in radio astronomy.

TECHNICAL DEVELOPMENTS

The first developments in receivers were mainly due to the work of Steinberg, Blum, and Le Roux, and for antennas to that of Arsac and Boischot.

Up to the early 1950s, the stability of the receiver gains was a major problem which was resolved firstly by developing stabilized power supplies (Blum et al. 1952), and then by applying the Dicke modulation process (Steinberg 1952). Improving the noise factor was the next step, in particular by using new metric wave tubes and by a comprehensive understanding of mixer operation (Le Roux 1955).

At the same time, important progress in interferometry was made, primarily by Arsac who, in the tradition of the French optical school (Duffieux, Françon, Maréchal, ...), proceeded with an in-depth analysis of the performance of an antenna network considered as a filter of spatial frequencies. This research led to the possibility of optimising the antenna distribution in the network while minimising the error factor in re-creating the images of a given source (Arsac 1956). The application of these theories resulted in the creation of an "Arsac Network" (Fig. 1) made up of four antennas, placed in positions 0, 1, 4, and 6, in a manner to supply six spatial frequencies of equal amplitude without redundancy.

Fig. 1. The "Arsac Interferometer" (1955) - Note the positions of the antennas (diameter: 1.1 m, frequency: 9350 MHz).

Later, other improvements in receiving methods were introduced, in particular, regarding the phase stability of long-distance transmission, improvement of mixers, better integration processes, use of correlation procedures (Blum 1959; Blum 1961), aperture synthesis (Kundu 1959), multilobe reception, etc.

SOLAR RESEARCH

Due to a lack of resolution, the first observations in France of solar radio radiation were mostly qualitative, statistical, and theoretical. However, with observations of others, they contributed to the identification and understanding of the various solar emissions: quiet Sun emission, slowly varying component, solar storms, bursts and flare outbursts, etc. These components are classical today, but in the 1940s were at the center of the observers' concerns, since it appeared that, in addition to the thermal emissions of the solar atmosphere, there were other emissions, proper to the radio domain, which were entirely new and for which the generating processes were still to be discovered (Denisse 1950a; Denisse 1950b).

During the 1950s, the emissions of the unperturbed solar atmosphere (quiet Sun) were measured, at Nançay, at several wavelengths and the comparison with theoretical models helped to determine the laws of variations of the density and electron temperature with altitude in the solar atmosphere. The slowly varying component was identified as solar condensations related to centers of optical activity and to the solar spot's magnetic field extensions into the solar atmosphere; the observations allowed measurement of the altitude of these emissive centers, and of their visibility as a function of their longitude, and of the frequency of observation (Denisse 1950a; Boischot & Denisse 1957). The influence on the E layer of the ionosphere of the X-rays emitted by the condensed regions of the solar atmosphere was also noted (Denisse & Kundu 1957).

In 1949, an expedition was organized to observe a total solar eclipse in Africa; the observations (Blum et al. 1952), made simultaneously at centimeter and meter wavelengths, brought forth quantitative measurements on active centers in the chromosphere and the solar corona. It should be noted that on this occasion, and probably for the first time, diffraction fringes were observed during occultation of a local solar source by the lunar limb. Unfortunately, these fringes were attributed to local interference and not recognized and exploited as they could have been. Other solar eclipses were also observed later (Laffineur et al. 1956).

Research on the mechanism of emission of non-thermal components stimulated theoretical studies: shock wave emissions (Denisse & Rocard 1951), Cerenkov emission (Denisse 1960), etc. These studies led, notably, to the publication of a book on waves that were likely to be excited in plasma (Delcroix & Denisse 1961), and to an in-depth study on synchrotron emission (Le Roux 1961).

Solar observations were also made at Meudon with a Würzburg reflector at 555 and 255 MHz (Laffineur 1953); these observations gave an indication of the directive properties of solar flares and their link with the morphology of associated flares.

Statistical studies permitted establishing the relations existing between the radio emissions of solar activity and the perturbations of the Earth's magnetic field (Denisse 1953; Simon 1956). This research showed fully the role that radio astronomy could play in obtaining a better understanding of solar-terrestrial relationships, and yielded a good method, still used today, for forecasting solar activity.

Parallel to these studies, work on interferometers with high resolution was progressing at Nançay.   One should note here that the ingenious concept of the "Arsac network" could not be used then: the small dimensions of the antennas limited sensitivity and the technology did not permit, at that time, the accurate phase adjustments which the system required.

Two other centimeter wavelength interferometers were built:
(1) in 1956, two 2 m antennas equatorially mounted which allowed aperture synthesis studies of the structure of the more permanent sources of solar activity, and measurements of the diameters of centimeter flare bursts (Kundu 1959) (Fig. 2);

Fig. 2. The two antenna interferometer (1955) used at 9350 MHz for diameter measurements of localised sources on the Sun.

(2) in 1957, 16 equidistant antennas (Pick & Steinberg 1961) which became the basic instrument for observing solar activity at low coronal altitudes (Fig. 3).   The two-antenna interferometer was also used (Kazès & Steinberg 1957; Kazès 1957) to study scintillation phenomena in the terrestrial atmosphere, as well as to measure the dimensions of atmospheric irregularities.

Fig. 3. The 16 antenna interferometer (1959) used at 9350 MHz for monitoring solar activity.

But the major event of the 1950s was the construction, begun in 1953, of the large interferometer at Nançay, designed by Blum in association with Boischot & Ginat. The setting up of this instrument, which had thirty-two 5 m antennas, on 169 MHz, aligned over 1600 m in the East-West direction (Fig. 4) stimulated all the members of the radio astronomy group; even now, 30 years later, our memory of that project is as a real adventure. The originality of this instrument was particularly in the use of 16 preamplifiers placed at the output of each pair of antennas in order to minimise the cable losses between the antennas and the central receiver (Blum et al. 1957). Later, the system was arranged to operate also at 408 MHz (Clavelier 1966) and was complemented in 1959 by a North-South network of eight 10 m antennas, using the multicorrelation procedure (Blum 1961); the instrument showed particularly good performance compared to similar equipment (Fig. 5).

Fig. 4. View of the east-west branch of the "Grand Interféromètre", used at 169 MHz for solar observations (1955). On the right is one of the two movable Würzburg interferometer antennas, used at 1415 MHz for fine structure observations of radio sources.

Fig. 5. The north-south branch of the "Grand Interféromètre".

This instrument enabled one to localise the sources of solar noise storms (Boischot 1958; Le Squeren 1963), and to measure their altitudes and dimensions. Its most significant contribution was probably Boischot's discovery of Type IV bursts (Boischot 1958) associated with the most intense chromospheric eruptions. These were identified from the onset (Boischot & Denisse 1957; Denisse et al. 1960) as synchrotron radiation from electron clouds accelerated to very high energy during flares and projected into the solar corona. This interpretation was supported by the close relation found by Avignon & Pick (1959) between Type IV bursts and the abnormal absorption observed in the vicinity of the polar caps (PCA's) and attributed to the incidence of protons of several tens of millions of electron-volts in the upper atmosphere of the Earth. Coming at the start of space research, when the first balloon observations detected, in situ, high energy particles produced by the Sun, the discovery of Type IV bursts raised great interest within the scientific community.

Another type of emission, the "continuum storm", was later identified (Pick 1961) and shown to result from the remnants of accelerated electrons trapped in the lower corona.

The routine observations obtained during that time with the Nançay interferometers (these observations continue today) represented an important contribution for many lines of research, in particular during the 1957-58 International Geophysical Year (relations with other forms of solar activity (Avignon et al. 1966), origin of solar cosmic rays, etc.).

The whole of these activities placed the Meudon group, by the end of the 1950s, among the most active in the field of solar radio astronomy, a fact which was demonstrated by the organisation of a symposium on radio astronomy in Paris in 1958 (Bracewell 1959)(Fig. 6).

GALACTIC AND EXTRAGALACTIC RESEARCH

The first galactic observations were made, at the beginning of the 1950s, with a Würzburg antenna at 900 MHz; at that time, the properties of radio sources and of the Galaxy at high frequencies were very little known. These observations led to the realisation of the first map of galactic emission seen at decimeter wavelengths (showing the high concentration of these emissions along the galactic plane), to measurements of the absolute flux of the most intense radio sources, and to the determination of an upper limit of 20° K for the sky temperature outside the galactic plane (Delannoy et al. 1957).

Fig. 6. Visiting Nançay in 1958, from left to right: J. Oort, L. Goldberg, M.J.G. Minnaert, and J.F. Denisse.

At the same time, a variable base interferometer composed of two equatorially mounted Würzburg reflectors was installed by Le Roux in 1959 at Nançay; the mirrors were movable along two railway tracks, one running East-West, and the other North-South (Fig. 4). The instrument was used at 1415 MHz, had a resolving power of a fraction of 1 arc min, and also incorporated several other original detection procedures, for instance, digital integration and detection by correlation. The observations carried out by Lequeux revealed the fine structure of the most intense radio sources (galactic center, supernova remnants, radio galaxies) and, for the first time, focused attention on the double source character of some radio galaxies; at the same time, other point sources remained unidentified and were to be recognized, only later, as quasars (Lequeux 1962).

Thus it became evident during the 1950s that the future of galactic and extragalactic radio astronomy had a high potential, both in high energy astrophysics (synchrotron radiation) and in the very low energy ranges.

At the Paris Institute of Astrophysics, the Laffineur group undertook building, at the Saint Michel Observatory, an interferometer with two fixed cylindro-parabolic antennas, equivalent to two 30 m reflectors one km apart. This 300 MHz apparatus began operating in 1959 and (Laffineur & Coupiac 1967) (Fig. 7) supplied a radio source catalog.

Fig. 7. One antenna of the Saint-Michel Interferometer of the Paris Institute of Astrophysics (CNRS) was used at 300 MHz for survey of radio sources (1959).

Meanwhile, the Meudon group built a more versatile instrument, one which has in fact supported radio astronomy activities in France for more than two decades. The original plan, to build a two-element (each of 25 m diameter) mobile interferometer on railway tracks, had to be abandoned due to the soft soil formation around Nançay which could not support the heavy mobile equipment. Instead, the Large Radiotelescope, still in use today, was built. Its construction was inspired by the system of semi-fixed antennas built by J.D. Kraus at Ohio State University; its advantage was to furnish a vast collective surface (about 6,000 m^2) usable at wavelengths as short as the 21 cm hydrogen line; furthermore, the instrument was achromatic, a property which later proved to be of great importance for the study of molecular lines.

Realisation of this instrument, to which Parise and Ginat made particular contributions, was already well advanced in 1963, and the first observations of galactic and extragalactic sources and of planets were obtained in the early 1960s with the incomplete instrument (Heidmann 1963; Boischot & Ginat 1963) (Fig. 8).

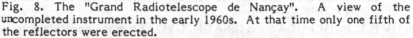

Fig. 8. The "Grand Radiotelescope de Nançay". A view of the uncompleted instrument in the early 1960s. At that time only one fifth of the reflectors were erected.

At the end of the period under review, the French radio astronomy group was considered one of the most valuable in the field of solar research; good progress had been made in galactic and extragalactic research and installation of adequate equipment was well under way for promoting further developments.

REFERENCES

ARSAC J. (1956). Etude théorique des réseaux d'antennes en radioastronomie et réalisation experimentale de l'un d'eux. Rev.Opt. 35, 65, 136,396.

AVIGNON Y. MARTRES M.J., and PICK M. (1966). Etude de la "composante lentement variable" en relation avec la structure des centres d'activité solaire associés. Ann.Astr. 29, 33.

AVIGNON Y. and PICK M. (1959). Relation entre les émissions solaires de rayons cosmiques et les sursauts de type IV. C.R.Acad.Sci. (Paris) 249, 2276.

BLUM E.J., DENISSE J.F., and STEINBERG J.L. (1952). Observations d'une éclipse annulaire de soleil effectuées sur 169 MHz et 9350 MHz. Ann.Astr. 15, 184.

BLUM E.J. (1959). Sensibilité des radiotéléscopes et récepteurs à corrélation. Ann.Astr. 22, 140.

BLUM E.J., BOISCHOT A., and GINAT M. (1957). Le grand interféromètre de Nançay. Ann.Astr. 20, 155.

BLUM E.J. (1961). Le réseau Nord-Sud à lobes multiples. Ann.Astr. 24, 359.

BOISCHOT A. and DENISSE J.F. (1957). Les émissions de type IV et l'origine des rayons cosmiques associés aux éruptions chromospheriques. C.R.Acad.Sci. (Paris) 245, 2194.

BOISCHOT A. (1958). Etude du rayonnement radioélectrique solaire sur 169 MHz à l'aide d'un grand interféromètre à réseau. Ann.Astr. 21, 273.

BOISCHOT A. and MOUTOT M. (1961). Etude du rayonnement thermique des centres d'activité solaire sur 169 MHz. Ann.Astr. 24, 171.

BOISCHOT A., GINAT M., and KAZES I. (1963). Observations des rayonnements des planètes Vénus et Jupiter sur 13 cm et 21 cm de longeur d'onde. Ann.Astr. 26, 385.

BRACEWELL R.N. (1959). Paris Symposium on Radioastronomy. Stanford, Ca.: Stanford University Press.

CLAVELIER B. (1966). Mise en service d'un nouveau réseau Est-Ouest sur 408 MHz à haut pouvoir de résolution destiné à l'étude du soleil. C.R.Acad.Sci. (Paris) 262, 228.

DELANNOY J., DENISSE J.F., Le ROUX E., and MORLET B. (1957). Mesures absolues de faibles densités de flux de rayonnement à 900 MHz. Ann.Astr. 20, 222.

DELCROIX J.L. and DENISSE J.F. (1961). Théorie des ondes dans les Plasmas. Paris: Dunod.

DENISSE J.F. (1950a). Contribution à l'étude des émission radioélectriques solaires. Ann.Astr. 13, 181.

DENISSE J.F. (1950b). Emissions radioélectriques d'origine purement thermique dans les milieux ionisés. Jour.Phys. et Rad. 11, 164.

DENISSE J.F. and ROCARD Y. (1951). Excitation d'oscillations électroniques dans une onde de choc. Applications radioastronomiques. Jour.Phys. et Rad. 12, 893.

DENISSE J.F. (1953). Sur le contrôle de l'activité géomagnetique par les taches solaires. C.R.Acad.Sci. (Paris) 236, 1856.

DENISSE J.F. and KUNDU M. (1957). Relation entre l'ionisation de la couche E de l'ionosphère et la rayonnement solaire radioélectrique. C.R.Acad.Sc. 224, 45.

DENISSE J.F., BOISCHOT A., and PICK M. (1960). Propriétes des éruptions chromo-
    sphèriques associées à la production de rayons cosmiques par le soleil. *In*
    Space Research, Proceedings of the First International Space Science
    Symposium (Nice, January 11-16, 1960), pp. 637-648, Amsterdam: North
    Holland Publishing Co.

DENISSE J.F. (1960). Les phénomènes radioéléctriques solaires et leur interpretation
physique. *In*Proceedings of the XIIIth URSI General Assembly, London.

HEIDMANN J. (1963). Etude de quelques galaxies normales ou peu émissives à 1430
    MHz. Ann.Astr. 26, 343.

KAZES I. (1957). Etude de la scintillation du soleil observée sur la longeur d'onde de
    3,2 cm. C.R.Acad.Sci. (Paris) 245, 636.

KAZES I. and STEINBERG J.L. (1957). Etude de la scintillation du soleil observée
    avec plusieurs antennes sur la longeur d'onde de 3,2 cm. C.R.Acad.Sci.
    (Paris) 245, 782.

KUNDU M. (1959). Structures et propriétés des sources d'activité solaire sur ondes
    centimétriques. Ann.Astr. 22, 1.

LAFFINEUR M. (1953). Contribution à l'étude du rayonnement électromagnetique du
    soleil sur ondes décimètriques. Contribution I.A.P. N° 169, Thèse, Paris.

LAFFINEUR M., COUPIAC P., and VAUQUOIS B. (1956). Observations radio-
    électriques de l'éclipse du 30 juin 1954. J.Atmosph.Terr.Phys. 6, 261.

LAFFINEUR M. and COUPIAC P. (1967). Un radio interféromètre méridien à liaisons
    radioélectriques. Contribution I.A.P. No 364.

LEQUEUX J. (1962). Mesures interféfomètriques à haute résolution du diamètre et de
    la structure des principales radiosources à 1420 MHz. Ann.Astr. 25, 221.

Le ROUX E. (1955). Mesures et recherches en radioastronomie; Analyses des résultats
    obtenus sur la longeur d'onde 33 cm; Etudes de certains spectres. Thèse,
    Paris.

Le ROUX E. (1961). Etude théorique du rayonnement synchrotron des radiosources.
    Ann.Astr. 24, 71.

Le SQUEREN A.M. (1963). Etude des orages radioélectriques solaires sur 169 MHz à
    l'aide de l'interféromètre en croix de la station de Nançay. Ann.Astr. 26,
    97.

PICK M. and STEINBERG J.L. (1961). Réseau à 16 antennes fonctionnant sur 9300
    MHz à la station radioastronomique de Nançay. Ann.Astrophys. 24, 45.

PICK M. (1961). Evolution des émissions radioélectriques solaires de Type IV et leur
    relation avec d'autres phénomènes solaires et géophysiques. Ann.Astr. 24,
    183.

SIMON P. (1956). Centres solaires radioémissifs et non radioémissifs. Ann.Astr. 19,
    122.

STEINBERG J.L. (1952). Les récepteurs de bruit radioélectrique. Onde Elec. 32, 445
    & 517.

BEGINNINGS OF SOLAR RADIO ASTRONOMY IN CANADA

Arthur E. Covington (retired)
Herzberg Institute of Astrophysics
National Research Council of Canada, Ottawa, Ontario

Thirty seven years have elapsed since I first thought about undertaking studies in cosmic noise, a large portion in the fifty year life of a new science. I had arrived in Ottawa in the summer of 1942, rather late in the development of the radar programs in progress at the Radio Field Station (RFS) of the National Research Council (NRC). By the fall of 1945 we were being asked to suggest projects needed for the transformation from war to peace. Before coming to the NRC, my colleagues and I had been students at university, so that at this time, when there was some hope of returning to our studies, there were many discussions concerning our future activities. It was in such an environment that I realized that the cosmic static survey of Grote Reber could be extended to microwavelengths by using radar radio receivers of greatly improved sensitivity. As a graduate student, while browsing through journals in a library sometime in 1939, I had learned of Reber's continuation of Karl Jansky's work. At least if I were to return to university, a simple experiment with more sensitive receivers could be quickly made at the RFS in a new region of the electromagnetic spectrum.

The possibility of observing cosmic noise from Sagittarius and determining its intensity by using the thermal noise from the sun as a known source was presented orally to W.J. Henderson, Microwave Section Head, sometime in the late fall of 1945 while he was attending to the vacuum system of the tube laboratory. He accepted the proposal for presentation to the Coordinating and Management Committee, and within a few days was able to say that my project had been approved, and that arrangements would soon be made for others to take over my field testing of an antenna operating at 1.25 cm. I doubt if such decisions could be made so quickly and easily nowadays when comprehensive programs have been established. By the end of 1946 a file for correspondence named "Research in Cosmic Noise" was opened; its first contents concerned the unexpected solar eclipse records of November 23, 1946.

The separation of galactic from similar sources of terrestrial noises depended upon Jansky's devotion to the scientific method in making various experiments over a number of years, and on his wisdom in interpretation. It was only when his observations were continued and analysed that he established that radio static originated in the direction of our galactic center, either from stars or interstellar gas. It was a puzzle that the sun, as our nearest star and most intense source of light, did not emit radio waves. If Jansky had been observing during sunspot maximum, hindsight suggests that solar radio emissions would certainly have been detected by the equipment he was using. This, in fact, was done later in 1935-36 by radio amateurs in the United Kingdom, although they did not fully recognize what they were detecting. They reported intense noises heard only during a sudden interruption of a radio signal, now known as a Sudden Ionosphere Disturbance (SID) (Ham 1975). Independently and in different circumstances, solar noise was discovered by J.S. Hey (1945), G.C. Southworth (1945) and Grote Reber (1944) during World War II, using highly directional antennas. I had learned about the work of Southworth while engaged in the NRC radar program; it was only after the war that the solar discoveries of Hey and Reber became known to me.

COSMIC NOISE OBSERVATIONS AT THE RADIO FIELD STATION
The first year-1946

A choice of wavelength for the cosmic noise program presented some difficulty since Henderson wanted to use a wavelength of 3 cm while I preferred 10.7 cm. In the end we proceeded with both. A design for a 30 ft diameter reflector good for 3 cm was commenced by the Engineering Design Section under H.E. Parsons, cousin to S.J. Parsons of Hey's group, and pioneer radio amateur. At the same time, a 4 ft parabolic reflector was polar mounted for immediate use with a 10 cm Dicke radiometer which I had designed as part of the maintenance program for the 10 cm radar sets at the RFS. I was assigned the assistance of two technicians, and we were joined by W.J. Medd upon his graduation from the University of Alberta. A paper on the declassified 10 cm Dicke radiometer was presented to the spring meeting of the Royal Society of Canada held in Toronto.

Sky noise. The temperature of the sky background at a wavelength of 10 cm had not been determined as far as I knew; so I decided its evaluation would be the first experiment. A small horn with a 30 degree beam

permanently pointing toward the zenith was connected to the radiometer,
which could be calibrated when needed by substituting a resistive
termination, initially at ambient temperature and later cooled by liquid
nitrogen. A zenith temperature of 50°K was measured. The possibility that
there could be large errors in the cool load precluded any serious
consideration of the origin of the emission, whether from the earth's
atmosphere or beyond. On Sunday, 5 May 1946, when interference was at a
minimum, a three hour record of sky noise showed a coarser texture of
fluctuations than thermal noise for a later period. This difference was
regarded as evidence of radio emission of unknown origin, the most likely
being water vapour in the earth's atmosphere. Within the next few weeks,
the passage of clouds through the beam was observed and found to be
unrelated to the coarse fluctuations. I then postulated the ionosphere as
a likely source, and obtained information on its activities for testing
such an origin from the recently formed Radio Propagation Laboratory under
F.L. Davies, and the Magnetic Division of the Dominion Observatory under
Glenn T. Madell. No obvious correlations were immediately apparent, and I
was left to learn about completely new fields. I soon found that a
standard text, Geomagnetism, by Chapman and Bartels, was a constant
companion.

Solar noise. In July of 1946, a naked eye sunspot aroused my curiosity
and stimulated me to use the small radio telescope, then lacking a drive,
as a transit instrument. The first drift curve was obtained on July 26th,
and showed a peak much higher than that calculated from a value of 6000°K
as the solar radio temperature. A second experiment was made a few
minutes later by advancing the telescope and rotating the dipole through
90 degrees when the peak of the drift curve had been reached. The N-S and
the E-W linearly polarized emissions were found to be equal. The two brief
experiments were significant, but were incidental to measuring galactic
noise in terms of the solar emission. The experiment would only be
completed when a drift curve through Sagittarius could be obtained. In
anticipation of evening observations, the antenna was set to a more
southerly declination. Upon arrival at the RFS at dusk, the southern sky
appeared unusually bright, and as darkness descended, colors became
evident as the center of activity moved northward. The possibility of
observing radio noise from such an aurora resulted in returning the
antenna to its zenith position and watching. The display became intense

as it moved northward, and then suddenly extended from the southern to the
northern sky with the formation of a remarkable overhead crown (Covington
& Covington 1946). The records of sky noise were unusually rough during
this display, and in the enthusiasm of the moment, the fluctuation which
occurred precisely when the rays passed overhead seemed unusually large.

During the next few days when the large spot was traversing the
solar disc, daily flux measurements were obtained from single drift
curves. When the spot was no longer visible on the disc, the daily level
dropped; and it was inferred that there were at least two components to
the total flux, one from the sunspot and another from the spotless solar
surface. The broadness of the beam, 7 degrees, naturally precluded any
inference about the structure of the radio emissive regions. Much later I
read that the geomagnetic storm accompanying the aurora was one of the
largest on record, and had been initiated by an intense flare on July
25th, one day before we had the antenna in operation!

When the auroral activity subsided a few days later, several
attempts were made to observe the noise difference between Sagittarius and
the zenith. None was found above the residual receiver noise. Before
disconnecting the radio telescope from the radiometer, I also pointed the
antenna toward Mars and Jupiter. Earphones were connected to the output
of the receiver as was the practice in identifying sources of local
interference, but nothing unusual was heard.

I was in a quandary. The observed value of the antenna
temperature of the sun, even when some allowance was made for the
existence of the radio emission from the sunspot, was much higher than
anticipated. While the shops installed a tracking motor and built a
synchronous detector, I made a review of the various formulas and
parameters, but the anomaly remained.

Eclipse. The Ottawa newspapers announced in early November that on the
23rd there would be a partial eclipse of the sun. The opportunities were
immediately apparent to me since ionospheric studies by an eclipse
expedition the previous year had been discussed by the scientific
community in Ottawa. At this time, the radio telescope was dismantled,
with parts yet to be finished. The installation was completed one day
before the eclipse in time to obtain the first tracking record, which
showed that noise bursts were not present as at metric wavelengths, and
that the intensity remained constant during the day. A second tracking

record was obtained the day after the eclipse, and the two were later averaged to provide the working reference for determining the changes in noise level as sources on the sun were covered and uncovered by the passage of the moon. The day of the eclipse, a Saturday, was clear and the radio astronomy project on the roof was probably watched by the whole establishment. A small optical telescope attached to the dish enabled the times of the moon's passage over various spots to be compared with the noise variations as they occurred. These showed qualitatively that a large fraction of the total flux originated from a large sunspot. A few days later, the record was shown to E.V. Appleton during his visit to the RFS. His interest in the qualitative evaluation was shown by the numerous questions he asked. Subsequently, he presented a survey of current work on the ionosphere which showed the global importance of the Canadian ionosphere network and of the detection of meteors and solar radio noise in the United Kingdom.

It was very fortunate that clear weather had provided the opportunity for the Dominion Observatory at Ottawa, under the direction of the solar astronomer Ralph DeLury, to obtain a series of photographs of the eclipsed sun. These provided the basis for analysing the variation of solar noise in relation to solar surface features; analysis showed that the brightness temperature of the radio emissive region associated with the large sunspot was 1.5 million degrees kelvin, and that of the solar background was 56,000 degrees. These high values were described in a letter to Nature (Covington 1947) at a time when only a thermal microwave emission, based on the optical temperature of the sun, was recognized.

The second year—1947

The second year commenced with the realization that further studies of spots by means of an eclipse would be infrequent, while an antenna with necessary resolution had yet to be built. Continuing studies of sunspots, differing as they do in many characteristics, could only be achieved through statistical studies of daily total flux measurements. Further, the detection of microwave noise during a solar flare, although extremely likely, nevertheless had yet to be achieved. Thus on February 14, 1947 daily observations were commenced. In the morning, upon arrival at the RFS, the sensitivity of the radiometer was determined by measuring ambient temperature of a "black box" which could be placed around the dipole at the focus of the dish, and by pointing the antenna

toward the zenith with its observed temperature of 50° K. On March 14 the
first burst was noted. When a call for papers for the Washington spring
meeting of the USA National Committee for URSI appeared, I presented a
description of the microwave solar noise observations taken at Ottawa
(Covington 1948). Shortly after the session a person introduced himself as
Karl Jansky; he had taken time from his work on amplifiers at the Bell
Telephone Laboratories to keep himself up to date on various aspects of
radio. Although not currently active in cosmic noise studies, he was much
interested in what had happened during the war. Soon Grote Reber
introduced himself with the comment that although Jansky and he had been
in correspondence, this was the first time they had the opportunity to
meet in person. They exchanged recent technical aspects of their work, and
Reber satisfied his curiosity that Jansky had been contacted by the same
person who wanted to use radio to reach a celestial home. Later at lunch
we learned of Jansky's illness which would lead to his untimely death in
1950. Reber had joined the National Bureau of Standards and was in the
process of moving his antenna from Wheaton to the Washington area.
Menzel's paper at the same meeting entitled "Wanted: New Indices of Solar
Activity" suggested to me that perhaps the 10 cm solar radio flux might be
a candidate, and I returned from the meetings with a glimpse of new
possibilities for solar noise observations and their bearing on solar-
terrestrial relationships.

GOTH HILL RADIO-OBSERVATORY

The radio telescope at the RFS was in the midst of operational
radar sets and other electrical equipment; interference often masked solar
events so that interpretation was impossible. We constructed two well
shielded radiometers, but this only showed that the leakage was through
the antenna side lobes, and that further reduction in interference would
require the removal of the radio telescope to a distant site. Henderson
and I investigated several possible sites near the RFS in late 1947, and
finally an elevated site some five miles away from the RFS on the property
of W.C. Goth was leased. Soon afterward, W.J. Henderson, as head of
Nuclear Physics and Radiology in the Division of Physics, became
completely involved in the construction of the electron cyclotron at the
RFS; the radio astronomy project was then administered by G.A. Miller in
the Microwave Section (Middleton 1981). In the spring of 1948, the
appointment of G.A. Harvey to manage the daily observations allowed me to

develop a polarimeter for the 4 ft reflector and a broad band radiometer with horn antenna for obtaining the solar spectrum from 10-30 cm. After the move to the new site, I began to envision our daily observations growing into a service not unlike the weather service. The correlation of solar radio flux with sunspot area had been shown, from a study by J.F. Denisse (1950), to be strongest at a wavelength of 10.7 cm rather than at 3 cm as observed by the U.S. Naval Research Laboratory (NRL), or at 62 cm as used by Reber (Figure 1). The possibility of deriving a radio index of solar activity was recognized early at the international meetings of the IAU and of URSI by appropriate resolutions. Invitations from C.W. Allen and M. Waldmeier prompted me to submit tabulations of the daily solar flux and distinctive events for publication in the <u>Quarterly Bulletin</u> on <u>Solar Activity</u>.

Figure 1. Scatter diagrams of sunspot Area versus Intensity of radio flux at 3, 10.7 and 62 cm show the highest correlation at a wavelength of 10.7 cm.

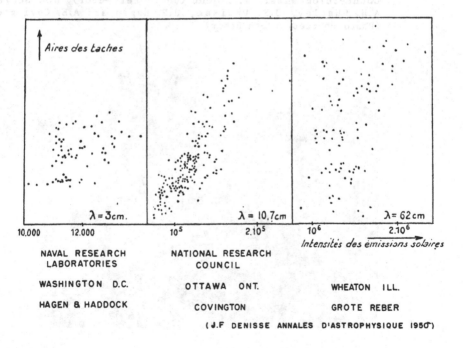

NAVAL RESEARCH
LABORATORIES

WASHINGTON D.C.

HAGEN & HADDOCK

NATIONAL RESEARCH
COUNCIL

OTTAWA ONT.

COVINGTON

WHEATON ILL.

GROTE REBER

(J.F DENISSE ANNALES D'ASTROPHYSIQUE 1950)

The 81st meeting of the American Astronomical Society (AAS) was hosted by the Dominion Observatory at Ottawa, and a number of papers on radio astronomy were given. Medd and I presented comparative solar observations on 200 MHz and 2800 MHz (Covington & Medd 1949). During the discussion C.S. Beals, Dominion Astronomer, raised the frequently asked question of just what was meant by the equivalent temperature of the antenna radiation resistance. I attempted an explanation of these theoretical concepts in terms of the collecting area of the antenna and the emission from a radio black box which was used each day to obtain the 10 cm solar flux. Later those who visited the observatory at Goth Hill had the opportunity to inspect an actual box, as well as the radio quarter wave plate used to measure the polarization (Figure 2).

Figure 2. Inspection of the 10.7 cm solar patrol radio telescope at the Goth Hill Radio Observatory during an AAS meeting held in Ottawa, June 1949. The radio black box and quarter-wave plate may be seen at the front of the platform. Counterclockwise: W.J. Medd (under reflector), C.L. Seeger, A.H. Shapley, J.F. Denisse, J.P. Hagen and A.E. Covington. (Photo courtesy Grote Reber)

Slotted waveguide array

The first phase in post-war radio astronomy adapted equipment from existing radar sets; its goal was exploratory and opportunistic. In 1950 a second phase emerged from the recognition that there was a new dimension to astronomy and that new observational results would be essentially determined by the performance of the new equipment. The design of the proposed NRC 30 ft reflector had been interrupted when the Air Force engineer in charge obtained his discharge, and was only revitalized by the completion of the 50 ft reflector at NRL and by reports of work in progress at other observatories. Since my radio astronomy program had used only the wavelength of 10 cm in the previous three years, the tolerances were relaxed to consider a much larger reflector, 120 ft in diameter. Even though costs were minimized by the use of a transit mount, they soon became prohibitive; a less expensive design was needed. In the end, a 150 ft parabolic cylinder with a slotted waveguide collector at the line focus was studied and accepted, in part because it could be constructed in two stages. The design was an extension of the 30 ft antenna used for the Microwave Early Warning radar with which the Radio and Electrical Engineering Division (REED) was familiar; the only uncertainties lay in the performance of an array of hitherto untried length and a design which would allow an increase in length at some future time. The non-resonant array made it possible to swing the beam a few degrees from the meridian by changing the wavelength of the received signal, thereby obtaining a number of drift curves on the same day. It was realized that field testing of this antenna's unprecedentedly narrow fan beam of 0.125 degrees (the far field was 25 miles away) could not be done. The only hope lay in expecting that every now and then a sufficiently intense and narrow radio sunspot would appear to serve as an external remote signal generator. The eclipse of 1946 had certainly shown that this did indeed occur, but the parameters and characteristics of radio sunspots in general were entirely unknown. It was also hoped that the discrete radio sources in our galaxy could be detected. Construction of the array was started in the summer of 1950; the first drift curves were obtained in October 1951, using on a regular basis a total power radiometer with a special IF amplifier designed by Broten. The drift curves showed emission from the vicinity of sunspots as point like sources superimposed upon a quiet sun background, and, unexpectedly, radio emission from a spotless calcium plage (Covington & Broten 1954).

Microwave solar patrol

From 1947 to 1949, data from the 4 ft radio telescope accumulated without any formal analysis while other programs were being set up. A single calibration of the solar patrol at local noon was found adequate for determining the level of total flux from the solar disc. During a month, these values change slowly, outlining what is known as the "slowly varying component", which is closely associated with the number of sunspots or the total sunspot area. The day to day variations arise chiefly from the rotation of the sun when there is a predominance of sunspots at one longitude, as well as from the intrinsic changes in radio emission associated with their evolution. The definitive study of this material (Covington 1951) was made only after observations at 150 cm were examined. In contrast to the steady flux level observed throughout the day at 10.7 cm, the metric wavelength record shows numerous bursts. When bursts could be intercompared--usually the more intense ones--the metric wave profile was generally more irregular and intense than the smooth rise and fall of flux observed at 10.7 cm. These differences were regarded as evidence of the physical conditions at the level of origin of the two waves in the solar atmosphere (the shorter wavelengths arising from lower levels), and separate analysis of the data obtained at the two wavelengths was warranted. Although equipment was being developed for metric wavelengths at Ottawa, this program was discontinued in order to emphasize the microwave observations.

Comparative studies of radio bursts and optical flares

Studies on burst profiles at 10 cm were made in February, 1953 through a visit by Helen Dodson and Ruth Hedeman of the McMath Hulbert Observatory. When they arrived at the RFS with a large selection of photographic material, the flare of May 19, 1951, with the ejection of a dark cloud of helium over a bright region, was the first one to be compared with the radio records. The comparison showed the start of an impulsive burst coincident with the optical brightening, quickly rising to a peak. However, the radio flux, instead of returning to the pre-burst or an enhanced post-burst level continued to decrease at a much slower rate before returning to the daily level. A small diminution of radio flux was recognized and was interpreted as the absorption of radio flux by the material from the jet lying over a lower bright emissive region. This type of event is infrequent and a new burst category was established only after a similar event was reported from the Toyokawa Radio Observatory, Japan (Tanaka & Kakinuma 1954).

    Continuation of the comparative study showed that the brightening of the flare and the onset of the radio event occur together, that a short duration impulsive radio burst is present during the rising part of the H-alpha emissions, while a longer gradual enhancement of flux which starts with the flare may last as long as or longer than the flare brightening. The name of gradual rise and fall (GRF) was introduced to describe the weak, long enduring event. It may occur separately or have other events superimposed upon it (Dodson et al 1954).

    Observations made with a Lyot filter at the Dominion Observatory (DO) have also been studied. The critical stages in the development of the limb flare of June 20 1961 shown in the sketches of Figure 3 are derived from the filtergrams and are to be compared with the 2800 MHz radio records, one at high sensitivity and another at low sensitivity. On the high sensitivity record before the flare, a GRF burst is associated with the maximum activation at 14:09 in a small loop prominence at region B. After return to a normal quiet level, there is a period of fluctuations as a precursor to the abrupt off-scale reading marking the onset of the impulsive burst. This event is associated with the various brightenings of regions A and B which lead to the ejection of material away from the sun (Gaizauskas & Covington 1962). Bursts of great intensity occur infrequently and, as in this case, often have two peaks. They are likely to be accompanied by major geophysical disturbances. A special category of "Great Burst" above a certain minimum level was created to draw attention to these events and to provide the means to alert those likely to be interested in their effects. Another study showed very clearly the simultaneous onset of the impulsive bursts and flare shock waves (Covington & Harvey 1961). With such examples in mind, a large block of data gathered in the 1950's was analysed with other synoptic data by Harvey (1965) to derive various statistical relationships.

Relative and absolute calibrations

    The initial calculations for the solar flux in 1946 used an antenna gain derived from the radar programs, and the preliminary sky background temperature of 50° K. At this time of sunspot maximum it was decided to concentrate immediately upon the development of a calibration

Figure 3. Flare-surge on western solar limb, July 20, 1961.
Left: Sketches from H-alpha filtergram of critical flare and
surge stages (dark areas are bright) (DO, Ottawa, Ontario).
Right: 2800 MHz solar noise burst (REED, Lake Traverse,
Ontario).

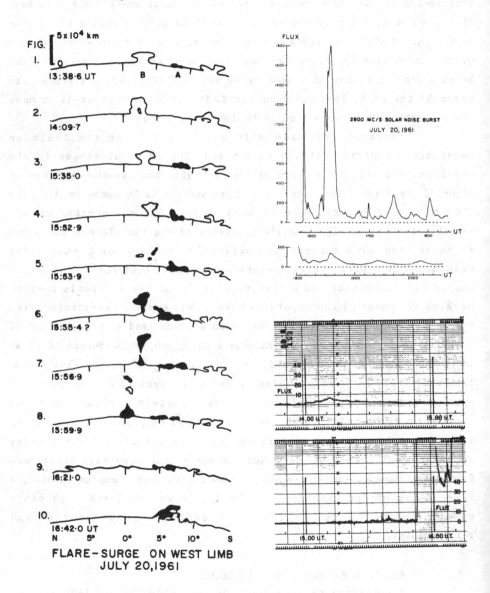

FLARE-SURGE ON WEST LIMB
JULY 20,1961

suitable for relative accuracy, and to postpone the more difficult
absolute determination for a later time. After a variety of heated
terminations had been constructed with varying degrees of success in
calibrating the radiometer, it was realized that the constancy of the
zenith sky temperature would provide one good reference level. A second
temperature was obtained by immersing the dipole at the focus in the
thermal emission from a radio black box, readily formed around the dipole
without disturbing it. Substitution errors and transmission line losses
were thus minimized.

In 1952, a second experiment was made to determine the zenith
temperature through the use of a framework containing an upward-pointing
horn and two waveguide terminations, one at ambient and the other at an
elevated temperature. It was known as the "swinging horn" experiment (Medd
& Covington 1958). The radiometer could thus be rapidly connected in
sequence to the three terminations, and the unknown sky temperature found.
The average value was 5.5° Kelvin with extremes of $+23^\circ$ and -18°. This
large range of values precluded any emphasis on the average non-zero
value. Later, A.A. Penzias and R.W. Wilson of the Bell Telephone Labs
made a much better determination of the "radio sky emission", and this is
of course now of major importance. At the time, I regarded the low
temperature as a trace of emission from the ionosphere or the heliosphere.

Compound interferometer

In January, 1954 the Carnegie Institution of Washington brought
together active workers from many countries and various disciplines to
review developments in radio astronomy. It was a time for critical
evaluation and new approaches. A dominant topic of the times was brought
out by a study of E.G. Bowen on the interferometer versus the big dish in
future radio astronomy. I reported on the solar results obtained with the
first stage of the slotted waveguide array. The design of this array with
its horn collector had been tested, but there still remained the mounting
of the unit at the line focus of a large parabolic cylinder, perhaps 20 ft
in diameter. Reflection on the developments presented at the conference
created second thoughts on how to proceed. Many of the accounts of
antennas either planned or under construction reported greater dimensions
or collecting area than the proposed cylinder, showing that it would no
longer be unique for new studies. However, if only solar noise studies
were attempted, it would be possible to develop an interferometer which

would make use of the linear array as one element and to construct one other antenna element. As with the slotted waveguide array, any small and intense spot emission would provide a means for making adjustments, testing the performance and determining some of the spot's physical parameters. After considering several different geometrical configurations, an interferometer formed by the array and another simple interferometer of identical aperture, adjacent to it and on the same axis, was finally selected as being most economical. This configuration became known as a "compound interferometer". The sharp lobe is produced by using a continuously rotating phase-shifter and synchronous-detector to combine the single lobe from the continuous array with the two interference patterns. The general principles had been outlined by Ryle and Mills. Modifications to the array and the construction of other components to make such an interferometer were begun in the fall of 1954.

Observations with the 2.4 arc min fan-shaped beam commenced in January 1956, and although more structure was seen in the drift curves, it was obvious that many individual solar regions were unresolved (Covington & Broten 1957). In 1958, the beamwidth was further sharpened by replacing the two element simple interferometer with a four element grating with gaps equal to the array length, thereby forming an aperture with an overall length of 600 ft. With north-south confusion absent, the resulting 1.2 arc min beam was adequate to separate and to study the structure of the larger spots. An aerial photograph of the Goth Hill Radio Observatory taken when the compound interferometer was being extended from 300 ft to 600 ft is shown in Figure 4.

URSI - INTERNATIONAL SCIENTIFIC RADIO UNION

Immediately after the war, before Canada became a member of URSI, the open door policy of the American National Committee for URSI and the Institute of Radio Engineers provided opportunities for me to exchange new ideas and current observations with others. In particular Commission 5 on Radio Astronomy studied topics in sub-committees largely described by their names: standards, nomenclature, frequency protection, and calibration. I was a frequent observer at many Washington spring meetings, from which many insights were gained. Although the National Research Council recommended the adherence of Canada to the Union as early as 1949, it was not until 1950 that a national Committee for URSI was formed under the chairmanship of D.W.R. McKinley, with R.E. Williamson from the

Figure 4. Aerial view of the Goth Hill Radio Observatory taken during the International Geophysical Year when the E-W aperture of the compound interferometer seen in the foreground was being extended from 300 ft to 600 ft by the addition of two parabolic cylinders of 8 by 10 ft aperture (shown pointing upward). The 150 ft slotted waveguide is at the bottom of a horn pointing toward the south. The 4 ft radio telescope is on a platform in front of the radiometer building on the hill; to the right is the 10 ft dish used for solar, lunar and galactic noise observations.

University of Toronto as first chairman of the Canadian Commission for
Radio Astronomy (McKinley 1952). In the fall of 1953, radio scientists
from both Canada and the USA were able to hold a joint meeting in Ottawa,
one of several which were to follow. The URSI General Assembly held in
1954 at The Hague accepted resolutions which recognized that solar noise
observations in the vicinity of 3000 MHz and 200 MHz would provide the
most suitable radio indices of solar activity. The next General Assembly
in 1957, held in Boulder, Colorado, reaffirmed these frequencies for the
needs of the International Geophysical Year.

VISITORS

Early visitors to the Radio Field Station and to Goth Hill whom
I can recall and who were much interested in the radio astronomy program
were: E.V. Appleton, W.R. Piggot, B.W. Currie, J.S. Hey, J. Ratcliffe,
J.G. Bolton, H.T. Friis, J.L. Pawsey and H.C. Van de Hulst. I was introduced to Pawsey during one of his early visits to the RFS by
W.J. Henderson; they had attended Cambridge at the same time (their
association is recorded in a group photograph appearing in Snow's book
entitled The Physicists, a Generation that Changed the World (1981)). When
Pawsey was being shown, sometime in 1949, the 10-30 cm broadband
radiometer with its horn antenna under construction for the absolute flux
determination, he told me about the 21 cm hydrogen line prediction and
wondered whether or not I could make, or would plan to make, any
observations for its confirmation. As it stood, the instrumentation was
hardly suitable. This was the first time that I had heard of the
prediction and is one occasion when I realized the magnitude of the
difficulties of switching from one promising area to another. I readily
gave a negative reply and realized that I would be continuing solar noise
work. The early determinations of the solar spectrum were limited but
sufficient to show that it was smooth. Almost twenty years later, other
investigators, including myself, formed a working group in URSI under the
chairmanship of Tanaka, to make acceptable and reliable evaluations at
discrete frequencies from 408 MHz to 17,000 MHz.

In the summer of 1949, Broten was among the first group of
students to be temporarily employed in the Laboratories of the NRC for
radio astronomy. Upon graduation the following year he was appointed to
the project, and R.M. Chisholm was a summer student. G.A. Harrower and
L.R. McNarry were also part of the student program, but in other projects.

In 1957, Harrower and Chisholm established a radio observatory of their own for Queen's University, while McNarry in 1960 commenced metric wave solar noise observations at ARO, later transferring to the Radio Astronomy group in the Radio and Electrical Engineering Division.

POSTLUDE - THE FUTURE IN 1960

Several events in 1960 mark a new era in astronomy in Canada: the opening of the Dominion Radio Astrophysical Observatory (DRAO) in Penticton, British Columbia was celebrated; the initial stage of setting up the 10.7 cm solar monitor at the Algonquin Radio Observatory (ARO) at Lake Traverse was completed; and Beals dedicated a 15 inch student telescope at Queen's University. At the dedication conference in Penticton, Chisholm expressed the hope that someday the two major observatories at Lake Traverse and Penticton would be linked by an interferometer. This hope was based upon interferometer experiments he had commenced in 1959 at Queen's University (Chisholm 1963). It was realized seven years later in a program including not only ARO and DRAO but also a radio telescope at Shirley Bay operated by the Defence Research Board outside Ottawa and another one at the Prince Albert Radar Laboratory in Saskatchewan. This was the first successful very long baseline interferometer using tape-recording techniques.

The solar noise program provided observations which have been useful in studying the propagation of radio waves through the earth's atmosphere as well as being of intrinsic interest in solar physics. Even though these studies were satisfying and more than enough to fully occupy several persons for a number of years, I had expectations of making at least one observation of cosmic noise. It would have provided a better answer in place of the upper limit I had to give in connection with the question often asked: "Have you measured cosmic noise?". Ultimately, at the close of the period assigned for this review, a 10 cm microwave travelling wave tube enabled Medd and Broten to use the 10 ft radio telescope to obtain observations of the moon, discrete sources, and the continuum of our galaxy at the Goth Hill Radio Observatory. The goal of the program set in 1945 had at last been achieved; all that remained to be done was the changing of a name on a still active correspondence file from "Research in Cosmic Noise" to the more appropriate "Research in Solar Noise".

SELECTED REFERENCES

Chisholm, R.M. (1963). Research in Communications and Antenna Theory at
    Queen's University, Kingston, Ont., 1959 to 1962. Kingston,
    Ont.,: Dept. of Elec. Eng., Queen's University.
Covington, A.E. & Covington, C.R. (1946). Aurora seen at Ottawa, Friday
    July 26, 1946. J. Roy. Astron. Soc. Can., 40, 275.
Covington, A.E. (1947). Micro-wave solar noise observations during the
    partial eclipse of November 23, 1946. Nature, 159, 405.
Covington, A.E. (1948). Solar noise observations on 10.7 centimeters.
    Proc. IRE, 36, 454–57.
Covington, A.E. & Medd, W.J. (1949). Observations of solar radio noise on
    1.5 metres and 10.7 centimetres. J. Roy. Astron. Soc. Can., 43,
    106–110.
Covington, A.E. (1951). Some characteristics of 10.7 cm solar noise,
    parts I and II. J. Roy. Astron. Soc. Can., 45, 15.
Covington, A.E. & Broten, N.W. (1954). Brightness of the solar disc at
    a wavelength of 10.3 cm. Astrophys. J., 119, 569–589.
Covington, A.E. & Broten, N.W. (1957). An interferometer for radio
    astronomy with a single lobed radiation pattern. IRE Trans. on
    Ant. and Prop., AP5, 247.
Covington, A.E. & Harvey, G.A. (1961). Coincidence of the explosive phase
    of solar flares with 10.7 cm solar noise bursts. Nature,
    192, 152.
Denisse, J.F.(1950). Contribution à l'étude des émissions radioélectriques
    solaires. Ann. d'Astrophys., 13, 181–202.
Dodson, H.W., Hedeman, E.R. & Covington, A.E. (1954). Solar flares and
    associated 2800 mc/sec radiations. Astrophys. J., 119, 541–63.
Gaizauskas, V. & Covington, A.E. (1962). Radio and corpuscular emission
    associated with the flare surge on the western limb of the sun
    on July 20, 1961. J. Geophys. Res., 67, 4119.
Ham, R.A. (1975). The hissing phenomenon. J. Brit. Astron. Assn.,
    85, 317–323.
Harvey, G.A. (1965). 2800 mc/s radiation associated with type II and
    type IV solar radio bursts and the relation with other
    phenomena. J. Geophys. Res., 70, 2961.
Hey, J.S. (1946). Solar radiations in the 4–6 metre radio wavelength band.
    Nature, 157, 47.
McKinley, D.W.R. (1952). Canadian National Committee of URSI for 1950–
    1952. Ottawa, Ont., Can.: Nat. Res. Council.
Medd, W.J. & Covington, A.E. (1958). Discussion of the 10.7 cm solar radio
    flux measurements and an estimation of the accuracy of
    observations. Proc. IRE, 46, 112.
Middleton, W.E.K. (1981). Radar Development in Canada. Waterloo, Ont.:
    Wilfred Laurier University Press.
Reber, Grote (1944). Cosmic static. Astrophys. J., 100, 279.
Southworth, G.C. (1945). Microwave radiation from the sun. J. Franklin
    Inst., 239, 285.
Tanaka, H. & Kakinuma, T. (1960). Sudden disappearance of a source of s-
    component of solar emission at microwave frequencies on Nov.
    30, 1954. Proc. of the Res. Inst. of Atmosph., Nagoya
    University, Toyokawa City, Japan, 7, 72.

DEVELOPMENT OF SOLAR RADIO ASTRONOMY IN JAPAN
UP UNTIL 1960

Haruo Tanaka
School of Electricity, Faculty of Engineering
Toyo University, Kawagoe-shi, 350 Japan

The start of radio astronomy in Japan was about four years behind the rest of the world.   However, considering the difficult social circumstances arising from the nation's defeat in World War II, its development was not so slow.   We were hungry, few materials were available, and the general power supply was desperately unstable —— frequency fluctuated between 54 and 61 Hz, and the voltage between 85 and 105 V, even in 1950.   My first work was therefore the design of an A.C. voltage stabilizer of 0.1 % accuracy, and I believe it greatly supported the development of observing facilities in Japan.

In this article, I will first introduce three topics of early Japanese research which have seldom been discussed, even in Japan. Then, I will briefly review how solar radio astronomy developed in Japan in those early days.   A precise review of scientific studies on radio astronomy in Japan until 1962 was made in Chapter 5 of 'Progress in Radio Science in Japan', September 1963, which was distributed at the General Assembly of URSI held in Tokyo.

1   TOPICS NOT WELL KNOWN

1.1   Solar radio noise was recognized in 1938 in Japan

Arakawa (1936) and Dellinger (1937) had reported that a kind of 'grinder' like noise sometimes appeared almost simultaneously with Dellinger phenomena in short-wave telecommunication receivers.   Dellinger phenomenon is the phenomenon of sudden decrease or fade-out of long distance telecommunication radio waves due to ionospheric disturbances caused by solar flares.   Nakagami & Miya (1939), of the International Telecommunication Co. Ltd., Tokyo, were interested in the origin of that temporary noise, and therefore tried to measure its incident angle, together with that of communication signals.

They built two horizontal half-wave dipoles, sufficiently

separated from each other to avoid mutual coupling, one $\lambda/2$ and the other
$5\lambda/4$ above the ground, and compared their outputs.   This arrangement was
designed to be suitable for measuring the variation of incident angles
less than 20 degrees, but by chance it was also capable of measuring
angles of more than 70 degrees.   It was awfully difficult to catch
Dellinger phenomena since no continuous recorder was available.   They
simply watched the output meters and wrote down the observed values in a
notebook every minute.   During their patient measurements over a half
year, April through September 1938, they succeeded in catching an example
of the temporal increase of noise during a Dellinger phenomenon, as shown
in Fig. 1.

In Fig. 1 it is seen that on 1 August the noise suddenly
increased to 40-50 dBμV as soon as the communication signal from station
PLJ at 14.6 MHz faded out.   The noise decreased rapidly in five minutes.
They were surprised that the noise received by the h=$5\lambda/4$ antenna was
more than 10 dB stronger than the one received by the h=$\lambda/2$ antenna,
which clearly showed that the incident angle was more than 70 degrees, as
plotted in the upper part of the Figure.   As the sun was then placed at
about 70 degrees in elevation angle, Miya believed naturally that the

Fig. 1. Nakagami and Miya in Tokyo received solar noise in
1938 (from Nakagami & Miya 1939).   At the onset of Dellinger
phenomenon communication signal faded out (ZAN), and the noise
from antennas increased remarkably to different levels corresponding to antennas of different heights.   This means that
the noise came from high up in the sky as shown in the upper
Figure.   PLJ is a station in Jakarta; JCST means Japan
Central Standard Time.

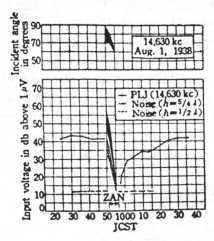

noise came directly from the sun.   However, his senior Nakagami was too
cautious to accept the young Miya's simple idea, and imagined that the
noise originated around the E-layer, connected with a Dellinger disturbance of the ionosphere.   In the end the possibility of direct noise
from the sun was not mentioned in their paper.

Dr. Nakagami passed away by accident about 10 years ago, but
Dr. Miya is now the President of International Telecommunications Installation Co. Ltd., Tokyo, and is my source for this story.

1.2 The original idea of the grating interferometer in Japan

The first experiment for receiving microwave (3300 MHz) solar
noise in Japan was made at Osaka University by Oda & Takakura (1951) in
November 1949, using an antenna as shown in Fig. 2.   They were physicists
who were interested in the origin of solar noise.   A hand-made horn was
mounted on a salvaged mount originally used for ship-borne searchlight.
Soon later, they moved to Osaka City University together with the antenna.
There they refined the antenna by replacing horn with a dish (Fig. 3), and
observations were made for about 15 months beginning in April 1950, for
two hours each day.   They noticed a linear relationship between solar

Fig. 2. First experiment to receive solar noise, at 3300 MHz,
by M. Oda (front) and T. Takakura (back) at Osaka University
in November, 1949.

Fig. 3. The dish version of Fig. 2, in 1950 at Osaka City
University.   Diameter of the dish was 1 m, which was equipped
with a rotary quarter-wave plate for circular polarization
measurements.   Drive was yet manual.

Fig. 4. The idea of the grating interferometer as proposed by
Oda's group in 1950.   Proposed wavelength was 7.5 cm and
diameter of horns, 50 cm.   Outputs from each row were
designed to pass through seesaw-driven trombone phase shifters
to shake the beam in vertical direction, but drift scan was to
be used in horizontal direction.   Chinese characters put on
the upper square means receiver; lower left, phase shifters;
lower right, master equatorial mount. From Ojio et al. (1950).

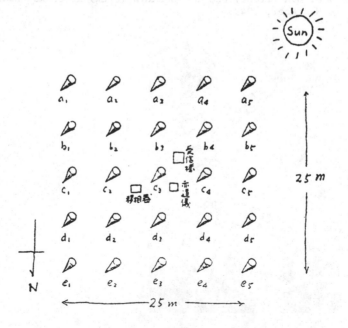

noise and sunspot number.

During the above experiments, Oda's group designed a grating interferometer for the localization of noise sources on the sun (Ojio et al. 1950). It consisted of 5x5=25 circular horns of 50-cm diameter for use at a wavelength of 7.5 cm (Fig. 4). This idea was presented in 1950 to the annual assembly of the Physical Society of Japan, but unfortunately was not realized. Prof. Oda belongs now to the Institute of Space and Astronautical Science in Tokyo, where he has applied a similar idea to X-ray observations, known as the modulation collimator or 'Oda Collimator'.

At that time I was an electrical engineer working with T. Kakinuma at Toyokawa (see below), and did not directly know of the above work. But we already knew the idea of the grating interferometer, probably through someone who was well acquainted with Oda and Takakura. After the completion of our first radiometer at 3750 MHz, we designed a one-dimensional grating interferometer and applied for funds for construction in 1951. The frequency of the interferometer was 4000 MHz (7.5 cm wavelength), which was by chance the same as Oda's design. The budget was partly approved in 1952, and the first 5-element interferometer was completed in March, 1953, as shown in Fig. 5 (Tanaka & Kakinuma 1953 b).

Fig. 5. The first grating interferometer at Toyokawa in April, 1953. Five 1.5 m dishes were driven by a common shaft. Wavelength was 7.5 cm. A wire-net antenna seen behind is the same as that in Fig. 12. The man is technician T. Takayanagi.

Fig. 6 is one of the first records taken by this interferometer.   This interferometer was extended to 8 elements in 1954 (Fig. 7).

It must be emphasized that the first Toyokawa grating interferometer was planned and built quite independently of the one developed by Christiansen at almost the same time (see the contribution by Christiansen in this volume).

Fig. 6. One of the first drift curves of the sun taken by the interferometer in Fig. 5, on April 25, 1953.   Time marks were put each one minute; 0 means meridian transit time, and time goes from left to right.

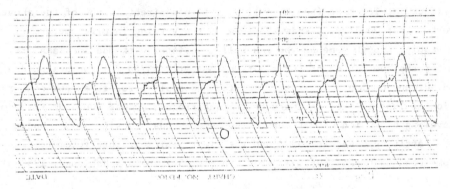

Fig. 7. 8-element grating interferometer on 7.5 cm and single frequency polarimeters on 30 cm, 15 cm, 8 cm and 3.2 cm, in 1957 at Toyokawa.

1.3 The sky temperature was reported as 0-5 K in 1951

Tanaka et al. (1951) found in 1951 that the background sky
temperature was 0 to 5 K during their extensive work on absolute calibration of the solar flux density at 3750 MHz.   In the original 1951 paper
only the abstract was written in English (Fig. 8); the full English
version was not published until later (Tanaka & Kakinuma 1953 a).
Unfortunately nobody recommended further study of this problem; fourteen
years later a general 3 K microwave background was discovered by Penzias
and Wilson.

2 EARLY RADIO TELESCOPES AND ACTIVITIES

2.1 Mitaka: Tokyo Astronomical Observatory

Activity in Japan on radio astronomy commenced in September,
1949 at Mitaka, and was directed by the late Prof. T. Hatanaka, who was
much interested in simultaneous observations of the sun with optical and
radio telescopes.   The other initial staff members were F. Moriyama and
S. Suzuki.   Development of radio astronomy at Mitaka was largely due to
the support of the Observatory Director, the late Prof. Y. Hagihara, who
had recognized the importance of radio observations in astronomy.   These
persons may be seen in Fig. 13.

The first 200-MHz telescope, 5 m x 4.5 m (Fig. 9), was constructed for use at Mitaka at the governmental Central Radio Observatory
(later called the Radio Research Laboratories).   100- and 60-MHz
telescopes were added in 1950.   A 10-m, equatorial-mount telescope (Fig.

Fig. 8. English abstract of the 1951 paper in Japanese wherein
Tanaka et al. measured an upper limit to the 8-cm background
sky temperature of 5 K.

ON THE SUBSTITUTION MEASUREMENT OF SKY
TEMPERATURE AT CENTIMETRE WAVES

Haruo TANAKA, Takakiyo KAKINUMA, Hidehiko JINDŌ
and Toshio TAKAYANAGI

The antenna toward the sky was replaced by a hot load, whose temperature was raised to
about 300°C.
When a square-law detector is employed, direct comparison is possible by reversing the phase
of either case in the low-frequency part of a radiometer. Uniform temperature rise of the hot
load, rejection of substitution error and the loss of transmission circuit are chief points of
discussion.
A result of measurement at 8 cm shows that the sky temperature is between 0° and 5°K.
(Pages 121-123)

Fig. 9. The first 200-MHz dipole-array antenna for solar
observations, started in September, 1949 at Mitaka, Tokyo.
Number of dipoles was 4 x 4 = 16, and the size 5 m x 4.5 m.
Reflectors were put behind the structure.   Drive was manual.
A dome seen behind is the 20-cm solar telescope for sunspot
observations, still in use.

Fig. 10. 10-m equatorial-mount solar radio telescope built in
1953 at Mitaka, Tokyo; the second largest microwave dish in
radio astronomy at that time.   This telescope was used mainly
for 200-MHz polarization observations of the sun, but was also
used for solar and lunar observations at 3000 MHz.   Mesh size
of the surface net was 4 mm.

10) was built in 1953; this in fact was the second largest microwave dish in radio astronomy in the world at that time.   Multi-frequency polariza-tion studies of the sun on meter wavelengths (Hatanaka et al. 1955) were extensively made by this telescope until 1963, when they had to be termi-nated due to man-made interference.

A 201-MHz, 4-element multi-phase interferometer (Suzuki 1959) built in 1958 was unique in that the source position of solar bursts could be determined with an accuracy of a fraction of an arc-minute.   Fig. 11 shows one of the antenna elements of size 8 m x 5 m.

Several other instruments were built at Mitaka, such as a 150-800 MHz spectrometer, 3000 and 9500 MHz polarimeters, etc.; by the end of 1950's about 10 radio astronomers were actively studying the sun.

2.2   Toyokawa: The Research Institute of Atmospherics, Nagoya University

The Institute was established in June, 1949 under the directorship of Prof. A. Kimpara, who was an expert on atmospherics and lightning.   The United States Air Force, which had occupied Toyokawa

Fig. 11. One of the four transit-type antenna elements for the 201-MHz interferometer built by Suzuki in 1958 at Mitaka, Tokyo.   5 x 4 = 20 dipoles are mounted over a 8 m x 5 m wire-net reflector.   On the right, a part of two-element Yagis may be seen, which were used for total solar flux observations at 100 and 67 MHz.

Naval Arsenal, a site completely bombed just before the end of the war, released a part of the site in April, 1947 for the study of atmospherics. This early release of the site was made in return for Kimpara's assistance in building a location network for lightning for the safety of trans-Pacific aviation. This is believed to have been the first peaceful use of wartime Japanese military installations.

Kimpara intended to observe the sun on centimeter wavelengths in connection with the ionospheric disturbances which affect radio communications and terrestrial radio noise. In 1949 he invited me to start a solar radio noise group; I had just terminated a 5-year post-graduate course at the Faculty of Engineering, University of Tokyo, and joined him at the end of 1949. T. Kakinuma, who was a young physicist, joined me in April, 1951.

Fig. 12 shows our first 8-cm radiometer, completed in April, 1951. At first we concentrated our efforts on observing the solar noise as accurately as possible. Fig. 13 is a picture taken in 1954 when the Japanese National Commission V of URSI met at Toyokawa for the first time. Most of the important persons in Japanese radio astronomy at that time are included.

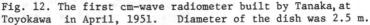

Fig. 12. The first cm-wave radiometer built by Tanaka, at Toyokawa in April, 1951. Diameter of the dish was 2.5 m.

Fig. 13. Japanese National Commission V of URSI held in 1954
at Toyokawa.　Back row, from right: T. Hatanaka, S. Suzuki,
H. Tanaka, T. Kakinuma, T. Takakura, K. Akabane and H. Jindo
(technician); front row, from right: Y. Hagihara and A.
Kimpara.

Fig. 14. Location of polarization sources by partial eclipse
observations on April 19, 1958.　Observations were made at
Toyokawa and Hachijo-jima (an island about 300 km south of
Tokyo).　See how the polarization sources coincided with the
strong magnetic field of sunspots.　Closed patches correspond
to N poles, and open areas, to S poles.

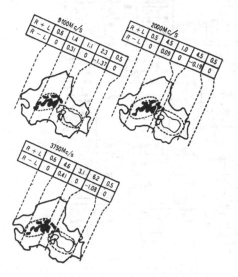

One of the most remarkable early findings at Toyokawa was that eclipse observations revealed that solar polarized emission comes from just over sunspots (Tanaka & Kakinuma 1958), as shown in Fig. 14. The finding of a narrow spectral feature on centimeter waves by simultaneous routine observations at 3750 and 9400 MHz (Fig. 15) encouraged our wide frequency-range observations on microwaves since the IGY (Tanaka et al. 1956). Our polarimeter antennas in those days are shown in Fig. 7. Later developments are outside the scope of this article.

At Toyokawa, 5 technicians, 4 assistants, and a few students from the Faculty of Engineering, Nagoya University, supported Tanaka and Kakinuma at the end of the 1950's.

Fig. 15. Simultaneous observations in 1956 of solar radio bursts on 8 cm (upper) and 3.2 cm (lower) revealed that some narrow spectral features exist even in the centimeter region (from Tanaka et al. 1956).

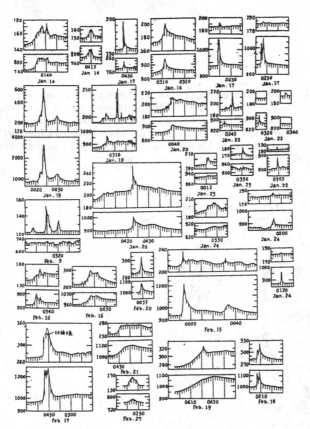

2.3 Hiraiso: Radio Research Laboratories, Ministry of Posts and Telecommunications

Besides supporting the Mitaka group at their beginning (see above), first tests for receiving solar noise at 61.2 MHz were made in 1950. Regular observations started in 1952 at 200 MHz (Fig. 16), mainly for the radio communications service. Observing frequency was later expanded.

So far, I have mentioned some selected topics and activities on radio astronomy in Japan mainly before the IGY. I am afraid my selection is far from complete, and not a few hidden topics have been left out. However I am especially happy to have had the opportunity to mention a few works in Japan in very early days, which otherwise will be buried forever.

Fig.16. The first 200-MHz equatorial-mount dipole array of 6 x 4 elements for routine observations of the sun at Hiraiso, Radio Research Laboratories in Japan; observations started in early 1952. Background is the Pacific Ocean.

References

Arakawa, D. (1936). Abnormal attenuation in short radio wave propagation. Rep. Radio Res. Japan, 5, no.4, 31-8.

Dellinger, J.H. (1937). Sudden disturbances of the ionosphere. Proc. IRE, 25, no.10, 1253-90 (see 1271).

Hatanaka, T., Suzuki, S. & Tsuchiya, A. (1955). Polarization of solar radio bursts at 200 Mc/s. I. A time sharing radio polarimeter. Publ. Astron. Soc. Japan, 7, no.3, 114-20.

Nakagami, M. & Miya, K. (1939). On the incident angle of short radio waves and high frequency noise during the "Dellinger effect". J. Inst. Elect. Eng. Japan, 59, no.608, 176 (in Japanese) ≃ Electrotech. J. Japan, 3, Sep., 216.

Oda, M. & Takakura, T. (1951). A study on the solar noise at 3300 m.c. J. Phys. Soc. Japan, 6, no.3, 202.

Ojio, T., Takakura, T. & Kaneko, S. (1950). A plan for the localization of noise sources on the solar surface by 5-column 5-row electromagnetic horns. Preprint for the 5th annual assembly of Phys. Soc. Japan, Nov. 5, 5C7, 1-8 (in Japanese).

Suzuki, S. (1959). Multiphase radio interferometers for locating the sources of solar radio emission. Publ. Astron. Soc. Japan, 11, no.4, 195-215.

Tanaka, H., Kakinuma, T., Jindo, H. & Takayanagi, T. (1951). On the substitution measurement of sky temperature at centimetre waves. Bull. Res. Inst. Atmospherics, Nagoya Univ., 2, no.2, 121-3 (in Japanese).

Tanaka, H. & Kakinuma, T. (1953 a). On the substitution measurement of sky temperature. Proc. Res. Inst. Atmospherics, Nagoya Univ., 1, 85-8.

Tanaka, H. & Kakinuma, T. (1953 b). Equipment under construction for locating sources of solar noise at 4000 Mc/s. Proc. Res. Inst. Atmospherics, Nagoya Univ., 1, 89.

Tanaka, H., Kakinuma, T., Jindo, H., Takayanagi, T. & Torii, C. (1956). The observation of solar radio emission at 9,400 MC. Bull. Res. Inst. Atmospherics, Nagoya Univ., 6, no.1-2, 61-6 (in Japanese).

Tanaka, H. & Kakinuma, T. (1958). Eclipse observations of microwave radio sources on the solar disk on 19 April 1958, Rep. Ionosph. Res. Japan, 12, no.3, 273-84.

SECTION FIVE: *Broader Reflections*

The final section consists of a group of articles addressing the early development of radio astronomy in various broader contexts. David Edge, a radio astronomer turned sociologist of science, has frequently used his former field as a case study for achieving new insights into the social forces which make science what it is. In his present contribution he summarizes a portion of this work as it applies to the contrasting styles of research found at Cambridge, Jodrell Bank, and Sydney.

William McCrea never himself became embroiled in the 1950s controversy over radio source counts and their cosmological implications, but he was there as it all happened. In his article he traces the development of cosmology in the twentieth century, in which he himself played a major role, and then discusses how the new radio data affected the cosmologist's perception of the Universe.

In the third article Hendrik Van de Hulst does not so much write about his own substantial contributions to radio astronomy, but rather examines how ideas in astrophysics evolve over a period of decades, that is, with frequencies in the nanohertz range. Although this frequency range is somewhat lower than radio astronomers normally consider, he argues that in fact such long-term changes, often overlooked, are vital to the proper understanding of the development of a science.

Owen Gingerich, trained as an astrophysicist and now an historian of science, closes the volume with a look at radio astronomy's overall effect on twentieth-century astronomy. Has it caused a "revolution" in the technical sense defined by Thomas Kuhn? In providing a cogent answer to this question, Gingerich leads us through the fantastic panoply of discoveries over the past 75 years in all areas of astronomy, but in particular those involving radio techniques.

Tony Hewish adjusting an aerial at the Grange Road site of the Cavendish Laboratory (circa 1949) (courtesy Bruce Elsmore)

STYLES OF RESEARCH IN THREE EARLY RADIO ASTRONOMY GROUPS

D.O. Edge
Science Studies Unit, University of Edinburgh

The development of the three pioneer radio astronomy groups at
Sydney, Jodrell Bank and Cambridge[1] offers historians and sociologists of
science an unusual opportunity to make a comparative study of concurrent
'research schools'.[2] What emerges is an intriguing story of the
interrelationship of technical strategy and commitments, social structure,
and scientific style.

The main advantage in studying a research school (as opposed
to, say, a more loosely defined 'specialty network' or 'invisible
college') is that any school is an unproblematic *group* - it consists of
specifiable people, working together. The three radio astronomy groups
also shared a number of other common factors, which simplifies any
comparative analysis. To begin with, they all embarked, essentially from
scratch, into a virgin area of research in which they shared the same
initial clues - namely, the original observations of Jansky, Reber and
Hey - solar and galactic radio emissions; the existence of discrete
sources, or 'radio stars'; and radar reflections from meteors. What is
more, since all the researchers had been engaged in wartime radar
development, they shared the same technical resources and insights, and
cannibalised or adapted the same equipment. Many of them had been active
colleagues. They understood the same patois of electromagnetic theory —
especially as it related to aerial design — and they all had access to,
and familiarity with, at least some elements of the crucial 'Fourier
transform' theory. The two British groups were both sections of
prestigious university Physics Departments, and both Martin Ryle
(at Cambridge) and Bernard Lovell (at Jodrell Bank) were given
effectively *carte blanche* support by eminent superiors — J.A. Ratcliffe
at Cambridge, Patrick Blackett at Manchester. The Sydney Group differed
significantly here. It was in a Government research establishment, and
J.L. Pawsey, the radio astronomy group leader, was at times overruled by

E.G. Bowen, the Director of the whole Radiophysics Division. Resulting
conflicts and competition for scarce resources seem able to account for
many details of the Sydney group's development. All three groups,
however, had no problems in recruitment, nor in access to publication
outlets.

And yet the three groups developed on very different lines.
Nor were these differences simply a matter of choosing distinct research
topics. Work on discrete radio sources emerged at all three centres —
and it is on this work that I wish to concentrate.

SYDNEY

At Sydney, by around 1950, the group consisted of essentially
autonomous research teams, working on physically separated sites. As one
Sydney pioneer put it :

> People tended to explode out first thing in the morning
> and you didn't see them until the evening. They tended
> to be quite independent groups, and competitive in some
> ways . . . and we would have this meeting once a week
> . . . this attempt to get all the information from one
> group to another, just because they were so relatively
> isolated during most of the week . . . A person like
> Pawsey knew the story well, and he would move from one
> place to another and kept a very close contact with the
> various groups. But everybody . . . was very involved
> with their own equipment and running it . . . and
> because they were responsible individuals they were given
> their heads to get on with what particularly interested
> them.

Two of these "responsible individuals" who formed teams at
Sydney for work on discrete radio sources were J.G. Bolton and
B.Y. Mills. At that time, since phase-stable cables, low-noise receivers
and massive computing facilities were not available, anyone attempting to
map the radio sky had to adopt a technical strategy which fell short of a
Fourier transform of a comprehensive sampling of the amplitude and phase
information in an incoming wavefront, and opt for a technique which
involved simplifying assumptions and compromise. Early experience at
Sydney, using initially a cliff-top 'Lloyd's mirror' technique, had
demonstrated the difficulty of analysing the confused records from simple
spaced-aerial interferometers. Mills was therefore convinced that the
compromise should mimic a 'pencil-beam' instrument, but aim for a
resolving power greater than could be obtained from any single 'Big Dish'.
He opted for a phase-switched 'Cross' array. This was designed to produce

unambiguous signals from point sources, but was, of course, much less
sensitive than a fully-filled aperture of comparable dimensions. At the
lower frequencies required for radio source observations, detailed
structure and weaker sources tended to be swamped by a combination of
receiver noise and the general galactic background. This, in turn, led
to problems of interpretation which, although different in kind from those
arising from interferometer records, proved almost equally troublesome.
Bolton, on the other hand, felt that a strategy built around at least
one large Big Dish was preferable — not only for its sensitivity, but
also for its ability to explore higher frequencies, and its adaptability
as a general-purpose instrument.

During the 1950s, Mills had the upper hand in this argument.
He built a major Cross array, and Bolton left Sydney for Caltech. But
Bowen later backed Bolton, who returned to CSIRO to build the 210-ft
Parkes dish; Mills then left for the University of Sydney, where he later
built a 'Super-Cross' array. The details of these movements are not
important for my argument here; much of the early work at Sydney is
discussed elsewhere in this volume by Bowen, Christiansen, Kerr, and
Mills. The point is that the social structure of the Sydney group, only
loosely coordinated by Pawsey, led to small autonomous (and competitive)
teams wedded to the exploitation of what are essentially *subsets* of a
general instrumental strategy — and hence committed to pressing these
specialised solutions, and their claims and results, to their limits.

JODRELL BANK

At first, there was strong social cohesion at Jodrell Bank :

There was an atmosphere of everything being new and
exciting about it, and Jodrell Bank was a rather
pleasant place. In the day there were only half a
dozen people there, and we were all tremendously
good friends, and there was sort of a huge great
personal friendship. And we used to go round there
weekends, Saturdays, Sundays, all night, all day.
We had one or two old ramshackle vehicles that we
managed to get the University to buy to transport
materials and people out; and we used to drive out
from Manchester . . . at all hours of the day and
night . . . If we'd done a whole day's work there
and were going home at night, we'd all pile into
the local transport and go round to the local pub
and have a drink, and then the transport would drop
people off at their homes . . .

Everyone at Jodrell Bank worked on one site. However, radar
research continued alongside the radio work until late in the 1950s.
This quickly introduced a clear differentiation, and the needs of both
sets of researchers effectively determined an early commitment to the
design and building of a 'multipurpose' Big Dish, which could be used for
both (radar) transmission and (radio) reception. Already, from the late
1940s, a 218-ft fixed reflector had been in use at Jodrell Bank for radio
work. By the mid-1950s, autonomous teams had developed at Jodrell Bank,
very much as at Sydney — but with the difference that they all used, or
planned to use, one of the Big Dish elements then becoming available.
In 1957, the 250-ft Mark I instrument came into use, and consolidated
this strategy. It was the first of a family of such multi-purpose
paraboloid reflectors at Jodrell Bank.

It is by now a truism in the sociology of science that one
common determinant of research problem choice by scientists is the nature
and scope of the instrumentation available to them. The point is well
illustrated here. It seems that, to Jodrell Bank researchers, this
family of multi-purpose instruments had something of the character of
unalterable 'givens' — fixed resources to be adapted, as far as they
allowed, to different ends. Here is one expression of what might be
called this more 'passive' approach to instrumentation :

> *Have you often had the feeling that you are limited*
> *by the available techniques?*
>
> Yes, to some extent. Often, at Jodrell, I would like
> to have had . . . well a spectrometer was one case,
> and I jolly well pushed ahead and got it eventually . . .
>
> *But you could have got it?*
>
> In some cases you can, but generally, I think, when the
> answer has been "No", you say, "Well, that's just too
> bad, but there are plenty of other things I want to get
> ahead with". I think your thinking and planning tends
> to be conditioned by what you have got on hand, and you
> can see interesting and important problems you want to
> tackle with the equipment you have got without putting
> effort into a wide number of other techniques or projects.
> I don't think we have ever gone through a phase where we
> have said, "This thing is a waste of time : let us build
> [a Cambridge-style array] at Jodrell" — quite apart
> from the politics of doing it, I don't think anyone would
> ever have contemplated that. There is so much that one
> felt was worthwhile to get ahead with, and we've never
> felt that there's been really a shortage of equipment or
> techniques to go ahead with a problem that you want to do.

Research topics were chosen, at least partly, in the light of the potential appropriateness of the available instruments to their exploration:

> There are quite a variety of topics where one really said, "Well, look, we've got a piece of equipment — what can we do with it?"

So radio source work at Jodrell Bank in the 1950s tended to concentrate on the development of long-baseline interferometry, with radio links to smaller satellite aerials — again, a subset of the ideal, Fourier transform approach (see the contributions in this volume by Hanbury Brown and by Lovell). But choices were also influenced by the team structure at Jodrell Bank, and by the alternately competitive and cooperative nature of scientific research. The process is neatly described by a radio astronomer who joined the Jodrell Bank staff in 1959 :

> I was interested in sources, and when I came here, I more or less joined with Y, who was at that time getting the Mark I going on source work. The first summer I came, I did spend quite some time working with Z on the long-baseline interferometer, but I think both sides realised this was a temporary measure when I first came, and I didn't want to continue with them as a permanency.
>
> *Why not?*
>
> Well, they had a going group. They had their own ambitions, they had their own techniques and ways of doing things, and you may fit in or you may not. I don't think there were any deep reasons for this, but I didn't want to muck in and become one of that gang . . . So X and Y and I fitted up the Mark I with 610 Mc/s and observed some sources at frequencies around there, and I think I would say the biggest break came when we got the offer of cooperation from A at Cambridge, with B [an American] . . . .
>
> *And you still work on sources?*
>
> Yes. At that time what interested me most was the spectrum . . . and this went on. We did some observations of other wavelengths here, because A came here . . . and we worked together, and well . . . it became apparent to me that this was becoming a fairly well-ploughed field, ploughed particularly by B, and I didn't just want to go on measuring the intensity of many, many and many and many, more sources. And we'd set up the instrument consisting of two telescopes here, the Mark I and Mark II, and it transpired that this is an exceedingly good way of measuring polarisation. So I changed over and transferred my major interest to polarisation of sources, and it has remained there ever since.

As at Sydney, the teams at Jodrell Bank tended to develop and 'own' their
subsidiary equipment :

> . . . there was rather a feeling that this was our bit
> of equipment, even to the extent that parametric amplifiers
> belonged to one group . . . If you change[d] from one
> experiment to another, then one group ripped everything out,
> and the other group had to install everything again . . . .

As the groups at both Sydney and Jodrell Bank differentiated
in this way, neither Pawsey nor Lovell could maintain an overall sense of
theoretical cohesion and direction. Ideas essentially originated in, and
were pursued by, the teams. Everyone retained a curiosity in what others
were doing : but, without social cohesion, theoretical (and often
technical) interplay was minimal. As a Jodrell Bank researcher
remarked :

> . . . I think there is a big contrast [here] between
> [Cambridge and Jodrell Bank]. But I think it's partly
> to do with the type of equipment we have. I mean, a
> big telescope like this [the Mark I] can have many jobs
> going and many specialties in hand. I really defy one
> man, you know, to unify line work and cosmology, on the
> one hand, and background polarization — the lot . . .
> A lot of astronomy goes on here . . . .

In such a situation, where theoretical cohesion and direction are lacking,
so is the main rational means for comparing competing bids for resources
and observing time. Such strategic decisions then become a matter for
broader, 'political' resolution, and group leaders are more exposed to
overt conflict.

CAMBRIDGE

The group at Cambridge, led by Ryle, stands in marked contrast
to the other two. At least until the early 1960s, it lacked any clear
and consistent internal differentiation. Throughout the 1950s, what
essentially united the group was the singlemindedness with which it
pursued the technical strategy of pressing Fourier transform principles
as far as the state of the art (and likely grants) would allow them —
taking advantage, as they became available, of technical advances in
phase stability, low-noise amplification, and computing. Mapping the sky
was conceived in terms of its most general principles, and these were
successively realised. In the process, the Cambridge researchers assumed
a more 'active' attitude to, and relationship with, their major
instruments. As we have seen, attitudes at Jodrell Bank can be

characterised as relatively 'passive' : the Big Dishes were treated as essentially *fixed* resources, 'givens', in whose design and construction the researchers had played little part. Their initiative was largely confined to the design of subsidiary equipment. But at Cambridge (as with Mills' team at Sydney) this initiative could encompass the design of the main arrays themselves, which could then be viewed as essentially *manipulable* resources. The nature of the technical task, and the way in which the Cambridge group organised themselves to tackle it, can together explain this difference.

The prime focus of the Cambridge group's work consisted of the planning, design, building and exploitation of a series of receiving arrays consisting of spaced aerials, on principles which eventually became known as 'aperture synthesis' (see the article elsewhere in this volume by Scheuer). These arrays were designed primarily to carry out comprehensive surveys of discrete radio sources, but they could be adapted later to other uses. The early arrays were complex inter- ferometers, with all the problems of confusion-limitation that their large secondary responses implied — which laid them open to constant criticism from Mills, at Sydney, with his pencil-beam commitment. But later arrays (designed, in principle, in 1953/54 — although the full theory was not published until 1960[3]) effectively overcame these problems, and cannot be so simply classified.

Ryle coordinated this technical work, bringing in other group members as appropriate, so that the senior researchers were all involved in the process. He supervised the building of the early arrays himself, without using outside consultants. He ran the group on broadly democratic lines, with all major decisions made by a small 'governing body', but it was clear that he was accepted as the intellectual and experimental leader. The whole group thus maintained a high level of social cohesion, and hence of theoretical discussion, and of the definition of shared scientific goals. This cohesion was supported, not (as in the early years at Jodrell Bank) by physical isolation and pub visits in ramshackle buses, but by such institutions as the weekly 'Range Day' (when everyone was expected to help with menial tasks, such as cutting grass and scraping paint off old aerials), and regular Saturday morning literature discussions. There were identifiable 'theorists' in the group, and the variety of work undertaken was, on the whole, viewed within one

integrating framework. Ideas tended to be treated as 'common property' :
there were a lot of them "sculling round the group". (F.G. Smith recalls
these days in his own contribution to this volume.)

In the mid-fifties, the Cambridge group, led by Ryle, started
to interpret its data in bold, cosmological terms, relating the various
research programmes in the group by explicit discussion of astrophysical
theory (see the article by McCrea elsewhere in this volume). In 1955, the
group published the over-ambitious 2C Survey of radio sources; this
listed some 2000 sources, of which all but 200 or so were later
discredited. (In 1959, the subsequent 3C Survey listed less than 500
sources.) However, such numbers invited statistical treatment, and the
slopes of graphs of the number-intensity relation ('Log N/Log S') of
samples of radio sources, backed by the independent analysis of the
deflections on the survey records ('P of D'), became major components in
these cosmological arguments. For a period, this approach dominated the
group's thinking : ". . . from the time that the counts were discovered,
then the cosmological problem, yes, filled the group". Or, as another
interview ran :

> *Cambridge took the lead in discussing Log N/Log S,
> and this defined a scientific strategy, as well as
> a technical strategy?*
>
> It did, yes.
>
> *And that was virtually unique to Cambridge?*
>
> That's right. But, of course, . . . [the] possibility
> of the cosmological significance on this sort of scale
> [only] came up at the time of 2C.
>
> *And became something that the group as a whole shared in?*
>
> Yes.
>
> *There weren't members of the group who were really
> sitting on the sidelines and saying, "Look here . . ."?*
>
> No, I think everybody was absolutely enmeshed in this . . .
> Previously to that time it had been quite clear that you
> could see a long way into the Universe, but the idea of
> ever finding out whether it bends, or whether there are
> more things, and so forth, I think didn't emerge until 2C.
>
> *Do you feel yourself, now, that there was a certain sort
> of overreaching . . . not just technical, but scientific
> too?*
>
> No, I don't. I see this as a real sort of scientific *grasp*,
> a sort of vision of what could come out of it. I mean, not
> presumption — daring, I would say — saying "By gosh, perhaps
> this is really the first time we can do some cosmology" . . .

Did this come predominantly from Ryle?

The first idea of counting radio sources to produce a
Log N/Log S was actually in [Smith's] thesis. But then,
of course, we thought that all the sources were galactic
. . . But certainly the cosmological thing came from
[Ryle]. But there again, you see, that is clearly another
idea sculling round the group, that by counting sources
and looking at populations you can find something about
their distributions in space.

*I have a feeling that the group really had made up its
mind about the Universe in an important sense.*

I think that's true.

At what point?

I think it was not until the P of D analysis had been done
. . . [When] the 2C attitude, if you like, had been sort of
vindicated by P of D, well, the details of 2C were wrong,
[but] then I began to realize that this was correct, and
meaningful.

When, in 1958, Ryle synthesised the range of available data in
his Bakerian Lecture to the Royal Society,[4] Cambridge group members were
impressed by his brilliance, but outsiders tended to shrug it off as
merely confirming positions they already effectively held. As one Jodrell
Bank researcher put it :

> . . . the main argument by this time was not so much what
> these things were — nobody was disputing they were
> extragalactic and high-powered, and so on — but people
> were still — still are, to a certain extent — disputing
> exactly what the Log N/Log S means . . . [With] the
> vigorous dispute over the reliability of the number counts,
> rightly or wrongly, [Cambridge] became identified with the
> Log N/Log S. It wasn't necessary.

What about P of D?

We were pretty impressed with it, more so than with the
number counts. But we always had these nagging doubts —
some of them were wrong — as to what happens with
extended sources, and what happens if we are dealing with
something we don't understand. We felt it was a pretty
big step to lump everything together and say "There you are,
this is [it], we have counted a few hundred sources" —
which had all been got wrong once, and if the slopes had
come down from 3 to 1.9, where are they going to come down
to next? You know, are we really going to solve the Universe
by counting 500 sources? . . . the Log N/Log S certainly
coloured people's attitudes. But I think possibly people
felt . . . that all these sources were a study in themselves,
and the Log N/Log S was a possibility beyond it. And it
always seemed that this was the end of the whole : this was
what it was all tending towards.

To a group with Cambridge's social cohesion and technical strategy, reaching for the "possibility beyond" seemed the natural thing to do. Indeed, so involved were the Cambridge group members in the task of *technical* development, that they are notably ready to view the *scientific* results of their work as a kind of fortuitous 'spin-off'. To them, the technical development had an autonomous character. These comments are typical :

> The point I would like to make very strongly is that from 1949 onwards the recognition that an interferometer provided one term of the Fourier transform of the sky led to an active and continuous development.

> I feel very strongly that the technical developments had (and still have) an active life of their own quite independent of the immediate demands set by astronomical problems. It is then just chance whether the necessary technique is to hand at the right time.

> The main feeling was a joy in pressing new techniques to the limit of their possibilities as we saw them . . . The [Cambridge] group was ignorant of, or impervious to, outside influences or examples.

It may seem paradoxical that a group so actively committed to treating their major arrays as manipulable resources should apparently develop such a firm belief in the *inevitability* of the steps they took. However, this version of 'technological determinism' (if that is what it is) emphasises the influence, not of specific instruments, but of the realisation of technical *possibilities*. Since the latter are conceptual in character (". . . as we saw them . . ."), and hence are socially mediated, the distinction is important. Those at Jodrell Bank see the influence of particular instruments (the Big Dishes) on their research strategies ; they therefore interpret the Cambridge group's development in similar terms :

> There's a programme that they've followed — mostly individual sources and so on — the techniques more or less force them to do it. They've got one piece of equipment, and they just follow this up and do work that this machine will do for them . . . I think, having one piece of equipment like that, there's not the possibility of diversification that we have with a telescope like this here [at Jodrell Bank].

But the inadequacy of an explanation of group differences in terms of this simple technological determinism was realised by an ex-Jodrell Bank worker :

. . . the Jodrell Bank telescope was used for what it was
intended, as a general-purpose, flexible instrument.
In Holland, using a moderately sized instrument, and at
Parkes [Sydney], single dishes were used in a more
singleminded way — in Holland for hydrogen line work,
and at Parkes where they *exploited* the field in source
position measurements.   Source positions *could* have been
done at Jodrell in the same way, but it was not so
specialised, and all the workers were not directed largely
to one single problem. You never had the whole group of
people working night and day on one single dominant topic.

'Exploitation' of this kind requires (and produces) social
cohesion. Technique alone is not sufficient.

DISCUSSION

One is tempted to say that at Sydney and Jodrell Bank
socially differentiated groups, composed of autonomous teams, coordinated
by essentially 'administrative' leaders, *therefore* produced *'fragmented
science'* — in the sense of a variety of scientific and technical
programmes, pursuing unrelated topics. Whereas at Cambridge a socially
coherent group, democratically led by an involved and active leader,
therefore gave rise to *'coherent science'* — and to results which many
outsiders found either premature or puzzling. Certainly, there seems to
be good evidence here of the close relationship between social structure,
technical strategy and scientific aim within a research school.

However, Nobel Prizes notwithstanding, one should resist the
temptation to look here for a 'recipe for success'. The singleminded
attempt at Cambridge to 'make everything fit' seems to have led them into
what would now be generally regarded as 'mistakes'. The initial
rejection by Ryle of the Sydney group's claimed radio identification of
the Crab Nebula,[5] and the odd tenacity, in the early 1950s, of Ryle's
belief that the majority of radio sources were *not* extragalactic,[6] are
two examples of claims which, it could be argued, were led astray by an
overemphasis on theoretical consistency. And, of course, the Cambridge
commitment to the success of their arrays led, in 1955, to the over-
confidence of their first major source catalogue (the 2C Survey), and to
the criticism led by Mills, with his alternative (if less dramatic)
commitments and claims. For a time, the Cambridge group was virtually
discredited. If (as is not entirely improbable) the verdict of history
on them in the end had been one of 'failure', then the story I am here
sketching would have to be read as a cautionary tale.

In its size, flexibility and cohesion; in the high level of
both its theoretical competence and the completeness of its in-house
expertise; in its competitiveness; and in the style of its leader —
in all these respects, the Cambridge group of radio astronomers seems to
resemble closely the Italian group of nuclear physicists, led by Enrico
Fermi in the 1930s.[7] Both groups adopted a strategy, in a competitive
situation, which entailed that their mastery of both theoretical and
technical 'frontiers' must be as rigorous as possible. This, in turn,
implies social cohesion and a high level of internal discussion. If
resources are to be concentrated, the risks entailed by possible errors
are larger. Everyone, therefore, has a stake in ensuring that the
intellectual exchange within the group is as efficient as possible :
coherence is encouraged.

In at least one respect, however, Ryle's group differs
markedly from Fermi's. Among astronomers, the Cambridge group is (or was)
notorious for the tightness of its boundaries, and for the limits to which
it would go to maintain 'secrecy' (for want of a better word) as to its
plans and results. Fermi, in contrast, was so keen to disseminate his
findings that he is credited with the invention of the preprint. Perhaps
the key difference here is that the Cambridge group was developing
(and therefore felt it necessary to protect) a *general principle* of
technical strategy; if this were to be grasped and realised elsewhere
(especially where resources could be more quickly mobilised), competitive
advantage might be lost. Fermi's group, on the other hand, was merely
bringing together techniques already established elsewhere, and responding
rapidly to a stream of fresh discoveries : to them, quick dissemination
of their results was the way to *assert* their competitiveness and *establish*
their lead.

Which leads naturally to one final point. Many people
(scientists among them), in seeking an explanation of the differences
between these research schools, trace them to differences between the
personalities of the leaders. Thus, for instance, the Cambridge group's
scientific achievements are 'explained' by Martin Ryle's 'genius', the
group secrecy by his alleged 'paranoia', and so on. I resist this kind of
'personality reductionism'. It can lead to distasteful excesses : but
much more important, it is too simple an approach. The detailed processes
of scientific advance involve a far wider range of factors than can be

attributed to (or encompassed by) any one 'personality'. Yet research
schools *are* established by *people,* and must reflect, in some measure,
their characteristics. One way of reformulating this difficulty is to say
that the kind of analysis I have sketched out here is 'another way of
talking about the personalities involved' — just as an X-ray diffraction
photo specifies a molecular structure as well as does an atomic model.
As I have written elsewhere[8]:

> Two major factors have emerged in explanation of the
> differences between the [Cambridge and Jodrell Bank]
> groups — the different leadership styles of Ryle and
> Lovell, and the different technical strategies adopted
> at the two centres. It is clear that these two factors
> are related. It is possible to argue that the second
> derives from the first and that the differences between
> the two groups can be adequately explained by postulating
> simply that Ryle and Lovell "are two different people".
> It is our view, however, that such a move is unnecessary.
> The sequences we have been describing have consisted of a
> series of scientific (i.e., cultural), technical, and
> social constraints *that have allowed two different styles
> of leadership to operate at the two centres.*

Although I must confess that I find this passage as puzzling
as when I first wrote it, ten years ago, its central point remains :
different styles of research are not *just* the product of the different
personal characteristics of those responsible for their emergence.

NOTES

1 Readers can find the full documentation of the study on which
   I am drawing in this article in Edge & Mulkay (1976),
   especially Chapter 9. (I refer to this book in these notes as
   AT.) This study concentrates on the development of the
   Cambridge and Jodrell Bank groups, with only incidental
   information on developments at Sydney : the latter therefore
   play a subsidiary role in this article. Quotations in this
   article are taken from interviews with radio astronomers
   conducted over the period 1971-73. For several reasons
   (including sociological convention) these interviews were
   carried out on the strict understanding that they could only
   be used anonymously : this undertaking cannot be relaxed for
   the quotations in this article.

2 For a survey of studies of research schools, and a statement
   of their importance, see Geison (1981).

3 See Ryle & Hewish (1960).

4 Ryle (1958).

5 See *AT,* 101-3.

6 See *AT,* 95-9.

7 See Holton (1974).

8 *AT*, 346.

REFERENCES

Edge, D.O. & Mulkay, M.J. (1976). *Astronomy Transformed : The Emergence of Radio Astronomy in Britain*. New York & London : Wiley-Interscience.

Geison, G.L. (1981). Emerging specialties and research schools. *History of Science*, 19, no.1, 20-40.

Holton, G. (1974). Striking gold in science : Fermi's group and the recapture of Italy's place in physics. *Minerva*, 12, no.2, 159-98.

Ryle, M. (1958). The nature of the cosmic radio sources. *Proceedings of the Royal Society*, A, 248, 289-308.

Ryle, M. & Hewish, A. (1960). The synthesis of large radio telescopes. *Monthly Notices of the Royal Astronomical Society*, 120, 220-30.

(l. to r.)   Tommy Gold, Hermann Bondi, and Fred Hoyle at the 1961 General Assembly of the IAU in Berkeley (courtesy Ron Bracewell)

THE INFLUENCE OF RADIO ASTRONOMY ON COSMOLOGY

W.H. McCrea
Astronomy Centre, University of Sussex, Brighton BN1 9QH
England

It is a privilege to participate - along with a number of those who played leading parts in the discoveries being commemorated - in this notable celebration. At the same time I do have the feeling of being dug up as a fossil theorist in order to illustrate one aspect of the developments concerned. Anyhow, my rôle is to try to bring alive the atmosphere of those years when radio astronomy first entered the field of cosmology with unexpected and exciting consequences. For other reasons as well, this was a time of crucial developments in cosmology. Since I have both to sketch in the background, as well as bring out highlights in the foreground, I necessarily give a picture without much detail. I begin with a review of developments in cosmology from the time of Hubble to the early 1950s, and then discuss the influence of radio astronomy. Such detail as I do give will be subject to much selection, largely because much of the action with which I am best acquainted took place in Cambridge and in the Royal Astronomical Society.

COSMOLOGY OF THE VISIBLE UNIVERSE

In 1929 Edwin Hubble began to publish his observational discovery of the expansion of the universe. I shall proceed on the assumption that the expansion is real in the most ordinary sense. Hubble found that for any remote galaxy

$$\text{recession speed} \overset{\sim}{\propto} H_0 \text{ x distance}$$

where for the given epoch of observation H_0 is a constant, the same for all galaxies. Hubble established his distance scale by a number of steps, all but the first and shortest employing standard candles of one sort or another. The quantity $T_0 = H_0^{-1}$ is then a time (Hubble time) characteristic of the observed universe. From the outset, it was natural

to suppose this simple linear relation to have the status of a first
approximation, and it appeared to be a good one. It had the commendable
feature that it must hold good in the same form for an observer on any of
the galaxies concerned. It was the first ever discovery of a concerted
behaviour of the universe as a whole. It had, however, been adumbrated by
earlier observations by V.M. Slipher, and by relativistic models discovered
by W. de Sitter, A. Friedman and G. Lemaître. So it was hailed also as a
predictive triumph for general relativity theory.

In due course Hubble looked for other empirical properties of
the universe in the large. The sort of observation that appeared likely to
be of interest in principle was subject to immense difficulties in practice,
e.g. any property likely to be of statistical significance would have to
apply to a very large region of the universe because the universe is
manifestly irregular up to a very considerable scale. Or again most
conceivable investigations demanded samples that for the purpose in hand
could be regarded as complete, which is in general the severest sort of
demand upon any set of observations. About all that Hubble could claim in
this context was: were it asserted that (a) an observer on any galaxy
would have statistically the same picture of the universe as an observer
on any other, (b) the universe as seen from our galaxy is statistically the
same in all directions, then available observations would not contradict
these assertions. For the rest, optical observations tended to concentrate
upon trying to find a second approximation (an acceleration term) in
Hubble's velocity law.

In Britain at any rate, part of the reason for the ready
acceptance of the notion of the expansion of the universe was that it
fitted in with ideas that A.S. Eddington was developing at the time about
the meaning of the constants of physics and about the harmonization of
quantum physics and cosmic physics. In his scheme the cosmical constant Λ
played a crucial rôle, and he saw in the expansion an empirical means to
evaluate this constant. Eddington had enormous prestige as a theorist, and
Hubble had very great prestige as an observer. If they agreed that the
universe is expanding, then it had to be expanding. Moreover, if they
agreed substantially about the rate of expansion, then that pretty well
had to be the rate. In this regard they played into each other's hands;
Hubble ultimately found a value of H_0 about 560 km/s/Mpc, and Eddington
predicted a somewhat greater value (between about 570 and 590). This
played a key part in the history of the subject. (Slater (1954) later

maintained that Eddington's own reasoning should have given him a value
near to Baade's first revision of Hubble's result, which will be mentioned
later. As a matter of history, however, the values just quoted were those
that mattered at the time). It may be noted that $H_0 = 500$ kms^{-1}Mpc^{-1} gives
$T_0 \overset{\sim}{} 2 \times 10^9$ years and $H_0 = 600$ gives $T_0 \overset{\sim}{} 1.6 \times 10^9$ years.

Almost any plausible interpretations of the observations
indicated that the observed expansion must have started about time T_0 in
the past. If $\Lambda > 0$, it could be inferred that the universe went through a
congested state about T_0 in the past, without trying to say much about what
might have happened before that. Here it should be recalled that in the
1930s there was the great debate among astronomers about 'short' and 'long'
time-scales for the past life of our Galaxy; some evidence indicated something between about 10^9 and 10^{10} years, other evidence appeared to require
about 10^{12} years. On the strength of Hubble's results, some astronomers
tended to think that the shorter time could be associated with the phase of
the observed expansion, while they had still to discover what had gone on
before that.

Other cosmologists, including most notably Einstein himself,
took the view that the most significant feature of the observed expansion
of the universe for relativity theory was that it allowed them to discard
the abhorrent cosmical constant. Einstein had introduced it only so that
the theory would admit a static universe; if the actual universe was now
seen to be non-static, then Λ was simply unwanted. In that case, simple
theory implies that the universe can never have been expanding more
slowly than it is now, so the age of the universe cannot be more than
T_0 - say about 2×10^9 years on Hubble's figure.

The evidence of radioactivity showed the Earth's oldest rocks
to be of that order of age. Up to the outbreak of World War II in 1939,
most astronomers were, I think, prepared to find that not too dramatic
revisions of various measurements would bring everything into agreement for
a universe not much more than 2×10^9 years old. And that is how the matter
remained for almost the next ten years. The one thing nobody ever actually
said in those days was that the observations might be badly mistaken. At
any rate I do not remember anybody saying this, but I certainly remember
continually wishing that our knowledge of practically the entire universe
did not have to depend upon one man's observations, or perhaps two - Hubble
and M.L. Humason together - no matter how skilled they might be.

NEW THINKING: STEADY-STATE COSMOLOGY

What triggered new thinking on the subject after the War was
the question of the origin of the chemical elements. Problems of stellar
structure, stellar energy-generation and stellar evolution were being
studied on a more realistically physical basis than before. It had become
generally accepted that a normal star lives mainly upon the conversion of
hydrogen into helium through the proton-proton reaction or through the
carbon-nitrogen cycle. So the raw material of stellar energy is hydrogen,
and it was fairly clear that normal stars still consist to a large extent
of hydrogen. So there were two very vital, and presumably closely related,
questions. Where did the hydrogen come from, and whence came any other
elements needed at the outset of a star's existence? George Gamow and his
colleagues in the United States had ideas on the subject which may be said
to have led ultimately to big-bang cosmology, and which about the time
concerned were predicting cosmic background radiation - as was so soon to
be forgotten! But in Cambridge in England in 1948 there was a different
approach. Fred Hoyle had the concept of modifying the field-relations of
general relativity by the introduction of a creation-term. This would
admit the continual creation of fresh matter, presumably in the form of
hydrogen atoms or of protons and electrons in about equal numbers. As the
simplest resulting possibility he looked for a steady-state solution in
which the old material expanding out of any region of the cosmos would be
continually replenished by new material created within the region. Working
at the same time Hermann Bondi and Thomas Gold started from the postulate
of the perfect cosmological principle, roughly speaking the hypothesis
that the universe in the large (or the smoothed-out universe) is the same
everywhere and everywhen. The only model satisfying this together with
relativistic kinematics is the steady-state expanding universe, which
obviously demands continual creation. In fact the two approaches led to
the same steady-state cosmology.

The unique feature of the resulting cosmological model was
precisly that it was unique. Apart from the scaling factors of the density
and the Hubble constant, there is only one steady-state universe. If
observation gives any property of the actual universe that disagrees with
a prediction of the model, the model must be rejected. In particular, if
observation shows that any intrinsic property of the actual universe varies
with redshift, i.e. with distance, then the model is falsified as a model
of that universe.

By 1948 it had become clear that the age of the universe given by relativistic cosmology ('evolutionary' cosmology) with Hubble's value of H_0 was absurd; it was less than the age of the solar system! I think it was not this that triggered steady-state theory, but in any case once that theory had been presented, I think for most astronomers the strongest argument in its favour was that they thought it offered a way of escape from such absurdity. This was because they jumped to the conclusion that if the universe is in a steady state it is infinitely old and so there must be all the time in the world for anything anybody needs. In fact this is not much more sensible than thinking that if you live in a very old city you should have more time to do things than if you live in a 'new town'. When it was proposed, steady-state theory actually made the situation more difficult; this can be seen as follows.

If t_0 is the age of an evolutionary model (with zero Λ) then it is easily shown that $t_0 < T_0$, the Hubble time. Nothing in the model is older than t_0, and so a fortiori nothing is older than T_0. But it is possible to have a model with t_0 very near to T_0, and so the possibility that, say, our Galaxy and all neighbouring galaxies have age nearly equal to the Hubble time is not excluded. On the other hand, in a steady-state model, it is easy to show (McCrea 1950) that in any region the fraction of galaxies with age greater than a, say, is $\exp(-3a/T_0)$; the mean age of the galaxies in the region is then $\frac{1}{3}T_0$. Thus if $T_0 \stackrel{\sim}{\sim} 2 \times 10^9$ years the mean age of the galaxies we see around us would be about 6×10^8 years, and if the age of our own Galaxy is about 1.6×10^{10} years as indicated by astrophysics it would be older than all but one in about $\exp(24) \stackrel{\sim}{\sim} 3 \times 10^{10}$ of those around it. With the Hubble value of H_0 the age problem was therefore actually more acute for steady-state cosmology than for evolutionary cosmology in general.

Looking back, it appears that the natural conclusion would have been: with the accepted value of the Hubble constant, evolutionary cosmology leads to an absurdity, and if we look to the drastic remedy of steady-state cosmology we find a worse absurdity; so there must be something wrong about the accepted value. There was indeed no need to mention cosmological models at all, or even any particular interpretation of the cosmological redshift. All that mattered was that there was a cosmological effect that appeared to be proportional to distance; so the cosmos must involve a characteristic distance; astronomers are accustomed to express distance in time (light-year), so this is equivalent to saying that the

cosmos must involve a characteristic time, the Hubble time. But a time of
about 2×10^9 years was not characteristic of the cosmos. What Hubble had
tried to measure was in fact distance. So Hubble must have got his
distances wrong.

It is surprising that so many physicists and astronomers so
quickly took a sympathetic view of steady-state cosmology after the first
publications on the subject in 1948. The main astronomical reasons appear
to have been (a) the actually mistaken belief that it offered a significant
escape from the dilemma of the short time-scale, (b) the fact that the
apparently greatest-ever discovery in astronomy - the expansion of the
universe - had accounted for nothing about the astronomical universe; but
this, as we now see, was because astronomers were looking for an effect at
an impossibly short time in the past, (c) the continual supply of new
hydrogen seemed to solve some problems in astrophysics, and (d) in a
steady-state universe, the problem of galaxy formation was that of forming
new galaxies in the presence of existing galaxies, whereas in evolutionary
cosmology there was the apparently far more intractable problem of the
formation of the first galaxies.

Many scientists have confessed, however, that they found an
aesthetic appeal in the steady-state concept as contrasted with a distaste
for the notion of the big-bang that has now replaced it. There appears to
be nothing rational about a preference for a little bang for every
elementary particle in the universe as against one big bang for the entire
universe; rationality appears to suggest that these are simply two extreme
possibilities for contemplation and that cosmologists might find it
interesting to look for intermediate cases.

NEW MEASUREMENTS

In 1952 Walter Baade startled the astronomical community by an
irrefutable demonstration that Hubble's measures of the distances of the
galaxies were wrong and that all remote galaxies are more than twice as far
away as Hubble had estimated. This was because what Hubble had taken to be
bright stars - which he used as such for 'standard candles' - in certain
nearby galaxies were not stars at all but bright HII regions. And, when
Hubble had believed he had confirmation of his inferred distances from
observations of variable stars, this was because he had wrongly assumed
the same period - luminosity relation to hold good for two different sorts
of variables. As a result of penetrating re-examination of the evidence

Baade came up - or came down - with the value of $H_0 \stackrel{\sim}{} 180$ km/s/Mpc which gives $T_0 \stackrel{\sim}{} 5.5 \times 10^9$ years.

As in the case of some other items in the history of modern cosmology the quantitative result was about as tantalizing as it could be. On the one hand, astronomers could not see how even the new value of T_0 would leave sufficient past time for the evolution of the universe. On the other hand, had Hubble found such a value in earlier years, they had to admit that steady-state cosmology would have made less appeal vis-à-vis evolutionary cosmology, even on the mistaken view that it is significantly less restrictive with regard to times.

The most significant thing about Baade's discoveries was undoubtedly the mere fact that the spell of Hubble's original quantitative results had been broken. The next most important thing was that the revision was substantial and was in the direction of making cosmology in general more plausible. I suppose the most immediate reaction from astronomers - apart from their realization as to how uncritically they had accepted Hubble's values - was that if a first revision was so large, there could well be subsequent revisions that would go much further. Baade himself was not dogmatic about his own numerical results; and we know that Allan Sandage, upon whom the mantle of Hubble fell, eventually proceeded to obtain H_0 values fully ten times smaller than Hubble's values.

Again as a matter of history, it can be said that Baade's work had almost no effect upon a decision between evolutionary and steady-state theory.

RADIO ASTRONOMY

From the account so far it is seen that in the early 1950s it was upon a decidedly misty universe that the dawn of radio astronomy had been coming up for some while. No one seemed to think then that radio astronomy might help to clear the mist. Indeed, anyone who considered it would have found good reasons to reject the idea. For J.S. Hey had detected radio emission from the Sun and, regarding the Sun as a typical star, this was consistent with the interesting but weak emission from the Galaxy as a whole discovered by K.G. Jansky and by G. Reber in the 1930s. It showed that a similar galaxy at the distance of the Andromeda Nebula might be only just about detectable, as turned out to be the case in 1950. So it looked as though any notion of doing serious extragalactic astronomy in radio frequencies should have been quite reasonably dismissed.

Astronomers who wanted to go outside the optical range expected
far more exciting discoveries from going into the ultraviolet and beyond
than from going in the other direction into radio frequencies - save for
the fact that high frequencies just do not get through the Earth's
atmosphere while radio waves do get through.

Looking back at all the non-reasons for going ahead with radio
telescopes, I now find it easier than ever to believe a tale that went
round at the time. When one group had almost completed making a radio
telescope carefully designed for a specific purpose, according to legend
one member woke up in the middle of the night in utter horror because it
had suddenly come to him that in calculating what signal they might expect
they had *multiplied* by light-speed c where they ought to have divided by c.
When he hastened to point this out to his colleagues they shared his
horror, but decided that having gone so far they might as well finish the
instrument. One of them is supposed to have said that in the whole universe
there might be something that would give a signal c^2 times stronger than
the one they had thought they were going to detect - and there was! I can
give no chapter and verse for any of this, but it seems as plausible an
explanation as any other as to how radio astronomy got started.

RADIO STARS

From the standpoint of our present topic the first great
discovery in radio astronomy was that of discrete sources, and more
particularly of 'radio stars' as those were called that had no extent
perceptible to early radio telescopes. Hey found the first in 1946,
several more were found over the next few years, and by about 1951 there
were a few fairly convincing identifications with optical galaxies, e.g.
Cygnus A identified with a faint double galaxy of redshift about 0.06, for
which a radio luminosity about a million times greater than that of our
Galaxy was inferred (see the contribution by Smith elsewhere in this
volume). This was most astonishing and astronomers had to practise hard
before breakfast at believing it - that two objects should have comparable
emission in one part of the electromagnetic spectrum but emissions
differing by a factor of a million in another part. Hitherto a galaxy had
been conceived of as an aggregate of stars with a relatively small amount
of interstellar material, but here was the indication that something could
be going on in a galaxy on a totally different scale from anything known
in, say, our own Galaxy. The first suggestion was that it was a case of

a collision between two galaxies (Baade and Minkowski 1954). This
hypothesis proved to be generally untenable, and it is mentioned here only
to emphasize that astronomers appreciated that they were presented with
something outside previous experience. However, in the present article we
shall not be concerned further with the physical character of these objects.

CATALOGUES

When any new class of object is discovered in the sky, the
first thing that astronomers think of doing is cataloguing the objects as
they are found. There are corresponding exercises in all empirical
sciences. In the first instance the catalogue may have no purpose other
than to ensure that, once found, a particular object may be found again.
So the catalogue must give for each entry the position and some assignment
of an apparent 'magnitude'. Of course, if the class turns out to be
interesting, other characteristics will soon find their way into the
catalogue, e.g. something about spectral features, variability, etc. In
the case of radio 'stars' the convenient measure of apparent 'magnitude'
was the flux density S at a particular frequency, measured at the observer
in watts per square metre per cycle; the 'flux unit' that proved useful
was $10^{-26} Wm^{-2} (c/s)^{-1} = 1$ Jansky.

An obvious feature of almost any catalogue in astronomy is that
it is never complete - rather like the list of friends to whom one sends
cards at Christmas: it is a pleasant selection, but it is not all the
friends you have. However, Martin Ryle soon appreciated - and presumably
others did too at about the same time - that, down to a flux density that
was about the limit of what instruments could detect at the time, it was
feasible to catalogue every source in the part of the sky that could be
properly observed. The number was not so large as to make the task
prohibitive, nor so small as to make a statistical interpretation illusory.

Log N - Log S GRAPHS

By about 1953 Ryle's group had built a radio telescope capable
of making a significant survey. By 1955 they had the 2C catalogue of 1936
apparent sources, believed to be complete for the accessible part of the
sky down to $S \sim 20$Jy at 3.7m wavelength. Of these sources some 30 were
extended and so inferred to be mostly within the Galaxy. Ryle & Scheuer
(1955) discussed the remaining 1906 sources - 'radio stars' in the
nomenclature of that time - those having no discernible extent. Adopting

what became common practice, they plotted log N - log S graphs, where for the part of the sky concerned $N = N(S)$ is the number of sources giving flux S or more. When they did this for each of several parts of the sky, they found no significant differences between those parts. Thus the sources were shown to be distributed around us with remarkable isotropy.

Figure 1. Reproduction of Figures 1,2 of Ryle & Scheuer (1955), with kind permission of the authors. Here I has the same significance as S in the present text.

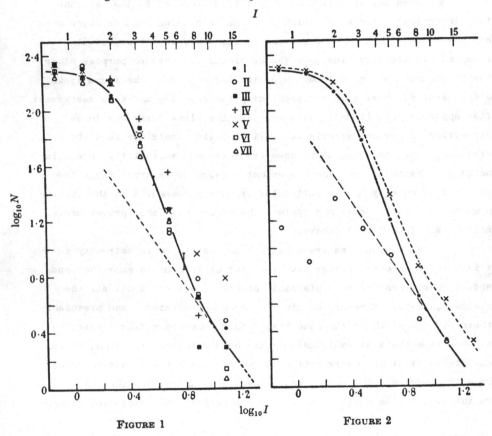

FIGURE 1   FIGURE 2

FIGURE 1. Plot of $\log_{10} N$ against $\log_{10} I$ for seven areas of sky. The full curve was obtained by combining the five non-equatorial areas, and its standard errors are shown by the solid vertical lines. The straight line corresponds to a slope of -1.5, such as would occur with a uniform spatial density.

FIGURE 2. Plot of $\log_{10} N$ against $\log_{10} I$ for the two equatorial areas (short dashes) and for the five non-equatorial areas (full line). The points obtained by plotting

$$\log_{10} (N_{\text{IV,V}} - N_{\text{I,II,III,VI,VII}})$$

are also shown (O).

Figure 1 of this article is a reproduction of Figures 1 and 2 of the paper quoted, together with the original captions. Reference may be made also to Ryle (1955, 1958, 1968).

From well-known arguments it then followed that the radio stars were either mostly local within the Galaxy or mostly very distant outside the Galaxy. The possibility of comparable numbers of both sorts was implausible. My recollection is that in the early days almost nobody concerned had unshakable conviction either way. The ultimate general conviction that the 'radio stars' were mostly very distant was the outcome of much careful balancing of evidence in the days of few optical identifications, and no one needed to be ashamed if it took some time to persuade him. But at any rate by the time that Ryle and his colleagues began to publish extensive results, they had inferred that they were indeed looking at the Universe on a very large scale. The story is in a general way reminiscent of William Herschel 150 years before mapping out the distribution of stars in the Milky Way Galaxy long before the distance from the Sun of any single one of them had been actually measured.

Before further discussion of the log N - log S results there are two general matters that it is important to mention here. The first concerns the possible ways of inferring the form of $N(S)$ from the observations. The radio telescope scanned some region of the sky and as it did so the flux of radiation received by it produced a response by its detector which was conveniently described as a 'deflection'. The plot of the deflections over the region constituted the 'observations' that were to be interpreted. In principle, the simple procedure was to recognize genuine sources in the record, to distinguish them from what might be called 'instrumental effects', and to have the system calibrated so as to derive from the amplitude of the recorded deflection the value of S for each such source. This, essentially, was the way in which Ryle's group was operating most of the time.

There was, however, an alternative procedure; again just in principle, it was the following. Knowing the characteristics of the telescope, it was possible to predict the record it would yield if sources appropriate to a given function $N(S)$ were scattered at random over a specified region of the sky. This produced the computed probability distribution of deflections corresponding to the assumed $N(S)$. The procedure was carried out for a range of functions $N(S)$. The statistics of the record of actual observations were compared directly with the

computed distributions and the $N(S)$ giving the best match was identified.
In this fashion information about $N(S)$ for the observed Universe could be
derived from the statistics of the raw observations. The information might
appear less precise than that got by counting sources, but it made more
comprehensive use of the observations, it was independent of any necessity
to distinguish between source effects and instrumental effects in interpreting the observations, it took automatic account of confusion effects,
and it took some account of faint sources beyond the resolving capability
of the telescope. On the other hand, the statistical information got in
this way was of no direct help in the business of cataloguing sources; in
that regard it supplied only an overall check. And so far as the 2C
catalogue was concerned, its authors did claim at the time that the
statistical method confirmed, at least qualitatively, the trend of $N(S)$
shown by the catalogue.

The foregoing is a crude attempt to give some general idea of the
'P of D'-probability of deflections - statistical analysis worked out by
Peter Scheuer. I am not in a position to say to what extent his detailed
quantitative results have stood the test of time, but I think that in due
course his work settled down to give a slope close to -1.9 for the main
linear part of the log N - log S graph, with which that ultimately got from
the source counts was found to be in good agreement. A more immediate
consequence of Scheuer's analysis was that he could show how a plausible
extrapolation of the $N(S)$ that Ryle found near the limit of his resolution
did indeed account for the properties of the observed radio background at
the wavelengths employed. (This is not to be confused with the microwave
background discovered much later by Penzias and Wilson.) This was
important not only because of the information it gave about $N(S)$, but also
because it indicated that there is no *different* population of sources
beyond those detected in obtaining the log N - log S results. It confirmed
that Ryle and his group could claim to be observing the whole (radio)
universe.

The statistical treatment played a part of real historical
significance around 1955 and in the next few years. It appears certain
that it helped to sustain the confidence of the Cambridge workers when, as
we see in the following section, their results attracted severe criticism.
There was a human touch as well that shows how the progress of science may
be influenced by the personal affairs of individual scientists. In the
midst of all this work upon the frontiers of the astronomical universe,

Scheuer had to lay it aside in order to complete his spell of military service. I remember Ryle telling me that the publication of some of their work would have to wait until after Scheuer got back to Cambridge - at the time he was apparently the only mathematician in the group.

The other general matter referred to above is to be mentioned only because it is one about which any astronomer inevitably enquires, particularly in regard to any statistical studies. It is the phenomenon of *clustering*. Do 'radio stars' occur in clusters of such objects, or in significant association with clusters of other astronomical objects? Questions like these were asked at the time. Interesting as they were, they had no major effect upon the history so far as we take it in the present sketch.

RADIO COSMOLOGY

About 1955, especially in consequence of the work leading to the 2C catalogue, everything pointed to the inference that radio astronomers were studying the Universe on a much greater scale than optical astronomers. Indeed, Ryle estimated that of the almost 2000 sources then believed to figure in that catalogue, only a few tens would be found to be identifiable with optical objects within reach of the 200-inch Palomar telescope. The discovery of the isotropy of the Universe on this scale was therefore of fundamental cosmological importance. We have seen that optical astronomy at best did not invalidate the hypothesis that the Universe is homogeneous and isotropic. But radio astronomy gave positive evidence that the Universe is isotropic on the largest scale on which it could be studied. If then there is nothing special about our situation, it followed that the Universe is also homogeneous on this scale. This was a tremendous advance in empirical cosmology. These general conclusions retained this validity even after the detailed interpretation of the 2C catalogue had been called in question.

Returning to the form of the $\log N - \log S$ graphs, qualitatively all the Cambridge plots were such that (see Figure 1):
a) for the largest S values, $\log N$ increased at first approximately linearly with slope about -1.5 for decreasing $\log S$, but this had not much significance because it depended on rather few points;
b) for S decreasing further, the main part of the graph was again approximately linear but with a considerably steeper slope;
c) for the least values of S for which sources were held to be resolved

the increase of log N became less and less steep, statistical analysis
confirming this trend even for smaller S.

Were the sources all of one standard power with uniform
distribution in Euclidean space, then the number N_0 giving flux not less
than S at a stationary observer would satisfy a relation

$$\log N_0 = -1.5 \log S + \text{constant.}$$

Were there some distribution of intrinsic powers, but with this
distribution the same throughout, then this relation would still hold good.
It serves as a useful comparison for the empirical log N - log S relation;
also some authors displayed the behaviour of $N(S)$ more clearly by plotting
log (N/N_0), instead of log N, with N_0 suitably normalized.

As time went on, radio astronomy was developing greatly in
various centres in various countries. Most of the groups did not equip
themselves specially for making source counts, but those that did were for
roughly a decade continually seeking to improve their capabilities. So the
history of this phase became rather complex and tangled. Each new survey
was of course a major undertaking in instrumentation and its application,
as well as in analysis of results. So the results themselves took a long
time to appear in full. While in the years concerned there was not much
coming and going directly between different groups, members of the groups
were apt to give interim reports at international conferences, large and
small, such reports being sometimes published and sometimes not. For these
and other reasons there is an 'internal' history of the developments as
seen by those immersed in them, and an 'external' history of the subject
as seen by an interested astronomer not actually involved in the work -
each such astronomer getting his own particular picture depending upon his
particular contacts with those who were involved.

The topic of this section is almost uniquely fortunate in
having been selected as the central feature of the history of radio
astronomy by Edge and Mulkay (1976), now regarded as a classic of such
'internal' history. Reference is made in particular to chapters 5 and 6
and their documentation.

Mills & Slee (1957) working in Australia at Sydney published
the first extensive source counts after those made at Cambridge; a fuller
account was given by Mills, Slee & Hill (1958) (see the contribution by
Mills elsewhere in this volume). Mills & Slee presented a devastating

challenge to the 2C results; where the two surveys overlapped in the sky
they declared the results to be "almost completely discordant". Not only
was there little agreement about even the existence of individual sources
in the common region, but also the two surveys disagreed about the slope of
the main part (b) of the log N - log S graph: 2C gave about -3, Mills &
Slee got about -1.8. But they were not at the time convinced that it was
significantly different from the comparison value -1.5, in which case the
result would be lacking in cosmological interest.

Radio astronomers in general then naturally became sceptical of
the 2C material. The Cambridge workers themselves were sustained by what
they regarded as largely independent support they had from the P of D
analysis, and by the results they were now gathering with their new
installation working at a shorter wavelength (Edge et $al.$ 1958). These
results became the 3C catalogue (Edge et $al.$ 1959). For a few years there
must have been tensions between some groups, but an onlooker at the time
saw little sign of it; he was aware simply of enthusiastic young men
getting on with the tasks they had set themselves and showing no particular
worry about what other such young men were up to elsewhere. It is easy to
exaggerate dissensions and rivalries in retrospect and, even if they were
unimportant at the time, they may seem to give piquancy to the story in the
telling.

Anyhow the telling has to be cut short because the story ended
in substantial consensus and some feeling of anticlimax. While it quickly
became apparent that the claims for the 2C work had been over ambitious,
it was evident that the participants had learned invaluable lessons from
it and had applied these to excellent purpose in the new 3C work. The
pendulum swung back: 3C cast doubt on many of the Australian sources; J.G.
Bolton (1960) working with a new group at Owens Valley in California in the
first considerable survey by a 'third party' did the same, while confirming
nearly everything in a test sample of 3C sources; from 1961, in Australia
itself Bolton et $al.$ (1964) began to get results in good agreement with
the Cambridge work. In particular, they found a slope of the log N - log S
graph of -1.85 ± 0.1 in good agreement with the then most recent Cambridge
value of -1.80 ± 0.1.

The three main inferences drawn by Ryle and his group from the
log N - log S graphs may be stated as:
(1) The diversion above the log N_0 line shows that the sources at the
distances concerned were more luminous and/or more numerous than those in

the cosmic neighbourhood of our Galaxy, while the diversion below this line
at smaller S shows that the sources at yet greater distances were less
luminous and/or less numerous than the latter. We have to say *were* to take
account of the look-back time. Over the whole look-back time concerned,
the Universe has therefore *evolved* to a very marked degree.

(2) Apart from the effects of evolution, the different Friedman-Lemaître
cosmological models predict different $\log N$ - $\log S$ relations. Naturally
it would have been of tremendous interest to be able to test whether any
such model is in good agreement with the radio observations in this regard.
From this standpoint it is unfortunate that the inferred evolutionary
effects largely mask any model-dependence of the interpretation of the
observations. From the standpoint of astrophysics, on the other hand, the
situation may be deemed hopeful in that something about the long-term
evolution of the system of sources may perhaps be inferred in an
effectively model-independent manner.

(3) The only cosmological model that can be tested conclusively is then the
one that expressly excludes evolution of the universe as a whole (not, of
course, of each individual galaxy). That is the *steady state* model. The
evolution inferred according to (1) was in outright contradiction of
steady-state theory.

        In any case, the $\log N$ - $\log S$ relation predicted by steady-
state cosmology is unique - like everything about the model. If the N_0
line is set to agree with this relation at the largest S values - which is
bound to be possible - then the relation gives a curve that falls below the
N_0 line for all smaller S, in complete disagreement with the radio results.
As Ryle pointed out, in the middle range of S the observations showed
values 10 or more times greater than the steady-state prediction, when
they are caused to agree at the largest S. (Two remarks are called for.
One is that at the time some writers seemed to suppose that the N_0 line
is the steady-state line. As we explained, the latter curves away below
the N_0 line with a slope that gets less and less steep than -1.5 with
decreasing S in the $\log N$ - $\log S$ plot. The second point is that a steady-
state theoretical $\log N$ - $\log S$ curve, because of redshift effects, cannot
be calculated precisely without a knowledge of the spectra of the sources.
That is why this curve did not usually appear in the pertinent discussions
of steady-state cosmology.)

        From 1955 Ryle had called attention to the conflict with
steady-state cosmology; his most definite presentation of the case - on

the basis of the 3C work - came when he read the paper Ryle & Clarke
(1961) to the Royal Astronomical Society in February 1961. Before the
meeting at least one London evening newspaper had got word of what was to
be reported, and astronomers on their way home after the meeting were able
to read all about the final overthrow of steady-state theory!

Of the three originators of steady-state cosmology only H.
Bondi was present at the meeting. However, after Ryle had spoken, Jayant
Narlikar gave a preliminary account of the paper Hoyle & Narlikar (1961)
in which they claimed to demonstrate that in steady-state cosmology a rare
happening like the occurrence of a radio source would show a statistical
distribution around an observer in agreement with Ryle's observations.
Most of those present were mystified by this claim. When the Hoyle-
Narlikar work appeared it seemed to me to require our Galaxy to be in a
preferred position, located in a fluctuation in the source distribution that
extends over practically the whole of the observable part of the Universe.
Also it seemed to be a matter of hypothesis and not of inference. In my
view this intervention served merely to confuse people into thinking that
Ryle's work after all had not invalidated straightforward steady-state
cosmology - which, granting the validity of the observations, most
indubitably it had (see McCrea 1963, especially pp. 196-7).

So much for a sadly sketchy 'internal' history. As to an
'external' history, since by definition this is a personal affair, it
seems natural that a writer should briefly describe his own. Now although
individuals came to be labelled 'cosmologists', nobody - certainly in the
years here concerned - spent his life working at cosmology. In practice a
'cosmologist' worked at whatever else was his ordinary avocation, but from
time to time followed up as best he could any idea that occurred to him,
or any observational evidence that came his way, which clearly looked as
though it could be of cosmological significance. I have to say this
because some colleagues at this present time are astonished at my
ignorance of, or apparent indifference to, what appear to them to have been
the stirring conflicts of two or three decades ago. So far as I can tell
now, the truth is that I was not wholly convinced of the general
cosmological significance of what was going on in radio astronomy, and to
the extent to which I was convinced on the general issue I was still
unconvinced about the certainty of any quantitative results. I had the
privilege of being consulted by Martin Ryle in 1955 about the presentation
of some of his source counts. I was intensely interested and never had

the slightest doubt that he was right in publishing them. But I was
woefully ignorant about the working of his telescope and I could do no more
than wonder in a general way about the extent to which the shape of his
resulting curves might be affected by confusion of sources and instrumental
cut-off. As to cosmological significance, I recognized that this would be
very important indeed if it could be established. But I took the naive
view that it could not be regarded as established until astronomers know
what the sources are and had some other observations, probably redshifts -
radio redshifts, were that possible - that would turn S values into
distances. So far as my own attitude was concerned, developments in the
next half-dozen years did little to change that, although obviously radio
astronomy itself was making immense progress. The Australian results of
1957 appeared in the Australian Journal of Physics which I did not see at
the time. I did know that they were highly critical of the Cambridge work.
But I was somehow aware that by then the Cambridge people knew they were
well on the way to having new results that would supercede both their own
earlier work and probably the existing Australian work as well. And then,
without knowing many details, I soon began to sense a trend towards general
agreement. I ought to have given more weight to this consensus. As it was,
I was bemused by the fact that successive surveys each appeared to find
that a good many of the 'sources' in a previous one were not sources at all,
and that they kept producing different slopes for the famous log N - log S
graphs. Having in mind the notorious case of the determination of the
Hubble constant, in this case of source counts I wondered whether history
was being repeated so that these revisions would be followed by a
succession of others. But the cases proved to be different in that for a
long time only one observatory was working on the Hubble constant, and that
since others entered the field they have been getting discordant results,
whereas there has long been good consensus about source counts - although
it came only towards the end of the period mainly concerned here.

When Ryle delivered his 1961 paper I came close to being
convinced by his cosmological claims. But I still wanted to know what
radio sources were before making up my mind. By then 'optical counterparts' had been proposed for still only about 20 sources. Of course, had
some sources been positively explained as objects for which an optical
counterpart would not be forthcoming, that would have had to be accepted.
No way of discovering the physical nature of the sources had, however, been
proposed other than via optical identifications. And at the juncture

this was particularly baffling because the first two or three quasars - all
of them 3C objects - were being observed. At the time astronomers were
most inclined to regard them as stellar objects within the Galaxy. This
shows clearly that the majority of astronomers had not yet accepted that
anything in the 3C catalogue was most likely to be at 'cosmological'
distance.

More trivially, I am sure that a contributory cause of
uncertainty was the presentation of the statistics in terms of the
cumulative $N(S)$ instead of the 'differential' $n(S)$, where $n(S)dS$ is the
number of sources in a region giving flux density in the interval $S, S + dS$,
so that $n(S) = -dN/dS$. Even better is, instead of N/N_0, to use n/n_0,
where n_0 is the suitably normalized differential parameter for a uniform
distribution in Euclidean space. Jauncey (1975) has discussed the
significance of this feature, which ceases to be trivial so soon as we
consider the problem of putting meaningful error bars on a log N - log S
graph.

CONCLUSION

In retrospect, in spite of confusing side issues, from about
1955 cosmologists would have been safe in accepting that radio astronomy
had shown the actual universe to be not in a steady state. Instead, they
waited until a decade later when the discovery of the microwave background
radiation had confirmed a positive prediction of big-bang cosmology.
Logically this was a less adequate reason for rejecting a steady-state
universe, since steady-state theory had made no prediction in this
particular regard, and so the discovery contradicted nothing in the theory.

Thus the direct cosmological outcome of the source surveys was
eventually somewhat meagre. To radio astronomers in many observatories -
notably Jodrell Bank - the predominating interest has been in the physical
nature and evolution of individual radio sources. But to those who study
these features the catalogues resulting from the surveys are all-important
both in providing a list of objects for study and in ensuring that the list
is representative. After such studies have proceeded sufficiently far,
cosmologists will nevertheless still have to come back and use these to
find a cosmological model that does account for the empirical log N - log S
relations. This persists as the main challenge bequeathed by radio
astronomers working a quarter-of-a-century ago.

REFERENCES

Baade, W. & Minkowski, R. 1954. Identification of the radio sources in
        Cassiopeia, Cygnus A and Puppis A. Astrophys. J. $\underline{119}$, 206-231.
Bolton, J.G. 1960. The discrete sources of cosmic radio emission
        (unpublished), quoted by Edge & Mulkay (1976) pp. 192,216.
Bolton, J.G., Gardner, F.F. & Mackey, M.B. 1964. The Parkes catalogue of
        radio sources, declination zone $-20°$ to $-60°$. Australian J.
        Phys. $\underline{17}$, 340-72.
Edge, D.O., Scheuer, P.A.G. & Shakeshaft, J.R. 1958. Evidence on the spatial
        distribution of radio sources derived from a survey at a
        frequency of 159 Mcs^{-1}. Mon. Not. R. Astr. S. $\underline{118}$, 183-96.
Edge, D.O., Shakeshaft, J.R., McAdam, W.B., Baldwin, J.E. & Archer, S. 1959.
        A survey of radio sources at a frequency of 159 Mcs^{-1}. Mem. R.
        Astr. S. $\underline{68}$, 37-60.
Edge, D.O. & Mulkay, M.B. 1976. Astronomy Transformed. New York: John Wiley
        & Sons.
Hoyle, F. & Narlikar, J.V. 1961. On the counting of radio sources in the
        steady-state cosmology. Mon. Not. R. Astr. S. $\underline{123}$, 133-66.
Jauncey, D.L. 1975. Radio surveys and source counts. Ann. Rev. Astron. &
        Astrophys. $\underline{13}$, 23-44.
McCrea, W.H. 1950. The steady-state theory of the expanding universe,
        Endeavour $\underline{9}$, 3-10.
McCrea, W.H. 1963. Cosmology - a brief review. Quart. Jl. R. Astr. S. $\underline{4}$,
        185-201.
Mills, B.Y. & Slee, O.B. 1957. A preliminary survey of radio sources in a
        limited region of the sky at a wavelength of 3.5m. Australian
        J. Phys. $\underline{10}$, 162-94.
Mills, B.Y., Slee, O.B. & Hill, E.R. 1958. A catalogue of radio sources
        between declinations $+10°$ and $-30°$. Australian J. Phys. $\underline{11}$,
        360-387.
Ryle, M. 1955. Radio stars and their cosmological significance (Halley
        Lecture). Observatory $\underline{75}$, 137-47.
Ryle, M. 1958. The nature of cosmic radio sources (Bakerian Lecture).
        Proc. Roy. Soc. A $\underline{248}$, 289-307.
Ryle, M. 1968. The counts of radio sources. Ann. Rev. Astron. & Astrophys.$\underline{6}$,
        249-66.
Ryle, M. & Clarke, R.W. 1961. An examination of the steady-state model in the
        light of some recent observations of radio sources. Mon. Not. R.
        Astr. S. $\underline{122}$, 349-362.
Ryle, M. & Scheuer, P.A.G. 1955. The spatial distribution and the nature of
        radio stars. Proc. Roy. Soc. A $\underline{230}$, 448-62.
Slater, N.B. 1954. Recession of the galaxies in Eddington's theory. Nature,
        Lond. $\underline{174}$, 321-2.

NANOHERTZ ASTRONOMY

H. C. van de Hulst
Sterrewacht, Leiden, Netherlands

Aim

This paper is an attempt at what with a sophistication
understandable to radio astronomers might be termed *nanohertz astronomy*:
the art of registering the coming and going of astronomical convictions
in periods of the order of 10^9 seconds = 30 years. This approach is
complementary to the common one, where the history of science is
described by focusing on the sudden discoveries and the rapid break-
throughs of insight. This complement is as necessary as are the added
measurements at very short spacings in the Fourier synthesis of extended
sources. Otherwise a broad underlying valley or elevation might be
misjudged and the basic structure misinterpreted. And – to continue this
metaphor – the historical development of science is indeed such an
extended source with a highly complex structure.

How can we 'measure' these low-frequency components? The
history of science that can be reconstructed from published papers is
far from complete. A substantial part of the development of scientific
knowledge is not documented this way. The relative importance of this
'hidden' part varies with the field: it is probably smallest in the
gathering of new data, moderate in the development of new techniques and
largest in the growth of theoretical insight and interpretation. In this
'hidden' part the ideas originate, grow (rightly or wrongly) into an
accepted theory, or (again rightly or wrongly) are discarded. The scene
is: sleepless nights, the private work room, the class room, visits,
phone calls, bull sessions, workshops, team meetings, some correspondence,
the hall where a symposium is being held, or the coffee shop across the
road. The record is scanty. Some correspondence gets published later, but
only if the subject is considered of outstanding importance. Some
symposium reports show a glimpse by publishing extensive discussions but
any chairman of a session or editor of a symposium volume knows how hard

it is to keep the weeds from overgrowing the flowers or even to recognize
the flowers. There have also been laudable initiatives to build through
interviews a permanent record of personal visions on the development of
ideas (Weart & DeVorkin 1981), but I had no time to consult these for the
present paper.

Consequently, the following remarks are not based on a
systematic study but on personal recollections and notes with some spot
checks from the literature. The comments found in *Classics in Radio
Astronomy* (Sullivan 1982) formed a welcome trigger. The choice of four
episodes is based purely on personal whim.

Dispersion in space

Shortly after the finite velocity of light was discovered,
Newton asked Flamsteed by letter (1691) to watch if the Jupiter
satellites briefly changed colour at occultation or immersion in the
shadow, as they should if light of different colours took different times
to reach us. I have this from the introduction to the paper in which
Tikhov (1908) summarizes the different methods leading to the discovery
of the Nordmann-Tikhov effect, which stated that such a time difference
did indeed exist in the observation of distant variable stars. Tikhov's
first long paper, published in 1905 in a rather unexpected place
(Communications of the Yekaterinoslav Mining Academy) derived, from an
analysis of observations by many authors, $\Delta t = 14 \pm 6$ minutes between
402 and 452 nm for the eclipsing variable β Aurigae. In his first
theoretical analysis in 1898 (he was 22 or 23 then; references may be
found in the biography by Suslov, 1980) he expressed the hope that this
dispersion in space, if ever discovered, would provide a practical means
for estimating distances in the galactic system. The alleged discovery
of this effect, by two independent methods, radial velocities of
spectroscopic double stars and photometry with colour filters, fulfilled
an ardent expectation. Nordmann (at Meudon) and Tikhov (at Pulkovo) both
received a prize from the Paris Academy in 1908 and Tikhov another one in
Russia. The *Astronomische Jahresberichte* for 1908 contains on pages 347-
352 dozens of references to papers, comments and reviews on this effect.

Among the comments were also sharp criticisms, the first from
Lebedev, who argued that, if interstellar matter was anywhere similar to
air or other matter known on earth, a dispersion of this magnitude should
be accompanied by an absorption rendering all stars invisible. Tikhov

in the 1908 paper cautiously left the door open to an astrophysical
explanation in the intrinsic properties of the stars, but he did not
accept the possibility that the measurements might be of insufficient
accuracy. Russell, Fowler & Borton (1917) confirm from a very thorough
review of new data that the measured differences are unquestionably real
in five of six eclipsing variables, rising to over 10 times the probable
error of about 1 minute. However, they discard the explanation by
dispersion in interstellar space because the needed differences in photographic and visual light velocity ranging from -2 to +5 meter/second are
too large to be physically acceptable. Soon after, the effect sank into
oblivion, except for an occasional cryptic reference like that of Shapley
(1930, see p.108) who says that 'it merits further careful study'. Any
living astronomer is excused for not having heard of the effect.

      What has this to do with radio astronomy? Nothing for a
historian focused on discrete discoveries. A lot for me trying to sense
the very-long-period waves of hopes, successes and frustrations in science.

      The first link is fairly casual. Presumably the same Charles
Nordmann, who in 1906 deposited with the Paris Academy a sealed document
on the dispersion of light in interstellar space, had in 1902 made an
attempt to detect radio waves from the sun (Nordmann 1902). It seems
ironic that he received a prize for what we now regard as a spurious
effect, and that he is remembered for an unsuccessful study. The initial
impression that these subjects are remote vanishes upon closer reading:
after all, the true excitement of both is to get to know something about
the propagation of extraterrestrial radiation. Jules Janssen was the
great inspirer. He invited Tikhov over to Meudon, told him to become a
member of the balloon club, which conditioned him also for mountain work,
among which were spectroscopic studies of the telluric absorption bands
on Mont Blanc. Jules Janssen also inspired Nordmann to do his 'radio
astronomy' experiment on the same mountain. The whole situation looks
almost modern: two young boys hunting for   big discoveries, fully aware
of the international scene, with a keen sense of adventure and a somewhat
petty selfishness protecting their personal credit.

      The second link is deeper. Let us remember the terminology. A
dispersion curve links the frequency with the wave number, or velocity
of propagation, or refractive index. The corresponding imaginary part
gives the absorption. Different modes of propagation may have different
dispersion curves. If these modes are linearly polarized, the difference

in refractive index is called 'birefringence', the difference in
absorption 'dichroism'. If they are circularly polarized, the difference
in refractive index usually is named by its effect, rotation of the plane
of polarization, and the difference in absorption 'circular dichroism.'

All of these phenomena form in principle one package, the
theory of which, as applied to laboratory situations, was a familiar part
of physics around 1900. What this package contained as applied to space
was unknown. The eagerness with which Nordmann and Tikhov tried to open
this package - at what we now know was the wrong end - would in principle
apply to any other way of opening it up.

The 70 years which followed gave indeed a number of surprising
discoveries which fit into this frame, but came from such unexpected
angles that one might easily miss seeing the connection. With no attempt
at a full or impartial account let me mention a few of them.

Interstellar absorption, now more correctly referred to as
extinction, was looked for and suspected by many from the beginning of the
century but proof came only around 1930. The wavelength dependence of this
effect has led to one of the commonest distance criteria, exactly as
Tikhov had hoped.

Interstellar dichroism, usually referred to as interstellar
polarization, came as a surprise discovery by Hiltner and Hall in 1948
and did a lot to make most astrophysicists aware that they must take
magnetic fields in the galaxy seriously. The proposal of synchrotron
radiation as the source of non-thermal radio emission in the years
immediately following (Sullivan 1982, papers 9, 10, 27) appears no
accident. Could interstellar polarization have been detected earlier?
Certainly. Several people may have stumbled upon it and discarded it as
impossible. Öhman once told me that he had been in that position. Even at
the time of Nordmann and Tikhov's work, Karl Schwarzschild by the technique of photographic photometry could quite well have detected interstellar polarization, had he tried to photograph the suitable stars
through a calcite prism.

Rotation of the plane of polarization - finally we come to
fullfledged radio astronomy - and its change with frequency in observing
non-thermal radio sources was found by Cooper & Price (1962), soon
yielding the *rotation measure* for many sources. This is by no means a
distance indicator but certainly it is a powerful tool, as Tikhov had
hoped to find, in the study of the physics of interstellar matter. Note

that the observed differences in these early papers are one or several full swings 2π, corresponding to differences in travel time between the two modes of one or several wave periods, i.e. fractions of a microsecond.

At last we return to straight timing. Tikhov at his time chose the widest interval in wavelength (a factor 1.12) and the sharpest cosmic clock (an accuracy of 5 minutes) that was available to him. How excited would he have been if he could have picked from the shelf a book on pulsars (Manchester & Taylor 1977) containing for 150 objects *dispersion measures* of great accuracy based on delay times in radio waves up to 1 second, measured with a precision of 10^{-4}. The pulses of the Crab Nebula pulsar have now been timed with an accuracy of better than 1 millisecond all the way from radio waves through the optical range to gamma rays over a wavelength range of a factor 10^{15}. Except for the dispersion in the long-wave radio domain these times coincide over the entire range. The only other detected gamma-ray pulsar, in Vela, however, does not show pulses at the same instant at all frequencies. The conclusion there reads as it should have read with the original Nordmann-Tikhov effect: the effect is intrinsic to the source and not interstellar.

Impossible temperatures

Wishful thinking in science is bad. Yet checking up on imaginative theories by taking new observations comes close to it and is a fully accepted method for making progress. The true art is to apply caution, care and scepticism as guards against wishful overinterpretation and still to keep the adventurous drive. These remarks form an introduction to the question I wish to address briefly in this section: Why did we not earlier live up to the existence of a million-degree solar corona and of non-thermal radio emission in the galaxy? I take these two episodes together because they were concurrent and initially presented themselves similarly, although the outcomes were quite different. I'll take the corona first. Again I apologize for producing little evidence besides recollections.

The true nature of the corona remained a riddle for a long time. About 1942, when I first heard about these problems in a graduate lecture course given by Minnaert, two independent phenomena, the scale height of the electron-scattering continuum and the state of ionization of the recently identified emission lines, led, *if* interpreted in terms of a thermal gas, to temperatures of about a million degrees. A third avenue,

the Fraunhofer lines in the corona, seemed to give conflicting evidence,
which was resolved only after the observations of the 1946 eclipse had
concluded the proof that the narrow lines arose from overlying zodiacal
light (Van de Hulst 1947).

But was it really a thermal gas? Alfvén (1941) argued that it
was, correctly emphasizing the strong thermal conductivity and the
relative thermal isolation. Edlén (1941) left the question open and even
in 1945 (Edlén 1945) talked about Alfvén's 'heating' mechanism in
quotation marks and concluded with classical caution: 'Before the various
suggestions have been more thoroughly examined it would be unwise to
judge in favour of the one or the other'. Minnaert showed the same
caution and Swings (1945) is equally noncommittal.

After all, during a good part of a century solar activity had
led from one surprising discovery to another, many of them clearly
associated with rapid gas motion or corpuscular radiation, so that
alternatives to a high temperature seemed abundant. Moreover, as I
showed later (Van de Hulst 1950) when returning to the subject with more
self-confidence, not all details in Alfvén's treatment were correct. This
explains why in 1945 (Van de Hulst 1945), not knowing that solar radio
emission had already been detected, I 'predicted' that it might soon be,
reckoning with a temperature of only 5000 K.

By 1947 the common opinion, including my own, had changed
completely and coronal temperatures of a million degrees had in 2 years
or less become a fully accepted fact. What did it? I think the full
credit goes to radio astronomy. Until then we only knew that the corona
looked quite different from one eclipse to the next, but that
spectacular changes *during* the 5 minutes or so of an eclipse were absent.
Radio astronomy not only coined the word 'quiet sun' but also for the
first time revealed the corona as a constantly present, quietly
radiating gas.

The next episode deals with two points off a curve. I refer
to Henyey and Keenan's calculation of the radio emission to be expected
from an interstellar gas layer of 10^4K (Henyey & Keenan 1940, Fig.1).
My own recalculation some 5 years later (Van de Hulst 1945, Fig.2) came
to the same result. The top Reber point at 160 MHz fits comfortably, but
the two Jansky points are higher than the computed curve by factors of
about 10 and 30. They are at such a low frequency that the curve is

already saturated, coinciding with the strongly sloping Rayleigh-Jeans
law. Hence, changing the density of the depth of the layer does not help;
only an increase of the adopted temperature by a factor 30 will give a
fit. Of course, with the keen interest with which many astronomers
examined these data at that time, this conclusion did not go unnoticed.
But why did we not take these points for what they are: the first clue
for the existence of non-thermal radio emission?

The answer may to some modern readers seem surprising. Again,
I have to work partly from memory, for doubts and misgivings are rarely
documented in the printed literature. First, an actual temperature of
300 000 K for the interstellar gas seemed an outrageous assumption. The
explanation of the forbidden lines in gaseous nebulae was already a
'classic' result and the recent detailed analysis by Menzel and coworkers
had shown that cooling and thermalization processes in such nebulae were
so efficient that a temperature of about 10^4K must result, almost
independently of the energies acquired by the electrons in the initial
ionization process. At the densities of the interstellar gas the cooling
and thermalization processes would be a factor 10^4 slower but still fast
enough. Moreover, the existence of dust and the measured excitation
temperature of the CH molecule pointed to even lower temperature.

The next possibility was that the measurements had been
wrongly calibrated. This seemed rather likely. From classical astronomy
I knew that it is notoriously difficult to avoid systematic errors in
absolute calibrations. Sometimes an author has his paper finished and
only then rapidly adds an absolute scale. (For that reason I tell my
students to read 'lazy units' when the figure is marked 'arbitrary
units.') I had not seen Jansky's original papers (see article by Sullivan
in this volume), nor did I have any knowledge of antenna theory and noise.
But the paper I did study said that Reber had recalibrated the Jansky
points. I admired Reber for his achievement but, having spotted that the
quoted width of his own antenna beam was off by a factor 3, I felt that
a serious calibration error could not be excluded. And there the matter
rested.

A year later the situation had changed markedly. I recommend
reading Reber and Greenstein (1947), not reprinted in Sullivan's
'Classics', presumably because it is only a review. They cite a number of
studies, including unpublished work by C.H.Townes, which appeared later

(Townes 1947), confirming the high brightness temperatures at low frequencies. In addition, the discovery of several discrete sources had proven the existence of extremely high brightness temperatures in some places. Two quotations from that review may serve to show where the front of radio astronomy was located at that time (15 September 1946): 'Large-scale electromagnetic interaction of moving and circulating charged particles must be included as a possible source of galactic radiation', and (addendum, 10 January 1947) 'If T_a should continue to rise at still lower frequencies, it may become necessary to ascribe the low-frequency component of cosmic static [= galactic radiation] to some such non-thermal radiation as has been suggested for the solar noise' [the active sun] .

These were the early, hesitant starts of high-energy astrophysics. The first quotation comes quite close to suggesting synchrotron radiation. The link with cosmic rays but not with synchrotron radiation was clearly pointed out by Ryle (1949); compare my historical comments (Van de Hulst 1972). Cosmic rays were well recognized as a population group of interstellar space but interactions with the more common population groups gas and dust had not yet been suggested except for a contribution to heating and ionization in Spitzer's yet unpublished work. The actual breakthrough of synchrotron radiation came several years later and is well documented in Sullivan's 'Classics' and in Ginzburg's paper in this volume.

The 'SE' of Maser

The last episode I wish to recall deals more with teaching than with research. Yet it illustrates well the way in which early radio astronomy was assimilated into general astrophysics. I do not remember when I first learned about stimulated emission (SE), but it belonged to the standard tools of astrophysics. Even if I had not known it before, I would have learned about it when redoing as a student's exercise in 1944 the Henyey and Keenan's computation of the thermal radiation cited above. Part of the fun in astrophysics is that familiar concepts may reappear in quite unexpected proportions. In this application SE was not a small correction but an almost fully canceling term with only 1 part in 10^6 remaining as the effective absorption.

Three follow-ups of this exercise are worth mentioning. First an immediate major worry appeared, which we did not find expressed in the

earlier publications. The near-cancellation arises generally from the
fact that at these very low frequencies the Boltzmann occupation numbers
(here represented by the Maxwell velocity distribution) of upper and
lower level of the transition hardly differ. The literature had convinced
us of the fairly rapid and complete thermalization of the interstellar
gas. But this could not be pretended to form a safeguard against
deviations of the order of 1 part in 10^6. So how well could the result
be trusted?

The answer was simple and conclusive. It was given in small
print (Van de Hulst 1945) because it was only a technical detail. Yet in
the present context it deserves to be recalled. The free-free radio
emission at any frequency is the superposition of many transitions arising
from a continuum of states in the electron-velocity distribution. Any
over- or underpopulation of one of these states works out with opposite
signs in the actual and the 'negative' absorption (= SE) *from* that state.
Hence, though in one transition *between* two particular states such a
deviation may spoil the near-cancellation, in the integral the deviation
will be compensated by the opposite deviation in an adjacent transition.
I formalized this reasoning (which I regarded as nothing more than a
trick) by considering instead of a continuum a discrete set of equi-
distant states permitting the transition $n \rightarrow n-1$ in emission and $n-1 \rightarrow n$
in absorption. Writing the occupation numbers (Boltzmann factors) as B_n
and the transition probabilities (here the Gaunt factors) as $G_{n, n-1}$, the
trick consists of shifting in the summation all the negative absorptions
by one place so that not the differences

$$(B_n - B_{n+1}) \, G_{n, \, n+1} \quad ,$$

but the differences

$$B_n (G_{n, \, n+1} - G_{n-1, n})$$

have to be summed. It is evident that in the second form minute
deviations in the B_n do not influence the result.

The second surprise came in teaching this subject in 1950. I
noticed that many astrophysics texts were unclear when introducing
stimulated emission. Some created the impression the SE was a kind of
quantummechanical correction to the classical theory of absorption, so
that it would vanish in the classical limit. The way it actually works
out is as follows:

process:	emission	stimulated emission = negative absorption	absorption
proportion:			
generally in any thermostat	1	$1/(e^{h\nu/kT}-1)$	$e^{h\nu/kT}/(e^{h\nu/kT}-1)$
example from optical astronomy $h\nu/kT = 2$	1	0.16	1.16
example from radio astronomy $h\nu/kT = 10^{-6}$	1	1000000	1000001

On each line in this table we may combine the last two terms in the form:

effective absorption = absorption minus negative absorption.

In the classical limit, which is needed in radio astronomy, the constant h vanishes, as it should, both from the radiation intensity (which becomes the Rayleigh-Jeans law) and from the effective absorption coefficient. It thus is the remaining small difference of two huge almost canceling terms which figures in Kirchhoff's law relating emission to absorption, formulated over 50 years before Einstein's transition probabilities.

The fact that 'negative absorption' is prominent in the classical limit can also be understood directly. If we exert on a swinging pendulum a resonant periodic force, there is a 50-50 chance, depending entirely on phase, that this will increase or decrease the amplitude.

Surprisingly, an even closer link with earlier problems emerged. In the quantummechanical description the near cancellation is made between the moves up and down *in the same transition*; stimulated emission starts from the upper level and absorption from the lower level of this transition. In the classical description of a swinging pendulum we compare instead moves up and down *from the same starting level*, i.e., when translated into quantum concepts, in two different transitions. The difference between one or the other way of formulating the near-cancellation was exactly the shift which I had devised as a mere 'trick' five years before. And the system with equidistant energy states, which I had used as the simplest manner to explain this trick, is of course the harmonic oscillator. The transition probabilities computed by quantum mechanics are well known. When converted into oscillator strengths in the usual manner the matrix reads:

```
                              to level n
                    n = 0   1   2   3   4   5  ..
                  ┌─────────────────────────────────
           m = 0  │       1
               1  │  -1       2
               2  │      -2       3
 from level m  3  │          -3       4
               4  │              -4       5
               5  │                  -5
              .. │
```

Zeroes must be read for all places not filled in. Positive is absorption,
negative is stimulated emission. The 'classical' subtraction occurs by
adding the two numbers on the same line, from which we see at once that
the effective oscillator strength for transitions from any level is 1.
On the other hand the 'quantum' subtraction occurs between the two
diagonal numbers which are equal and opposite so that the full burden of
proof that there is a net positive absorption falls on the occupation
numbers.

This story by itself is funny enough to be recalled, but it
gains some depth by adding the next chapter. It is obvious that the 'trick'
does not work in a spectrum with only one transition or with a variety
of lines at different frequencies. So if someone would have asked if such
spectra were safeguarded against strange results, or even inversions, in
the net absorption arising from inconspicuous deviations in the
occupation numbers, the answer would have been a clear 'no'. In other
words: astronomical masers might have been stated then to be an obvious
possibility. The actual history is that the question was neither asked
nor answered, for the prospect to observe separate lines had not even
emerged. Almost immediately after the 21-cm line was detected in 1951
(see article by F. J. Kerr in this volume) an ad-hoc discussion started
in which Purcell showed (after a number of unconvincing attempts by
others) that there was sufficient thermalization. Full details were
published much later (Purcell & Field 1956). Masers were realized as a
laboratory instrument in 1954, and were first used as a radio astronomy
amplifier in 1958, but were not recognized to exist in space until 1965.

Conclusion

Looking over these four episodes what have we learned?
Certainly, the popular way of science writing, to schematize the history
into separate discoveries, each popping up suddenly and then leaving in
a few years a wake of inferences and elaborations, is too simplistic.
There are over considerably longer portions of a life time, say, in the
1-10 nanohertz domain, long-wave undercurrents which stimulate or
discourage such discoveries. Some conclusions which might have been drawn
were not. Other conclusions that were drawn could not really, at least
in the responsible judgment of a large number of astrophysicists of the
time, be accepted.

One might well ask at this point whether there are particular
factors determining on which side of the fence individual scientists find
themselves in such dilemmas. I am convinced there are and have tried
elsewhere to sketch what factors of personality, education and cultural
climate may be important (Van de Hulst 1980).

Let me try a metaphor. The history of astrophysics runs like
the spiral arm of a galaxy. At first sight it impresses us as fitting
into a coherent design. Closer inspection draws us into desperation: the
outlines are ragged and at times we wonder if it is just a collection of
odd events. But still further study brings us to the conviction that there
is a dynamical pattern underlying it all and that this pattern may be
unraveled, provided we are prepared to drop a too simplistic attitude. I
hope that the notes presented in the preceding pages may help some
students of astrophysics, or of its history, to get an open eye for this
underlying dynamics.

Acknowledgements

I acknowledge with pleasure the help of Woody Sullivan and
of many of my colleagues at Leiden for removing errors and obscure points
from an earlier draft. I am grateful to Heleen van Herk-Kluyver for
translating the relevant pages of Suslov's book.

References

Alfvén, H. (1941). On the solar corona. Arkiv för Matematik, Astronomi och Fysik 27A, no.25, 1-23.

Cooper, B.F.C. & Price, R.M. (1962). Faraday rotation effects associated with the radio source Centaurus A. Nature 195, 1084-85.

Edlén, B. (1941). An attempt to identify the emission lines in the spectrum of the solar corona. Arkiv för Matematik, Astronomi och Fysik 28B, No. 1, 1-4.

Edlén, B. (1945). The identification of the coronal lines. Monthly Notices Royal Astronomical Society 105, 323-33.

Henyey, L.G. & Keenan, P.C. (1940). Interstellar radiation from free electrons and hydrogen atoms. Astrophysical J. 91, 625-30 (= Sullivan 1982, paper 8).

Manchester, R.N. & Taylor, J.H. (1977). Pulsars. San Francisco: Freeman.

Nordmann, C. (1902). Recherche des ondes hertziennes émanées du Soleil. Comptes Rendus Académie des Sciences Paris 134, 273-75 (= Sullivan 1982, paper 14).

Purcell, E.M. & Field, G.B. (1956). Influence of collisions upon population of hyperfine states in hydrogen. Astrophysical J. 124, 542-49.

Reber, G. & Greenstein, J.L. (1947). Radio-frequency investigations of astronomical interest. Observatory 67, 15-26.

Russell, H.N., Fowler, M. & Borton, M.C. (1917). Comparison of visual and photographic observations of eclipsing variables. Astrophysical Journal 45, 306-47.

Ryle, M. (1949). Evidence for the stellar origin of cosmic rays. Proceedings Physical Society 62, 491-99.

Shapley, H. (1930). Star Clusters, p. 108. New York: McGraw-Hill.

Sullivan, W.T. (1982). Classics in Radio Astronomy. Dordrecht: Reidel.

Suslov, A.K. (1980). Gabriil Adrianovich Tikhov. Leningrad: Nauka.

Swings, P. (1945). The line spectrum of the solar corona. Publications Astronomical Society of the Pacific 57, 117-37.

Tikhov, G.A. (1908). Deux méthodes de recherche de la dispersion dans les espaces célestes. Mitteilungen der Nikolai-Hauptsternwarte zu Pulkowo, 2, No. 21, pp. 141-83.

Townes, C.H. (1947). Interpretation of radio radiation from the Milky Way. Astrophysical J. 105, 235-40.

Van de Hulst, H.C. (1945). Radiogolven uit het wereldruim. II. Herkomst der radiogolven. Nederlands Tijdschrift Natuurkunde 11, 210-21 (= Sullivan 1982, paper 34).

Van de Hulst, H.C. (1947). Zodiacal light in the solar corona. Astrophysical J. 105, 471-88.

Van de Hulst, H.C. (1950). On the polar rays of the corona. Bulletins Astronomical Institutes of the Netherlands 11, 150-60.

Van de Hulst, H.C. (1972). Cosmic-ray electrons. Quarterly J.Royal Astro-
        nomical Society 13, 10-24.

Van de Hulst, H.C. (1980). Style of research. *In* Oort and the Universe,
        ed. H.van Woerden, W.N.Brouw & H.C.van de Hulst, pp.165-171.
        Dordrecht: Reidel.

Weart, S.R. & DeVorkin, D.H. (1981). Interviews as sources for history of
        modern astrophysics. Isis 72, 471-76.

*(l. to r.)   Jan Oort, Bill Erickson, Jan Högbom, and Ben
Hooghoudt consult on the design of the Benelux Cross in 1962
(courtesy Sterrewacht Leiden)*

RADIO ASTRONOMY AND THE NATURE OF SCIENCE

Owen Gingerich
Harvard-Smithsonian Center for Astrophysics, Cambridge, Mass.

The decade of the 1970s saw the four-hundredth anniversary of
Kepler, the quinquecentennial of Copernicus, the tercentenary of the
Greenwich Observatory, and the Einstein centennial. Now, at the beginning
of the 1980s, we are celebrating the fiftieth anniversary of the
beginnings of Karl Jansky's work in what was to become known as radio
astronomy. The difference between those earlier anniversaries and the
present one is that most of us have been, if not active participants, at
least interested bystanders as the discoveries at radio wavelengths have
made their impact upon astronomy. What we lack in historical perspective
is perhaps compensated by the immediacy of our own experiences.

RECOLLECTIONS OF EARLY RADIO ASTRONOMY

J.S. Hey, in his 1973 book The Evolution of Radio Astronomy,
has singled out the years 1950-51, as particularly crucial. In 1949 the
Australian group of J.G. Bolton, G.J. Stanley, and O.B. Slee had given the
first three optical identifications for discrete radio sources: Taurus A,
the Crab Nebula; Virgo A, the elliptical galaxy M87; and Centaurus A,
the peculiarly distorted galaxy NGC 5128. In 1950 H. Alfvén and
N. Herlofson introduced the idea that the observed radio emission from the
discrete radio sources was synchrontron radiation, K.O. Kiepenheuer
interpreted the radio emission of the Milky Way as synchrotron radiation
from cosmic ray electrons spiraling in the interstellar magnetic field,
and M. Ryle, F.G. Smith, and B. Elsmore published their preliminary list
of fifty so-called radio stars in the northern hemisphere. In 1951
H.I. Ewen and E.M. Purcell at Harvard detected the 21-cm line from neutral
galactic hydrogen, and in the other Cambridge F.G. Smith used
interferometer methods to find the precise positions of several radio
sources, positions that subsequently allowed W. Baade and R. Minkowski

(1954) to establish the optical identification of Cygnus A with a distant
and peculiar extragalactic object, and Cassiopeia A with an unusual
collection of gaseous knots and filaments. (All of these developments
have been discussed earlier in this volume.)

In those same years I was an editorial assistant at <u>Sky</u> <u>and</u>
<u>Telescope</u> magazine, and in 1951 an entering graduate student at Harvard
College Observatory. My classmates were David Heeschen, Edward Lilley,
Thomas Matthews, Charles A. Whitney, and Richard E. McCrosky. The fact
that half of the class entered radio astronomy indicates the influence of
Professor Bart J. Bok, who was quick to see the potentialities of the new
technique and prompt to integrate these innovations into the Harvard
curriculum. In the summer of 1952, Bok organized a series of fourteen
seminars on the structure of the Milky Way. Radio astronomy appeared
twice in Bok´s series. Heeschen and Lilley´s thorough review of galactic
and extragalactic radio astronomy still sticks in my mind as one of the
most memorable of all of the seminars. By that time the term "radio star"
was already suspect, and they systematically used the expression "point
sources." There were sufficiently few that they were able to tabulate
individually the fifty sources from the first Cambridge survey plus the
eighteeen observed by the Australians. In discussing the origin of the
radiation, they described both the 1937 paper of F.L. Whipple and
J.L. Greenstein and the 1940 paper of L.G. Henyey and P.C. Keenan, which I
mention here to show that the theorists had not entirely neglected
Jansky´s and Grote Reber´s early work.

The second of the 1952 seminars in which the results of radio
astronomy made a strong impact was Bok´s final summary, on the spiral
structure of our galaxy. Radio astronomers remember the competition
between the Dutch, Harvard, and Australian groups to detect the 21-cm
radiation, but from the astronomical perspective of 1952, the race was to
find the spiral structure of the Milky Way, and it was narrowly won by the
optical astronomers. Several decades of counting stars had been to no
avail in detecting the spiral pattern, and not until Baade´s
conceptualization of the stellar populations was the necessary clue in
hand for finding this structure at visible wavelengths. The idea of using
the luminous O and B stars as spiral tracers enabled W.W. Morgan,
S. Sharpless, and D. Osterbrock to announce the detection of the three
spiral arms nearest the sun. Their paper proved to be the sensation of

the December 1951 Cleveland meeting of the American Astronomical Society,
but most astronomers probably first learned of this discovery through the
January, 1952 Sky and Telescope, and similarly, most astronomers learned
of the radio confirmation of these spiral arms using the 21-cm line
through the July issue in which Otto Struve reported briefly on the work
of J.H. Oort and C.A. Muller. There seemed to be some poetic justice that
the optical astronomers found the spiral arms first, since such an
enormous and largely futile effort had been expended trying to find the
spiral arms by optical methods. But by the summer of 1952 we already
recognized the wave of the future: within a few years the radio
astronomers would track the spiral arms across the entire galaxy through
regions unpenetrable by ordinary light and the optical astronomers would
have to take the back seat.

        Important as radio astronomy may have been on the Harvard
scene in the early 1950s, it did not make much impact on the programs of
the American Astronomical Society until some years later. By the early
1960s the boom in astronomy in general had forced the Society to adopt
simultaneous sessions, and there were by then enough papers on radio
astronomy almost to fill up some entire sessions. I remember discussing
this situation at the 1963 Tucson meeting with George McVittie, then
secretary of the Society. I argued that radio astronomy was not an
independent discipline, but merely one of many techniques for learning
about the universe. I believed that it would make more sense for all of
the papers on a given object, regardless of the technique of the
observation, to be grouped together in the same session. At the business
meeting, McVittie put the suggestion to the membership, where it was
enthusiastically endorsed, and thereafter the radio astronomers were no
longer segregated into their own sessions.

IMPACT ON MODERN ASTRONOMY

        My assignment, however, is not to reminisce about the earlier
days of radio astronomy, but to try to assess its impact on astronomy
itself. It occurred to me that some indication of the significance of
radio astronomy could be gleaned by examining the 132 selections in A
Source Book in Astronomy and Astrophysics 1900-1975, edited by Kenneth
Lang and myself; full citations to the papers mentioned below can be
found in the Source Book. These selections, illustrating the progress in

astronomy since the turn of the century, were chosen in consultation with
astronomers in a large variety of subdisciplines, so I believe that they
accurately represent the march of astronomy during this era. The
pervasive impact of radio astronomy across the board is well illustrated
by the fact that six of the eight major topical divisions include articles
made possible by the new radio astronomy techniques.

   Altogether, how many of the 132 selections can be considered
radio astronomy? The difficulty of answering this question again reflects
the degree to which radio astronomy techniques have now been fully
accepted as part of the mainstream of astronomy. There is no question but
that the pioneering papers by Jansky and Reber belong to radio astronomy,
as do the papers by B.F. Burke and K.L. Franklin (1955) on the Jupiter
radio noise, A. Hewish, S.J. Bell, et al. (1968) on the discovery of
pulsars, S. Weinreb, A.H. Barrett, M.L. Meeks, and J.C. Henry (1963) on
the discovery of the hydroxyl lines, A.A. Penzias and R.W. Wilson (1965)
on the 3 K background radiation, or the classic papers of J.S. Hey,
S.J. Parsons, and J.W. Phillips (1946) on the discovery of a discrete
radio source in Cygnus and of Bolton, Stanley, and Slee (1949) on the
optical identifications of radio sources. But should the papers by
T.A. Matthews, J.L. Greenstein, A.R. Sandage et al. (1961) on the first
true radio star (which of course turned out to be a quasar) be classed as
radio astronomy? Or what about the optical identification of Cygnus A by
Baade and Minkowski (1954)? And do the papers on the discovery of a
candidate black hole, Cygnus X-1, get counted because this object is also
a radio source? The count is fuzzy, but including all of the above and
two papers on radar astronomy, radio astronomy impinges strongly on at
least 23 of the 132 selections.

   It next occurred to me to inquire as to what other techniques
had a comparable impact. What about the advent of large telescopes, with
apertures greater than two meters? I count exactly the same number of
selections, and this again includes the Baade and Minkowski paper, as well
as the Matthews, Greenstein, Sandage et al. contribution. A high
proportion of the papers on galaxies and quasars are included. These
papers come in two main bursts, the first in the decade following the
completion of the 100-inch telescope on Mt. Wilson, and the second
following the completion of the 200-inch Palmoar telescope. It would be
unfair to say, however, that the 100-inch telescope produced dramatic

results for only a decade in the hands of H. Shapley, W.S. Adams,
E.P. Hubble, and others, and then afterwards had little impact. Among our
132 selections is Carl Seyfert's 1943 paper on nuclear emission in spiral
nebulae, and Baade's 1944 paper on the resolution of M32, NGC 205, and the
central region of the Andromeda Nebula; in both of these the Mt. Wilson
telescopes played a critical role.

Another instrument that has had an enormous influence in
shaping contemporary astronomy is the electronic computer. Curiously
enough, there is no selection in the Source Book in which a high-speed
computer was the tool centrally involved in achieving the results,
although Sandage and M. Schwarzshild's 1952 study of the evolutionary
tracks of red giant stars relied heavily on desk calculators; by today's
standards, the computing power then was primitive indeed. Of course, much
of radio astronomy including very-long-baseline interferometry and
aperture synthesis would be unthinkable without modern computing power.
Similarly, the spacecraft required for the research reported in eight more
selections in the Source Book could hardly have been launched without the
assistance of electronic computers.

Now let me examine another group of articles, this time the 30
contributions that depend on new theoretical knowledge, namely, the
twentieth-century development of quantum mechanics and nuclear physics.
Modern understanding of the basis of spectra has enabled astronomers to
analyze the chemical composition of the stars, to establish the physics of
the million-degree solar corona, to unmask the mystery of nebulium, to
estimate the temperature of interstellar matter, and to predict the
existence of the 21-cm line. The Source Book includes contributions on
these topics by M. Saha, C.H. Payne, H.N. Russell, B. Edlén, I.S. Bowen,
G.E. Strömgren, L. Spitzer, M.P. Savedoff, H.C. van de Hulst, and others.
Similarly, our knowledge of the nuclear atom has enabled astronomers to
understand stellar evolution, including neutron stars and even the origin
of the elements themselves. The Source Book includes articles by
R.d'E. Atkinson, C.F. von Weizsäcker, H.A. Bethe, G. Gamow, E. Öpik,
E.E. Salpeter, E.M. and G.R. Burbidge, W.A. Fowler, F. Hoyle, and others.
We can easily say that without the twin theoretical structures of quantum
mechanics and nuclear physics, we would know very little about stars and
their evolution.

HAS RADIO ASTRONOMY CREATED A REVOLUTION IN ASTRONOMY?

This question is posed within the framework of Thomas Kuhn's model of scientific change as delineated in his The Structure of Scientific Revolutions. He describes how a crisis state leads to the overthrow of one major paradigm for a new one. There is general agreement that something of revolutionary proportions happened to science in the early Renaissance, beginning with Copernicus and culminating with Newtonian physics. There is also a widespread feeling that two theoretical developments of the twentieth centry have so altered our ideas of time and space, and of the world of the small, that they, too, are revolutionary in their nature. I refer, of course, to Einstein's relativity and to quantum mechanics.

In contrast with these epoch-making changes, it seems to me that radio astronomy has only altered our previous conceptions of the universe, not overthrown them. It has not made the universe older or younger, larger or smaller. It has, at least not yet, destroyed our confidence in the conservation laws of physics. If the pulsars had turned out to be signals from little green men, perhaps radio astronomy would have radically revolutionized our conceptions of the universe; instead, we have accommodated neutron stars into the existing scheme of stellar evolution. But even if radio astronomy has not so much destroyed our older astronomical viewpoint, it has enormously enlarged and enriched it. It is like that magical moment in the old Cinerama, when the curtains suddenly opened still further, unveiling the grandeur of the wide screen. Optical astronomy in the 1950s, on that narrow, central screen, offered a quiescent view of a slowly burning universe, the visible radiations from thermal disorder. But then the curtains abruptly parted, adding a grand and breathtaking vista, a panorama of swift and orderly motions that revealed themselves through the synchrotron radiation they generated—the so-called violent universe.

Such moments are not without precedence in science or in astronomy. They transform the scene; new problems and projects suddenly compete for attention. Armed with the new techniques, the researchers blaze a trail into the unknown. But this is not a revolution, at least not in Kuhn's sense. It is the horizon of knowledge expanding into a vacuum where no paradigm had existed, not a revolutionary overthrow of a well-established world view.

In its incremental enlargement of the exisiting astronomical scene, a parallel can be drawn between the impact of radio astronomy and the influence of large telescopes, or perhaps even more closely with the introduction of astronomical spectroscopy, which began with the interpretations of G.R. Kirchhoff in 1859. By artificially producing an absorption spectrum, Kirchhoff discovered the secret of the Fraunhofer lines in the sun and stars, and thus the key to the chemical and physical nature of the stars. Within a few years of that dramatic announcement, William Huggins in England, Father Secchi at Rome, and Henry Draper and Louis Rutherford in New York began the investigation of stellar spectra. About a decade after Kirchhoff's announcement four remarkably placed eclipses, in 1868, 1869, 1870, and 1871, allowed the discovery of what turned out to be the chromospheric helium line, of coronal lines, of the flash spectrum and the reversing layer, and of the f corona. This glorious succession of eclipses essentially inaugurated the study of solar physics, promoted particularly by Norman Lockyer in England and Jules Janssen in France. Their work laid the foundation for what S.P. Langley called "the new astronomy," and within a few years George Ellery Hale would champion the name "Astrophysics" to characterize this new endeavor. Astrophysics had not overthrown classical astronomy, but spectroscopy had transformed and enriched the possibilities for interpreting the sidereal universe. By 1900 no major observatory could be considered in the mainstream of research without a spectrograph.

Some of the most important spectroscopic discoveries were made in the decade following Kirchhoff's interpretations, but by no means were all the major discoveries skimmed off within the early years. On the contrary, as spectrographs were improved and as the theoretical understanding caught up with the observational data, an undiminishing stream of discoveries was made by deciphering spectral lines. We still have scientists who might be most appropriately   called spectroscopists, but a great many of the spectrographic advances are made by general astronomers who simply use spectroscopy as their tool. Similarly, the techniques of radio astronomy are today scarcely monopolized by those who call themselves radio astronomers.

But let me return to my central question about whether radio astronomy has wrought a revolution within astronomy. Granted that we now know a great deal more about the universe than five decades ago when radio astronomy stepped over the threshold, but has this process been revolutionary? There have, to be sure, been unexpected surprises, but I would say this was all in the day's work, normal science, "puzzle solving" if you like Kuhn's rather apt term. Yet the result seems much grander than anything implied by the word <u>normal</u>. As Philip Morrison recently remarked to me, it was radio astronomy that woke up astronomy.

New knowledge, by its sheer proportions, can seem revolutionary, but that does not necessarily constitute a scientific revolution within Kuhn's definitions. Perhaps we should turn our examination inside out and ask whether his analysis really illuminates the proliferation of scientific knowledge brought about through radio astronomy. If you believe in the picture of scientific development implied by his model, then you would expect such giant steps to involve a major change in the basic paradigm, and such an alteration, I maintain, has not been the case for radio astronomy. The undeniably enormous progress of astronomy is not so much the overthrow of a previous paradigm as the explosive increase of knowledge in areas where essentially no knowledge existed and where there was not even a paradigm to overthrow. Ironically, in the end it seems that my central question, concerning whether radio astronomy brought about a scientific revolution, tells us more about the deficiencies of Kuhn's theory of the growth of scientific knowledge than about the inadequacies of radio astronomy.

Science and technology now march closely hand in hand. And so it is in modern astronomy. Technique, whether it is Joel Stebbins with his primitive selenium cell or George Ellery Hale entrepreneuring a giant telescope, goes step by step with theory, particularly with the revolutionary concepts from quantum mechanics and relativity. Perhaps, in some fundamental fashion, only changes in theoretical structure map into Kuhn's model of scientific revolutions, and hence radio astronomy is ruled out by its very nature. At least one conclusion seems clear: if we wish to understand the evolution of modern science, we cannot fail to consider the deeply intertwined nature of technology and theoretical science. The undeniable impact of radio astronomy tells us as much.

References

Alfvén, H., & Herlofson, N. (1950). Cosmic radiation and radio stars. Physical Review 78, 616. Reprinted in Lang & Gingerich (1979), p. 781.

Baade, W., & Minkowski, R. (1954). Identification of the radio sources in Cassiopeia, Cygnus A, and Puppis A. Astrophysical Journal 119, 206–14. Reprinted in Lang & Gingerich (1979), pp. 788–91.

Bolton J.G., Stanley, G.J., & Slee, O.B. (1949). Positions of three discrete sources of galactic radio-frequency radiation. Nature 164, 101–2. Reprinted in Lang & Gingerich (1979), p. 778.

Ewen, H.I., & Purcell, E.M. (1951). Radiation from galactic hydrogen at 1420 MHz. Nature 168, 356. Reprinted in Lang & Gingerich (1979), p. 635.

Henyey, L.G., & Keenan, P.C. (1940). Interstellar radiation from free electrons and hydrogen atoms. Astrophysical Journal 91, 625–30.

Hey, J.S. (1973). The Evolution of Radio Astronomy. New York: Neale Watson Academic Publications, Inc.

Kiepenheuer, K.O. (1950). Cosmic rays as the source of general galactic radio emission. Physical Review 79, 738–9. Reprinted in Lang & Gingerich (1979), p. 679.

Kuhn, T.S. (1970). The Structure of Scientific Revolutions. 2nd ed. Chicago: University of Chicago Press.

Lang, K.R., & Gingerich, O., eds. (1979). A Source Book in Astronomy and Astrophysics, 1900–1975. Cambridge, Mass.: Harvard University Press.

[Morgan, W.W., Sharpless, S., & Osterbrock, D.] (1952). Spiral arms of the galaxy. Sky and Telescope 11, 138–9. Reprinted in Lang & Gingerich (1979), pp. 640–2.

Ryle, M., Smith, F.G., & Elsmore, B. (1950). A preliminary survey of the radio stars in the northern hemisphere. Monthly Notices of the Royal Astronomical Society 110, 508.

Whipple, F.L., & Greenstein, J.L. (1937). On the origin of interstellar radio disturbances. Proceedings of the National Academy of Sciences 23, 177–81.

Graham Smith overhauling "Jerry" cables (from German wartime production) at Grange Road, Cambridge circa 1950 (courtesy Ken Machin)

BIOGRAPHICAL NOTES

EDWARD G. ("TAFFY") BOWEN was born in Wales and obtained his Ph.D. from The
University of London in 1933. He soon became involved with
Sir Robert Watson Watt in developing the air warning radar
which formed the basis of the coastal radar chain guarding
Britain during World War II. During the war he worked on airborne radar and soon afterwards became Chief of the CSIRO Division of Radiophysics, Sydney, a post which he held for a
quarter of a century. He has received numerous awards for his
engineering and scientific achievements, and is a Fellow of
the Royal Society.

RONALD N. BRACEWELL was associated with the Radiophysics Laboratory, Sydney
from 1942 to 1955, including a leave of absence to obtain a
Ph.D. at the Cavendish Laboratory, Cambridge from 1946 to 1950.
After working with J. L. Pawsey to write *Radio Astronomy* (Oxford, 1955) he spent a year lecturing on radio astronomy at the
University of California, Berkeley at the invitation of Otto
Struve. Since 1955 he has been Professor of Electrical Engineering at Stanford University. His contribution was written
during sabbatical leave spent at the Radiophysics Laboratory
at the invitation of Dr. R. H. Frater, Chief of the Division.

WILBUR N. ("CHRIS") CHRISTIANSEN was born in 1913, graduated from Melbourne
University, and after an unsuccessful attempt to find a job in
astronomy became a research engineer in an industrial radio
organization. Here he obtained a number of patents on antennas
and transmission lines while modernizing the Australian international radio telegraph circuits. He joined the CSIRO Division
of Radiophysics in 1948 and developed several new radio telescopes of high resolving power while working in solar radio
astronomy. In 1960 he left CSIRO to become head of the Department of Electrical Engineering at Sydney University. He retired
in 1979 and is now a Visiting Fellow at Mount Stromlo Observatory, Foreign Secretary of the Australian Academy of Science,
and *Président Sortant* of URSI.

ARTHUR E. COVINGTON was born in Regina, Saskatchewan. He graduated in 1938
from the University of British Columbia with a B.A. in Mathematics and Physics, and in 1940, after presenting a thesis on
the electron microscope, received an M.A. Following postgraduate studies at Berkeley, California, he joined the radar program of the National Research Council of Canada at Ottawa in
1942. In 1978 he retired from the Radio Astronomy program at
the Herzberg Institute of Astrophysics of NRC, and is now adding
radio and radio astronomy archival materials to the McNicol and
Riche-Covington collections at the Douglas Library, Queen's
University, Kingston, Ontario.

JEAN-FRANÇOIS DENISSE, born in 1915 and trained as a plasma physicist,
joined the group of radio astronomers at the Ecole Normale
Supérieure in Paris in 1946. After several years spent in

Africa as professor of physics and at the U.S. National Bureau
of Standards, he was appointed in 1953 to the Paris Observatory
where he led the Radio Astronomy Department and the Nançay
Station up to 1963, when he became director of the Paris Observatory. He was Director of the Institut National d'Astronomie
et de Géophysique (1967–70), head of the French space program
as President of the Centre National d'Etudes Spatiales (1967–
1973), and responsible for research activities in French universities (1976–1981). He was also Chairman of E.S.O., the
European Organisation for Astronomy (1976–80), and President of
COSPAR (1978–1982).

DAVID O. EDGE is Reader in Science Studies and Director of the Science
Studies Unit at the University of Edinburgh, Scotland. He graduated in Physics at Cambridge in 1955, and gained his Ph.D. in
radio astronomy there in 1959. Before taking up his appointment
in Edinburgh, he was for six years a BBC science talks radio
producer. He is Editor of *Social Studies of Science*; co-author,
with M. J. Mulkay, of *Astronomy Transformed* (Wiley-Interscience,
1976); and co-editor, with Barry Barnes, of *Science in Context:
Readings in the Sociology of Science* (MIT Press and Open University Press, 1982). He is a Fellow of the Royal Astronomical
Society, and a member of the Council of the Society for Social
Studies of Science (4S).

OWEN GINGERICH is an astrophysicist at the Smithsonian Astrophysical Observatory and Professor of Astronomy and of the History of Science at
Harvard University. As an astronomy graduate student in the
early 1950's, he took several courses with B. J. Bok, but later
turned to research in model stellar atmospheres. Since 1971
Gingerich has worked primarily on the history of astronomy,
especially on Copernicus and Kepler, but recently he has been
editing the 20th-century volume of the *General History of
Astronomy* (Cambridge University Press), and has co-authored
with Kenneth Lang *A Source Book in Astronomy and Astrophysics,
1900-1975*. He is currently chairman of the Historical Astronomy
Division of the American Astronomical Society and head of the
U.S. National Committee for the International Astronomical Union.

VITALY L. GINZBURG, born in Moscow in 1916, is currently Head of the I.E.
Tamm Department of Theoretical Physics at the P.N. Lebedev
Institute of Physics of the USSR Academy of Sciences. In 1942
he obtained his doctorate in physics at the P.N. Lebedev Institute and Moscow State University. He has worked over the years
in a wide variety of fields of theoretical physics, including
radio physics, solid-state physics, superconductivity (for which
work he received the Lenin Prize in 1968), and astrophysics.
Since 1966 he has been a full member of the USSR Academy of
Sciences.

JESSE L. GREENSTEIN is the Lee A. DuBridge Professor of Astrophysics,
Emeritus, at the California Institute of Technology, and form-
erly staff member at the Palomar, Mount Wilson and Owens Valley
Observatories. Born in 1909 in New York City, he obtained his
A.B. from Harvard in 1929, and his Ph.D. in 1937; he was on
the faculty of the University of Chicago, at the Yerkes

Observatory, until 1948, when he became a professor at Caltech.
His early interests were in interstellar matter, but changed
towards chemical compositions as related to the theory of nuclear evolution and nucleosynthesis and study of faint stars at
end stages of stellar evolution, such as the hot subdwarfs and
white dwarfs. He participated in the first interpretation of
the optical spectra of quasars. He has had many organizational
responsibilities over the years, for example helping to found
the Owens Valley Radio Observatory, and chairing the National
Academy of Sciences study "Astronomy and Astrophysics for the
1970s."

ROBERT HANBURY BROWN was born in Aruvankadu, India in 1916. After graduating in Electrical Engineering in 1930 he did post-graduate work
at the Imperial College (University of London) and in 1936 joined the original team developing radar in the United Kingdom.
For several years he worked on the development of airborne radar
and from 1942-46 worked on radar at the Naval Research Laboratory in Washington, D.C. For two years he worked with Sir Robert
Watson-Watt as a partner in a firm of consulting engineers, and
in 1949 he joined Sir Bernard Lovell at Jodrell Bank where he
worked in radio astronomy until 1962. At that time he moved to
Australia to build and operate an optical stellar intensity
interferometer at Narrabri Observatory in New South Wales. He
is Emeritus Professor of Physics (Astronomy) at the University
of Sydney, a Fellow of the Royal Society, and President of the
International Astronomical Union.

FRANK J. KERR was born in England in 1918, but grew up in Australia. He
has a Doctor of Science degree from the University of Melbourne.
He was a member of the radio astronomy group at the CSIRO Division of Radiophysics in Sydney from 1940 to 1966, and has been
Professor of Astronomy at the University of Maryland since 1966.
From 1973 to 1978 he was Director of the Astronomy Program and
since 1978 has been Provost of the Division of Mathematical and
Physical Sciences and Engineering. He has been President of
Commission 33 ("Structure and Dynamics of the Galactic System")
of the International Astronomical Union, and a Vice President
of the American Astronomical Society. His scientific interests
have included VHF radio propagation, radar echoes from the Moon,
the Magellanic Clouds, the structure and kinematics of our
Galaxy, and interstellar matter.

ALFRED CHARLES BERNARD LOVELL was born in 1913 and obtained his Ph.D. in
the University of Bristol, working on thin metallic films
deposited in high vacuum. He was appointed Assistant Lecturer
in Physics in the University of Manchester in 1936 and worked
with P.M.S. Blackett on cosmic rays. During World War II he
was in the Telecommunications Research Establishment in charge
of the group which developed the centimetric H_2S blind bombing
device and anti-submarine radar. At the close of the war he
founded the Jodrell Bank Observatory. He was elected a Fellow
of the Royal Society in 1955 and knighted in 1961. In 1981 he
retired from his post as Professor of Radio Astronomy and
Director of Jodrell Bank.

WILLIAM HUNTER McCREA is Professor of Theoretical Astronomy (Emeritus) at
the University of Sussex. Along with the late E.A. Milne, he
invented Newtonian cosmology in 1934 and wrote some of the first
papers on observable relations in relativistic cosmology. He
has written on mathematics, mathematical physics, relativity,
cosmology and cosmogony, theoretical astrophysics, and history
of astronomy, including a short history of the Royal Greenwich
Observatory (Her Majesty's Stationery Office, 1975). He was
President of the Royal Astronomical Society in 1961-63 and
received its Gold Medal in 1976.

BERNARD Y. MILLS was born near Sydney and educated at the King's School and
Sydney University where he graduated B.Sc. in 1941 and B.E. in
1943. He was immediately attached to the CSIR Division of
Radiophysics and was employed in the development of various
radar systems. After the war he spent some time developing a
linear accelerator system for an X-ray tube, and in 1948 joined
the newly formed radio astronomy group; his work in this group
is described in the present contribution. In 1960 he took up
a Readership in the University of Sydney and, with financial
support from the University's Nuclear Research Foundation and
later from the U.S. National Science Foundation, led a group
which constructed the "Molonglo Cross", an instrument of one
mile aperture operating at 408 MHz. This instrument was completed in 1967 and operated for a decade, when it was converted
in the period 1978-1981 to a novel form of synthesis telescope.
Mills was elected a Fellow of the Australian Academy of Science
in 1959 and a Fellow of the Royal Society in 1963, and appointed
Professor of Physics (Astrophysics) in the University of Sydney
in 1965.

GROTE REBER was born in Chicago, Illinois in 1911 and graduated from the
present Illinois Institute of Technology in 1933. He worked
as an engineer for several radio manufacturers in Chicago over
the period 1933-47 during which he conducted his initial researches in radio astronomy. From 1947 to 1951 he was employed
at the National Bureau of Standards, but beginning in 1951 he
became self-employed with his research in Hawaii and Tasmania,
Australia being supported by the Research Corporation. His
interests included mapping of the background radiation at
$\sim$1-2 MHz, cosmic ray studies, dating of Aboriginal sites, and
botanical studies. He received an honorary Sc.D. degree from
Ohio State University in 1962, the Bruce Medal of the Astronomical Society of the Pacific in 1962, and the Jackson-Gwilt
Medal of the Royal Astronomical Society in 1983.

ALEXANDER E. SALOMONOVICH was born in Moscow in 1916 and graduated from the
Physical Faculty of Moscow State University in 1939. In 1949
he obtained his Candidate of Science (radio physics) degree and
in 1964 the Doctor of Science (radio physics), both at the
P. N. Lebedev Physical Institute, Academy of Sciences of the
USSR. In 1953-1959 he was chief scientist responsible for
creation of the Lebedev "RT-22" 22 meter precision radio telescope and until 1964 the chief scientist for operations with
this instrument. Since 1964 he has been head of the Submillimeter Department of the Laboratory of Spectroscopy of the

Lebedev Institute. He is the author of more than 120 papers
on radio astronomy investigations of the sun, moon and planets
in the millimeter wavelength range, and on theory and design of
radio telescopes, particularly of submillimeter ones for observations from artificial satellites.

PETER A. G. SCHEUER was born in 1930 in Germany and studied physics at
Cambridge University. In a lecture to final year undergraduates,
Martin Ryle showed him an interferometer record of numerous
point sources of radio emission of totally unknown nature, and
these proved an irresistible mystery. His dislike for electronics was accepted when it turned out that his first preamplifier could only be opened with a tin opener, but he showed
some understanding of Fourier transforms and was allowed to
continue. He has remained at the Cavendish Laboratory throughout his career, except for the period 1960-1962 when he was a
Visiting Scientist at the CSIRO Radiophysics Division in Sydney.
In the mid-1960's he decided it was time to start thinking about
the physics of radio sources - too early, as it turned out -
and is still trying to understand them.

F. GRAHAM SMITH was born in 1923 and read physics at Cambridge University,
obtaining his Ph.D. in 1952. He remained in the Cavendish
Laboratory until 1964 when he became Professor of Radio Astronomy at Jodrell Bank, University of Manchester, subsequently
being appointed Director in 1981. For the period 1976-81 he
was Director of the Royal Greenwich Observatory. His researches have included a wide range of interests in instrumental
techniques and astrophysics, including the study of radio sources
and pulsars. In 1977 his book *Pulsars* was published. He was
elected a Fellow of the Royal Society in 1970 and has been
Astronomer Royal since 1982.

WOODRUFF T. SULLIVAN, III was born in 1944 and obtained his B.S. in Physics
at the Massachusetts Institute of Technology and a Ph.D. in
Astronomy at the University of Maryland in 1971. His research
interests have included galactic microwave spectroscopy, galaxies, the search for extraterrestrial intelligence, and history
of radio astronomy. He is the compiler of *Classics in Radio
Astronomy* (Reidel, 1982) and is currently writing a monograph
on the worldwide history of radio astronomy before 1960. He
is presently Associate Professor of Astronomy at the University
of Washington, Seattle.

HARUO TANAKA was born in 1922 near Tokyo and was a post-graduate in the
Faculty of Engineering of the University of Tokyo. From 1949
to 1976 he was on the faculty of Nagoya University, including
five years as Director of the Research Institute of Atmospherics. From 1976 to 1982 he was associated with the University
of Tokyo as Director of the Large Radio Telescope Project,
later Nobeyama Radio Observatory. He was Chairman of Commission J ("Radio Astronomy") of URSI for 1978-81 and is currently head of the Japanese National Committee for URSI. He is
presently conducting research on radio telescopes at Toyo
University.

HENDRIK C. VAN DE HULST, born in 1918 in Utrecht, the Netherlands, is
   Professor of Theoretical Astrophysics at Leiden University.
   He has written *Light Scattering by Small Particles* (1957) and
   Multiple Light Scattering (1980), and his 1951 mimeographed
   A Course in Radio Astronomy was an important forerunner of
   early textbooks in the field. His work first centered on the
   solar corona and on interstellar matter and shifted later to
   radio astronomy. He has been active in many functions promoting international cooperation in space research, and acted as
   the first President of COSPAR for 1958-62.

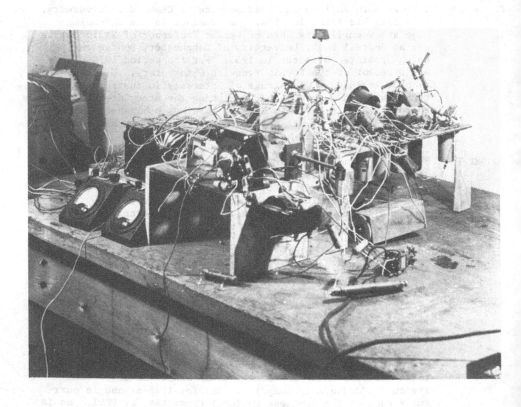

*Bench prototype for the "Cosmic Radio Pyrometer" (a VHF
receiver) developed by Ken Machin at Cambridge circa 1946-47
(courtesy Cavendish Laboratory)*

NAME INDEX

SUBJECT INDEX